Ludomir M. Laudański

Between Certainty and Uncertainty

Intelligent Systems Reference Library, Volume 31

Editors-in-Chief

Prof. Janusz Kacprzyk
Systems Research Institute
Polish Academy of Sciences
ul. Newelska 6
01-447 Warsaw
Poland
E-mail: kacprzyk@ibspan.waw.pl

Prof. Lakhmi C. Jain
University of South Australia
Adelaide
Mawson Lakes Campus
South Australia 5095
Australia
E-mail: Lakhmi.jain@unisa.edu.au

Further volumes of this series can be found on our homepage: springer.com

Vol. 11. Samuli Niiranen and Andre Ribeiro (Eds.)
Information Processing and Biological Systems, 2011
ISBN 978-3-642-19620-1

Vol. 12. Florin Gorunescu
Data Mining, 2011
ISBN 978-3-642-19720-8

Vol. 13. Witold Pedrycz and Shyi-Ming Chen (Eds.)
Granular Computing and Intelligent Systems, 2011
ISBN 978-3-642-19819-9

Vol. 14. George A. Anastassiou and Oktay Duman
Towards Intelligent Modeling: Statistical Approximation Theory, 2011
ISBN 978-3-642-19825-0

Vol. 15. Antonino Freno and Edmondo Trentin
Hybrid Random Fields, 2011
ISBN 978-3-642-20307-7

Vol. 16. Alexiei Dingli
Knowledge Annotation: Making Implicit Knowledge Explicit, 2011
ISBN 978-3-642-20322-0

Vol. 17. Crina Grosan and Ajith Abraham
Intelligent Systems, 2011
ISBN 978-3-642-21003-7

Vol. 18. Achim Zielesny
From Curve Fitting to Machine Learning, 2011
ISBN 978-3-642-21279-6

Vol. 19. George A. Anastassiou
Intelligent Systems: Approximation by Artificial Neural Networks, 2011
ISBN 978-3-642-21430-1

Vol. 20. Lech Polkowski
Approximate Reasoning by Parts, 2011
ISBN 978-3-642-22278-8

Vol. 21. Igor Chikalov
Average Time Complexity of Decision Trees, 2011
ISBN 978-3-642-22660-1

Vol. 22. Przemysław Różewski, Emma Kusztina, Ryszard Tadeusiewicz, and Oleg Zaikin
Intelligent Open Learning Systems, 2011
ISBN 978-3-642-22666-3

Vol. 23. Dawn E. Holmes and Lakhmi C. Jain (Eds.)
Data Mining: Foundations and Intelligent Paradigms, 2011
ISBN 978-3-642-23165-0

Vol. 24. Dawn E. Holmes and Lakhmi C. Jain (Eds.)
Data Mining: Foundations and Intelligent Paradigms, 2011
ISBN 978-3-642-23240-4

Vol. 25. Dawn E. Holmes and Lakhmi C. Jain (Eds.)
Data Mining: Foundations and Intelligent Paradigms, 2011
ISBN 978-3-642-23150-6

Vol. 26. Tauseef Gulrez and Aboul Ella Hassanien (Eds.)
Advances in Robotics and Virtual Reality, 2011
ISBN 978-3-642-23362-3

Vol. 27. Cristina Urdiales
Collaborative Assistive Robot for Mobility Enhancement (CARMEN), 2011
ISBN 978-3-642-24901-3

Vol. 28. Tatiana Valentine Guy, Miroslav Kárný and David H. Wolpert (Eds.)
Decision Making with Imperfect Decision Makers, 2012
ISBN 978-3-642-24646-3

Vol. 29. Roumen Kountchev and Kazumi Nakamatsu (Eds.)
Advances in Reasoning-Based Image Processing Intelligent Systems, 2012
ISBN 978-3-642-24692-0

Vol. 30. Marina V. Sokolova and Antonio Fernández-Caballero
Decision Making in Complex Systems, 2012
ISBN 978-3-642-25543-4

Vol. 31. Ludomir M. Laudański
Between Certainty and Uncertainty, 2013
ISBN 978-3-642-25696-7

Ludomir M. Laudański

Between Certainty and Uncertainty

Statistics and Probability in Five Units with Notes on Historical Origins and Illustrative Numerical Examples

Springer

Author
Ludomir M. Laudański
Rzeszow Technical University
Rzeszow
Poland

ISSN 1868-4394 e-ISSN 1868-4408
ISBN 978-3-642-25696-7 e-ISBN 978-3-642-25697-4
DOI 10.1007/978-3-642-25697-4
Springer Heidelberg New York Dordrecht London

Library of Congress Control Number: 2012930478

Printed on acid-free paper

Springer is part of Springer Science+Business Media (www.springer.com)

Acknowledgements

Bertrand Russell (1872–1970), who left about 40 books for the coming generations, in a preface to one of them confessed that they were completely ready in his mind before he sat down to write them. But this is not the case even with a single one of my books whose number does not came to one third of Russell's books. Most frequently the preparatory procedure of my book resembles a mosaic where the general concept becomes the opening stage of the design process. Here I must mention the pleasure stemming from a suitable selection of the pieces used to make successive parts of the mosaic. On the other hand, my books have been predominantly the result and product of my lectures and of the stimulating processes of my students' reception of them. As Czesław Miłosz once said after giving a lecture: "Again I was lucky enough to draw a rabbit from the hat". Therefore my first words of acknowledgement are addressed to my students. On a more personal note – I commence by mentioning my best American colleague and friend, the late Professor Frederick O. Smetana (1928-2011), whose assistance over several decades was an important stimulus in all I did – despite thousands of kilometres dividing us – he lived in Raleigh and I in Rzeszow (even the first letters of our home towns are the same). I also have profited, which a careful reader might note in the book, from the assistance of Professor Antoni Smoluk, now emeritus of Wrocław University of Economics and long-time Head of the Mathematics Department. Next I would like to thank Professor Stephan M. Stigler of Chicago University, an internationally recognized figure, the author of histories and stories on statistics and probability. His help with respect to literature for my book was exceptional: I can confess that each and every of my calls for help, even during his summer holidays, was answered and appropriate assistance was provided. Lecturing for two weeks at Yasar University, Ismir, Turkey, I had the pleasure to met Professor Giuseppe Burgio of La Sapienza, Rome, and our further contacts became fruitfully stimulating for this book. I have also received friendly support from Professor Anthony W. F. Edwards, Cambridge. The true Godfather of this book was Dr. Thomas Ditzinger from Springer Verlag, Heidelberg; I would also like to thank Helger Schaepe from the same place, whose friendly help I greatly appreciate. I received assistance from a very distant place, Chennai, India and here I must mention Ms. Suguna Ramalignan, the leader of the team which patiently and thoroughly took the book through its final printing stage. Let me also express my gratitude to professor Daniel Simson who gave his permission to reproduce a drawing by Leon Jeśmanowicz. My daughter Maria Klara Laudanska did the final proof reading of the book. To all those persons, named and unnamed, I would like once more to express my deep appreciation, gratitude and sincere thanks.

Rzeszow
May 2012

Ludomir M. Laudański

Contents

Polish Probabilists

Leon Jeśmanowicz (1914-89) : "Stochastic Processes, Wrocław 1952". © Daniel Simson, UMK, Toruń

Leon Jeśmanowicz's drawing – usually taken for a caricature – depicts the members of the Symposium on Stochastic Processes which was held in Wrocław at the end of September 1952. This post-war meeting gathered eminent mathematicians not necessarily probabilists. All the mathematicians depicted are carefully listed in the Index with their dates of birth and death for the readers' convenience given in italics. To say a few words about the time when the Symposium was held, it may be mentioned that the first book by Joseph Leo Doob (1910-2004) - "Stochastic Processes" was published by John Wiley & Sons in 1953. Today numerous biographies of the Polish mathematicians on the Internet are accompanied by drawings of Leon Jeśmanowicz. There is also a special album to commemorate Leon Jeśmanowicz – entitled "Drawings and Caricatures", edited by the University of Mikołaj Kopernik, Toruń, 2005, containing 96 pages and 142 caricatures and drawings – mostly – of professors of this University during its first 20 years of operation.

Prologue

S Y L L A B I 1. What are “statistical data”?; two parallel subdivisions – based on the dimension & based on the order; first order statistics dimension one – descriptive statistics – compression: position and dispersion; preliminaries of the Cartesian Geometry – a point and a line; linear transformations, universal statistics *z-scored* statistics. **2.** Second order statistics dimension one – grouping data; qualitative and quantitative statistical data; ordering qualitative statistics by using combinatorial rules - factorial, binomials, Pascal’s arithmetical triangle, basic combinatorial schemes; rules of grouping; grouping variables; frequency histogram & cumulative curve; compression – direct method and coded method to derive the mean and variance; two kinds of percentiles. **3.** Descriptive statistics dimension two – linear regression based upon Descartes's geometry; grouped data dimension two – great correlation table. **4.** Tracing the binomial distribution and its historic origins – Pascal’s arithmetical triangle, Newtonian symbol, Newtonian binomial; practising binomials; Poisson and Bortkiewicz's contributions. Negative binomials. **5.** Towards normal distribution – from de Moivre & Laplace to the law of the large numbers; practising two theorems of de Moivre and Laplace.

Logistics – per se

Let us commence with a definition according to *The Oxford Companion to Philosophy*, Cambridge 1995, edited by the renowned philosopher Ted Honderich [1]:

“A postulational method of constructing formalized logical system by specifying one’s symbols, recursively defining the well formatted formulae, and lying down an economical set of axioms and inference rules for proving theorems. Such a procedure is axiomatic”.

The above suggested approach does not fully coincide with the common meaning of this term such as is to be found for instance in Wikipedia:

Logistics is the management of the flow of the goods, information and other resources in a repair cycle between the point of origin and the point of consumption in order to meet the requirements of customers.

The specific and possibly the oldest branch of Logistics indicates its military origins:
The New Shorter Oxford English Dictionary [13] defines Logistics as *"the organization of moving, lodging and supplying troops and equipment"* also as *"the detailed organization and implementation of a plan or operation..."* mentioning some well known people: *"The historical leaders Hannibal Barca, Alexander the Great, and the Duke of Wellington are considered to have been logistical geniuses."*

The classic nature of statistical logistics is illustrated by the well-known "*travelling salesman problem,*" for which the Internet provides quite satisfactory references. One of the earliest pioneers in the field is surely the Irish genius Sir William Rowan Hamilton (1805-1865) – to whom we most likely owe thanks for the first mathematical treatment of the matter. On a personal note, the author of this book about 15 years ago was under pressure from the Polish Airline "LOT" to start cooperation by working out the best network connecting Warsaw with all LOT destination, including other Polish airports. This was to be done by using a doctoral thesis of a German student from Brauschweig who apparently prior to that time had been cooperating with Lufthansa for the same purpose. Right from the start it was completely clear that such an adventure would require a change of employer i.e. leaving Rzeszow TU for PLL "LOT" because the suggestion of LOT that *part time employment* would be completely enough to do the job *was greatly exaggerated* - to quote Mark Twain. The project did not take off. One day my computer HDD refused to work so decisively that all its content became inaccessible, including the thesis from Braunschweig – and after a while also LOT as the Polish government carrier experienced similar irreversible circumstances.

Dice Players

ALEA IACTA EST

Gauske, Briccius – period of activity 1476-1495 . Old master – painter, mason and architect, acting in Goerlitz (now Zgorzelec), Kutna Hora and Breslau (now Wrocław). The sculpture shown above is a part of the bay window on the southeast corner of the Wrocław City Hall and was created in the years 1476-1488. It depicts the famous phrase of Iulius Caesar spelled out on 10 January 49 BC while crossing with his army the river Rubikon (Northern Italy). It was a symbolic act initiating the civil war against Pompeius. This phrase was the first time quoted by Svetonius in his history of Roman Caesars – originally as *iacta alea est.* Plutarch wrote in his "Pompeius" that this phrase shouted Ceasar in Greek and it could mean the order to cross the river. Thinking how this phrase could be understood regarding the book on statistics and probability – we would suggest their meaning as a moment of a suspension – expecting this what will happen after a while. Contemplating how ingeniously the Old Master Briccius Gauske, could depict it.

BOOK ONE
THEORY

Chapter 1
Descriptive Statistics

Francis Galton and his obsessive statistical habits. Rules for compressing small statistics: the basic definition of the mean - deriving the basic mean and variance, their main properties; linear transformations – z-scored statistics – introduced with the elements of the Cartesian geometry

Considering the First Lecture as the best opportunity to give the audience a short account of the subject of the studies – we shall follow a very effective technique invented many years ago by a Polish writer Witold Gombrowicz (1904-1969): a dialogue between the Student and the Author.

1.1 A Dialogue

Student – Can you tell me why develop a special branch of Statistics dealing with Logistics as its main field of applications?

Author - As far as I know – there is no need to have such a particular branch. Even an extended study of the numerous Internet examples of using Statistics instructs us that the variety of statistical data sees Logistics as a receiver of statistical data. They primarily document the state of the art of a particular branch of Logistics – for instance in traffic planning, in the control of mechanical means of transportation (buses, trains, air planes), and in the management of stores and transporting stored goods. They serve, for instance, to run Police Logistics Department which collect and store statistics of traffic accidents, criminal events, and so forth. Then statistical data serve as the means of controlling ongoing processes – helping in their rational planning and/or amending their implementation.

S. – Does it mean that the lectures in this book are not going to be in any way different from other known courses which are accessible on the market of books on Statistics?

A. – The answer is both yes and no. Let us explain that somewhat. On the one hand the book has to do with the fact that the teaching hours of subjects related to mathematics are being reduced, which is very disturbing but real and of growing concern. It is evident in Poland at the secondary school level and later at the

L.M. Laudański: Between Certainty and Uncertainty, ISRL 31, pp. 9–36.
springerlink.com

university level as well. Those subjects also include Statistics. So, we are pressed to look for some attractive disguise to potentially broaden the boundaries of those disciplines. In this course the student will find more historical information concerning the origins of the basic components of the subject and the people in Statistical Science, here presented in close connection with Probability. The subdivision of the book includes five main units that approximately correspond to five teaching units of three teaching hours each. In other aspects it corresponds to typical timing for part-time Statistics studies at universities with economic and management profiles. The units related to the lectures are complemented by tutorial and laboratory hours. This task is reflected in the content of the second part of this book.

S. – Yes – I get it – but I think you may also say more regarding those differences which you consider specific for this book in comparison to other similar books on the same subject, and to your own previous books on Statistics.

A. – Well, let us try to be more specific in this respect! First there is something called frame of reference. Thanks to it normal distribution may be introduced without any introductory steps on the theory of probability, simply like statistical data undergoing analytically given recipe. Seeing such an example in [11] and knowing that it is a very popular approach in US student textbooks, I also originally decided to follow the same concept (see [2]). In my classes it has been a standard procedure for a decade. Now I have decided to change it. Therefore after initial and classic three steps (in Chapters 1-3) there comes Chapter 4 on *Binomial distribution*, and only then Chapter 5 about *Normal distribution.* To recall one of the major arguments for such an approach we should take a look at an introductory book on probability that is very little known even in Poland. It is by Witold Pogorzelski (1895-1963) and it is entitled *An outline of probability and errors theory* [3].

In this modest book, the renowned specialist on the theory of integral equations presented an approach starting with binomial distribution, then by showing its limiting property in view of normal distribution, went on to the limiting theorems, and in the end considered the law of large numbers. As this introduction is rather informal, let me mention my personal fondness for that great man dating back more than half a century to a time when I was a student of Aeronautical Engineering at Warsaw TU and attended Mathematics lectures by Prof. Witold Pogorzelski. My grade book features a “B” with his signature and date of 26 June 1957. It was the end of the Spring semester of 1956-1957 when I took an oral examination before professor Witold Pogorzelski - an aged, dignified, charming man of gentle manners... To quote a poet, let me say: "Where are flowers of those times?!" or "Where are the snows of yesteryear?" Does anyone know this line attributed to François Villon (1431-1463), a French poet, thief, and vagabond?
S. – I get the impression that you, Professor, have exhausted the topic so I will return to my seat and start listening to your first lecture.

1.2 Defining the Subject

To begin with, we recommend examining the Logistics stage of Statistics which has to be considered at the beginning of some future statistical investigations. It comes in the shape of Table.1.1 from [4], a book by "Two Muses" [Makać-Urbanek]:

Table 1.1 Statistical procedures and their stages

<table>
<tr><td rowspan="9">Logistics of investigations</td><td rowspan="2">1</td><td colspan="6">Aim of investigations</td></tr>
<tr><td colspan="3">general</td><td colspan="3">partial</td></tr>
<tr><td rowspan="2">2</td><td colspan="6">Field of investigations</td></tr>
<tr><td colspan="6">Choice of the particular population</td></tr>
<tr><td rowspan="3">3</td><td colspan="6">defining the range of the investigations</td></tr>
<tr><td colspan="6">pointing out the particular entries</td></tr>
<tr><td colspan="3">attributes</td><td colspan="3">variables</td></tr>
<tr><td rowspan="2">4</td><td colspan="6">the choice of the scale of investigations</td></tr>
<tr><td colspan="2">full scale investig</td><td colspan="2">representative invest</td><td colspan="2">partial investigations</td></tr>
<tr><td rowspan="4">Collecting data</td><td rowspan="2">5</td><td colspan="6">the choice of the storing data</td></tr>
<tr><td colspan="2">complete registration</td><td colspan="2">partial registration</td><td colspan="2">random choices</td></tr>
<tr><td>6</td><td colspan="6">revision of the results</td></tr>
<tr><td>7</td><td colspan="6">classifiying results of registrations</td></tr>
<tr><td rowspan="3">Documenting collected data</td><td>8</td><td colspan="6">grouping statistical data</td></tr>
<tr><td>9</td><td colspan="6">preparing statistical tables of the grouped data</td></tr>
<tr><td>10</td><td colspan="6">preparing graphical representation of the grouped data</td></tr>
<tr><td rowspan="3">Analysis</td><td>11</td><td colspan="6">structural analysis - determining means and percentiles</td></tr>
<tr><td>12</td><td colspan="6">regression and correlation analysis</td></tr>
<tr><td>13</td><td colspan="6">statistical indexes - time analysis</td></tr>
</table>

Among other things, Table.1.1 provides an opportunity to turn our attention towards the terminology that will be used. Not always, but quite frequently, it follows that used by U. Yule [20], who used the term "attribute" to refer to the main designate of statistics. Retaining the term, we propose to include two distinct types of attributes – qualitative attributes (in short "attributes"), such as gender, eye colour, marital status (widowed or married), and quantitative attributes, which will be called "variables" and lead to numerical statistics.

This is a good place to ask the following question: How should the subject called "Statistics" be understood? In general – from the point of view of the most common meaning – the answer is: "Statistics it is a collection of statistical data." This explains why Table.1.1 does not include mathematical statistics. Therefore, to suggest the direction of a further discussion, one can ask the question of the terse Latin adage: *cui bono*? That is: Who may benefit from statistical data? Without great risk, one may say that there is no single answer to such a question, even from the historical point of view. In an attempt to find an answer, a king may be recalled who requested statistical data to find out how much he possessed. However, this is immediately followed by another question: what did he want to know it for? In light of the definitions of Logistics given in the Prologue, statistical data serve to perform logistic tasks, enabling control of their progress. Thinking about the consequences of such prospects, consider an example of quite unusual possibilities. It is reported that the birth of Christ was significantly affected by the census of a particular Roman province in the Mediterranean, even though that was surely not the intent of the Caesar who called for the census. Turning back to the main subject of our considerations, the second stage, which transformed Statistics significantly, must be considered the moment this new branch of knowledge meets the developing new field of Probability. For instance, probability provided an opportunity to supplement statistical analysis of very real collections of data with similar analyses of new sets containing infinite members – and not only countable but also uncountable continuous sets, using the theory of numbers to include real numbers. The example of first such sets was the binomial distribution, and the second was the normal distribution. Therefore, this second stage of development of Statistics as a branch of knowledge acquired a good amount of abstraction, unimaginably so, making it a branch of abstract mathematics which can only be further developed by mathematicians. This can be considered the third stage – which at least in part is called mathematical statistics. Statistics also went back to considering small samples, but now with highly sophisticated tools and advanced mathematical models. After such a long reasoning, the answer to the above *cui bono?* question, is that it is the *knowledge of science and humanities* that benefits. The development of knowledge becomes the primary task of *Homo Sapiens* species. This leads to the appearance of the developer called the *scientist.* Here we can go back to Witold Gombrowicz, who saw around him two species of people: the *dumb* and the *bright.* Much earlier Erasmus (Desiderius Erasmus Roterodamus, b. 28 Oct. 1466 d. 12 July 1536, known as Erasmus of Rotterdam) made similar comments in his famous book *Laus Stultitiae* (or *Μωριαζ Εγκωμιο*). In a science-fiction book entitled "*Out of the Silent Planet*" by C. S. Lewis [4], Martian population has three distinct *strata.* Only the first caste, the *scientists* (sages, and philosophers) is the productive part of the Martian population. It seems a pity that the development of Cosmonautics has not left room for such interesting speculations now.

The above account should help in distinguishing the classification of statistical data types. They naturally fall into two groups. The first differ with respect to dimensions as they are known in geometry: one-dimensional, two-dimensional, three-dimensional data. The second we suggest to divide into four orders. Statistics of the first order are called *descriptive statistics.* A common feature is their small number, usually not less than six and rarely more than sixteen. Such statistical sets can be seen at a glance. Here in this Chapter, the leading example involves statistics expressing the ages of *five* children in one family - expressed by numbers: 14, 14, 15, 17, 20 . Chapter 2 is entirely devoted to statistics of the second order which we call *grouped data*. Numerous examples of such statistics fill *statistical annals*. By third order statistics we mean statistics having distributions with mathematical shapes, such as the countable binomial distribution, and the uncountable normal distribution. These statistics are in this book considered in Chapters 4 and 5, respectively. This kind of statistical data belongs to the first abstract class of statistical data, and as such cannot be an example of any goods stored in any warehouse. Maybe the cosmos can be the example of where this kind of statistics occur? Statistics of the fourth order are of the kind of sample statistics – and they are the subject of mathematical statistics not presented in this book. Chapter 3 in this book is related to the statistics of the first two orders/types – always of dimension two. Therefore it traditionally refers to *relationships* – being restrained to linear regression and linear correlation.

1.3 Descriptive Statistics Dimension One

1.3.1 Mean Value – Definition and Significance

As a matter of practice, descriptive statistical samples usually do not include less than 6 and/or more than 15-16 entries. Without great risk one may consider them as a *testing ground* for Statistics – to demonstrate, and define the basic tools of Statistics. *Mean value* becomes the opening concept. Some ambiguity is associated with this term demanding the beginners' attention. From one point of view – it is what we can call *the basic mean*. But more generally it is a concept spanning a variety of possible *means*. Also, the reader should be warned that quite frequent practice offers a variety of *means,* possibly of undesired origin, such as *arithmetic mean, geometric mean, harmonic mean.* For clarity we propose to make use of the synonym "average." So for the purpose of all the Statistics presented here we propose to use only a single term "*mean,*" but if we are forced to adapt our habits to frequent practice we rather shall say *arithmetic average, geometric average, harmonic average.* Such an attitude underlines the fact that Statistics applying the Probability Theory defines a unique *mean* composing vertical hierarchy of *means* united under a single general idea. Therefore the proposed definition of the *mean* has not been dressed in the guise of mathematics – and retains the heuristic idea:

To get the mean value of the statistics under consideration – add all its entries altogether and then divide the obtained sum by the number of those entries - the obtained number gives the desired mean value.

The first example of making use of the above definition involves descriptive statistics. Descriptive statistics not only offers finite statistics but also guarantees free access to every entry. This procedure is formalised by using a simple algebra with indexed symbols – denoted by x_i -- with the auxiliary notation stating that

$i = 1, 2, \ldots, N$ in which the symbol N denotes the number of entries *of the entire statistics*. With these in mind we can define *the basic mean value* – denoted by $\overline{x}$ – by the following formula:

$$\overline{x} = \frac{1}{N}\sum_{i=1}^{N} x_i \tag{1.1}$$

Before turning our attention to the first example of using formula (1.1), it can be noted that: the descriptive statistics under attention can represent either *ordered statistics* or *disordered statistics.* Assuming that the second case can be understood intuitively let us define the concept for the first case. From the formal point of view it is necessary to make use of the procedure which uses either the symbol $\geq$ or the symbol $\leq$. As a result, it leads either to the statistics where all the entries are arranged from the greatest to the smallest or to the contrary situation of statistics ordered from the smallest to the greatest element. It should be assumed that the indicated symbols have to be inserted between two successive entries. The unique result can be obtained only while proceeding from the disordered statistics to ordered statistics.

Descriptive statistics given in the second column of Table 1.2 may be interpreted as the age of children in a family. As a consequence, the statistics have been ordered in a natural way. Putting in the same order the weight of the children (or their height), the most probable statistics would be an example of disordered statistics.

The resulting number obtained by the formula (1.1) is not affected by whether the statistics are ordered or (remain) disordered – changing the order of the terms does not change the basic mean of the statistics. By using simple Algebra, the above given formula (1.1) can be transformed into (1.2) as follows:

$$0 = \frac{1}{N}\sum_{i=1}^{N}(x_i - \overline{x}) \quad \rightarrow \quad \sum_{i=1}^{N}(x_i - \overline{x}) = 0 \tag{1.2}$$

The conclusion from (1.2) is: the *mean value of the statistic deviations from the basic mean is zero.* It is also justified to state: *the basic mean is a unique value such that the mean deviation from it is zero.* An impatient student is awaiting the calculation of the *basic mean* for the statistics given in Table 1.2.

Let us confirm that in this case $N = 5$ so, formula (1.1) leads first to:

$$\sum_{i=1}^{5} x_i = 80 \qquad \text{and then we get} \qquad \overline{x} = 16 \tag{1.3}$$

The result (1.3) can be used to start a discussion regarding the basic statistical procedure–i.e. the "*compression of statistical data*". The basic mean indicates the *positioning* of the given statistics with respect to the axes of real numbers, which is considered the main reference for all statistical data. It can be also considered as a result of the compression of the entire statistics. Recalling secondary school Physics - in particular its sub-branch of Mechanics – there is clearly an analogy between the basic mean and the centre of gravity. We are particularly going to make use of the *Archimedes' lever*. The *basic mean* – with a help of Fig. 1.1 can be interpreted as the fulcrum position for the statistics given in Table 1.2. To catch such an idea one has to use simultaneously the result (1.2) together with the values given in the last column of Table 1. 2.

Table 1.2 Statistics of Deviations.

i	x_i	$\overline{x}$	$x_i - \overline{x}$
5	20	16	4
4	17	16	1
3	15	16	-1
2	14	16	-2
1	14	16	-2
			$\sum(x_i - \overline{x})=0$

Let us focus our attention on the numerical values shown in the last column of Table 1.2. They are the same as the positions indicated on the plank at the bottom of the Fig. 1.1. Moreover the same sequence of numbers can be understood as the arms of the "weights" put on the weightless plank in the indicated positions. It is obvious that such a situation corresponds to the balance of the lever.

From one viewpoint it is possible to say that Table 1.2 and Fig. 1.1 may be considered as an *illustration* or a *picture-story* whose meaning could be left for the student to decipher. On the other hand there is a temptation to comment that this particular situation has some special purposes. The upper section of Fig. 1.1 suggests that the weights representing statistical entries *hang* above the lever. Therefore the lever itself represents now nothing more but a uniform scale numbered as shown in two lower pictures of Fig. 1.1. This scale can be *shifted freely in either horizontal direction.* The initial position indicates the scale showing numbers from 14 to 20. They depict the original statistics (as given in the second column of Table 1.2). There is no number "16" – but we know that it corresponds to the fulcrum position - put below the *basic mean*.

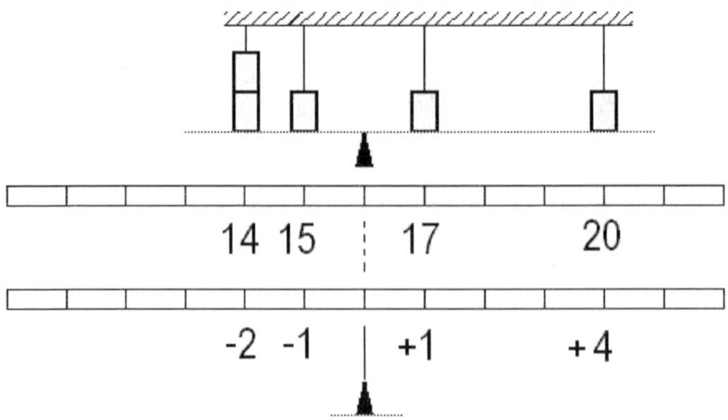

Fig. 1.1 Archimedes' lever – a balance with respect to the fulcrum placed under the mean

Then, by moving the lever from the left to the right by sixteen units – the new fulcrum position indicates "0" – shown in the lowest part of Fig. 1.1, we can read the *shifted statistics* entries as numbers from -2 – to +4 (also given in the last column of Table 1.2). The equilibrium of the shifted statistics is especially easy to check. Values in Table 1.2 confirm the *zero sum* of these statistics – while the lever equilibrium follows from the resulting moment of all the weights about the fulcrum point, which is equal to zero. Therefore – concluding the presented *picture-story* – we have a rather unusual proposition in the end.

The self-explanatory meaning of Fig. 1.1 encourages us to call this situation *the Divine Proof* of the Theorem expressed formally by (1.2). The idea comes from the obvious understanding that *God does not need proofs – God sees the Truth*. To close the above consideration – the description beneath Fig. 1.1 should have the form of a unique imperative expressed by a single word: *Vide*!

The above (somewhat provocative) proposal was born some time ago during a discussion of a paper [6] by A. Smoluk. The discussion ended with the outcome shown here in Fig. 1.2.

Having in mind that the Pythagoras Theorem is now familiar to a wide range of educated people all over the world, it may well serve as a leading example of a *proof* that can be *SEEN* by contemplating the content of Fig. 1.2.

1.3.2 Variance, and Variability

It is quite obvious that statistics differ one from the other by differences in variability – which in technical applications is called *scatter*. Possibly the first scientific discipline in which scatter occurs is in the quantitative differences among observations of stars and planets to determine their positions. In a natural way it leads to the concept of *errors*. Nowadays we consider that the famous book by Galileo Galilee "*Dialogo i dui massimi* ..." [7] written in the Italian language (which was published almost simultaneously in Florence in 1632 and in Leyden in 1638) laid the foundation for the future *Theory of Errors*. But it is also

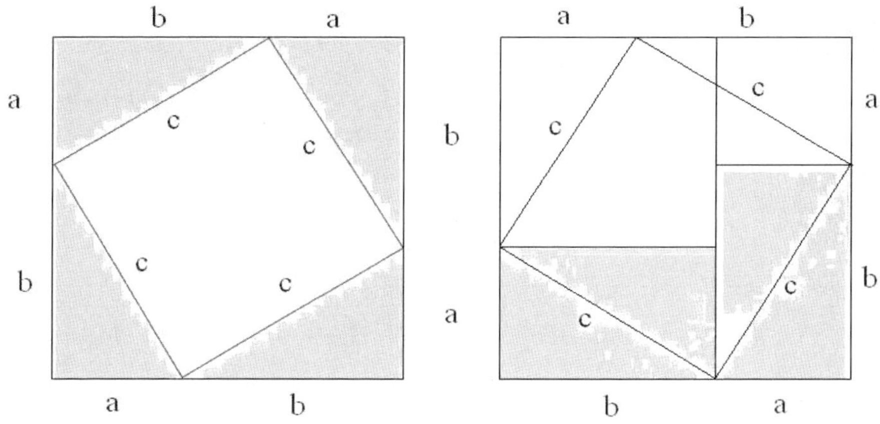

Fig. 1.2 God's Proof that $a^2 + b^2 = c^2$

commonly agreed that we owe to F. Gauss the scientific development of this discipline in his "*Theoria motus …*" [8] written in Latin in 1809 - (see also [12]). We acknowledge here these facts mainly having in mind that *Logistics* – especially as a discipline satisfying Engineering expectations lies closer to the *Theory of Errors* than *Statistics* - naturally satisfying the goals and expectations of *Economics* incorporated especially into the branch of Econometrics.

Table 1.3 Squared Deviations

i	x_i	$\overline{x}$	x_i^2	$x_i - \overline{x}$	$(x_i - \overline{x})^2$
5	20	16	400	4	16
4	17	16	289	1	1
3	15	16	225	-1	1
2	14	16	196	-2	4
1	14	16	196	-2	4
			$\sum x_i^2 = 1306$	$\sum (x_i - \overline{x}) = 0$	$\sum (x_i - \overline{x})^2 = 26$

There is no doubt that the statistics of deviations from the mean value is a way of reflecting the variability of any statistical data. However, the mean of the deviations cannot serve as a measure of variability, since it is always zero. Thus *squared deviations* (see Table 1.3) become the simplest useful concept. The mean of these squared deviations is always guaranteed to be positive – and this positive outcome is called the *variance.*

The formal definition (1.4) may be viewed as the *second basic mean*, but it has a common name, in numerous languages, of *variance*:

$$\sigma_x^2 = \frac{1}{N}\sum\left(x_i - \overline{x}\right)^2 \tag{1.4}$$

If we literally read (1.4) it says that *variance is the mean squared deviation away from the mean*. Direct application of (1.4) to Table 1.3 gives: $\sigma_x^2 = 26/5 = 5.2$. Together with the procedure using the definition in (1.4) there is another algorithm to derive the variance. The following successive equivalent formulas lead to the final outcome:

$$\sigma_x^2 = \frac{1}{N}\sum_{i=1}^{N}\left(x_i^2 - 2x_i\overline{x} + \overline{x}^2\right) \rightarrow \frac{1}{N}\sum_{i=1}^{N} x_i^2 - \overline{x}^2$$

and finally $\sigma_x^2 = \overline{x^2} - \overline{x}^2$ (1.5)

For a student who wants to see the omitted step, the below should do:

$$-2\frac{1}{N}\sum_{i=1}^{N}\left(x_i\overline{x}\right) = -2\,\overline{x}\,\frac{1}{N}\sum_{i=1}^{N} x_i \rightarrow -2\overline{x}^2$$

What (1.5) says is: *variance is the difference between mean square and squared mean.* By using numerical values given in Table 1.3 the *mean square* is 261.2 and the *squared mean* is 16 * 16 = 256. Their difference gives $\sigma_x^2 = 5.2$. Exactly the same as the previous result.

The algorithm based on (1.5) has an important property known as *additivity*. Before we comment on it there is a remark related to the major difference between using the definition in (1.4) and the property in (1.5). The point is that in numerical practice the stage of calculating the *deviations* (depicted in the two last columns of Table 1.3) increases significantly the volume of arithmetical operations leading to the desired outcome. The above-mentioned property of *additivity* is used when implementing the procedure (1.5) into calculators. This "philosophy" enables calculations to proceed step-by-step; entering into the calculator memory successive terms belonging to the statistics under consideration, and enabling one to trace the current volume and the current mean and standard deviation. Usually it is called the "standard deviation procedure" and built-in for all *scientific calculators* on the market. The details related to this are left for the student. Instruction manuals attached to the mentioned products offer sufficiently accessible guidance.

Assuming that the user of either of the two practical ways described above resulting in numerical variance overlooks the units appearing in the statistical data under attention, then probably a characteristic *flaw* of this measure of variability may escape his attention. The point is that the variance units are *squared* units of the entries which occur in the considered statistics. Sometimes it may cause some confusion with respect to the considered subject. For instance: the unit of the statistics which is the subject of analyses in Table 1.2 and Table 1.3 is *time* – which with respect to the people's ages is measured in years. But what about the units of the variance for this case? Help comes from the concept of the *standard deviation* – sometimes called *dispersion.* Returning to the example under consideration:

$$\sigma_x \equiv \sqrt{\sigma_x^2} \cong 2.28 \qquad \text{or, using more digits,} \qquad \sigma_x \cong 2.28035085$$

Recalling the concept of *compression,* which is mentioned from time to time, and anticipating future findings, we now offer the student a proposal in the shape of some information:

$$\bar{x} \pm \sigma_x \qquad \bar{x} \pm 2\sigma_x \qquad \bar{x} \pm 3\sigma_x \tag{1.6}$$

The formulas in (1.6) are standard for the Theory of Errors, telling us that the mean of some measured quantity is not an accurate value of the statistic. However, complete understanding of the common use of (1.6) will require the concept of *normal distribution* (which gives a 99.7% likelihood of the statistic lying within the range indicated by the third formula). Nevertheless it is interesting to look at these banded values for the statistics in Tables 1.2 and 1.3. They are:

$$(13.72,\ 18.28) \qquad (11.44,\ 20.56) \qquad (9.16,\ 22.84)$$

It is seen that the second range of the bounding values is enough to cover the entire statistics in the set –that there is no value bigger than 20.56 and there is no value smaller than 11.44.

1.3.3 Linear Transformations

Geometric prerequisites: points and lines in Cartesian geometry. The subject introduced here will be developed further in Chapter 3, but the material considered here is sufficient to fulfil the needs of Chapter 1.

The history of Geometry commences not with science, but with practical needs, some of which have now been taken over by Geodesy and related disciplines. It is interesting that the scientific origin of Geometry is credited to a Greek – a man whose name, Euclid, is known to all, but whose life time is known only approximately as close to the year 300 B.C. There is a wide range of beautiful books on geometry – we recall here a book by H. R. Jacobs [9] and encourage the Student to make wide use of it. Some of our pictures below will reflect this book. The last chapter of the book (Chapter 17) has the title "*Coordinate Geometry*". Jacobs, like most authors of books on Geometry, prescribes the use of a coordinate system which he attributes to Rene Descartes (1596-1650) or Renatus Cartesius in Latin. This is not exactly correct and we shall discuss the topic below. Classic Euclidean Geometry used a compass and a ruler to draw and to construct geometric figures. But what is the most profound constituent of Euclidean geometry is what is hidden behind simple practice: it is its *deductive*, or *axiomatic structure.* Throughout centuries there were notable efforts to imitate the axiomatic approach of Euclid also in the field of probability. As it is denoted in the title of this sub-chapter, everything begins with points and lines [9]. To draw a line – we have to chose two points as seen in Fig. 1.3.

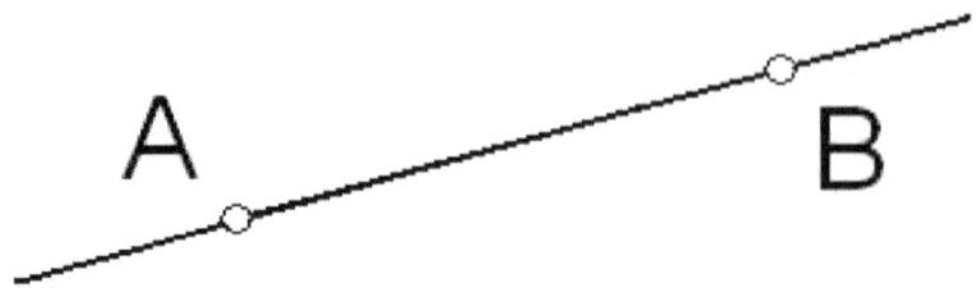

Fig. 1.3 Two points determine a line

Three *noncolinear* points determine a plane, see Fig. 1.4. Combining the two given statements, the plane is determined by a line and a point which is not on the line. The plane can also be determined by two intersecting nonparallel lines. When we think about the past we may feel greatly astonished: How could such highly developed civilizations as Babylon, and then Egypt go through a period of time longer than a millennium using geometry entirely based upon utilitarian practical rules without any attempt to attach them to the abstract Geometry invented by Euclid?

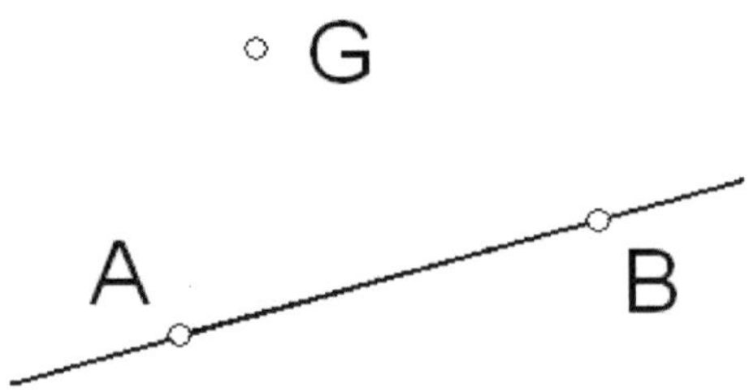

Fig. 1.4 A point and a line determine a plane

Almost two thousand years later, long, long after Euclid, a French philosopher, sage, mathematician (and also at times a mercenary soldier) Cartesius (most often cited as the father of rational philosophy, related to the Latin sentence "*Cogito – ergo sum*") invented coordinate or analytical geometry. In all popular accounts there is always reference to two perpendicular axes called "Cartesian coordinates" – although this attribution actually follows the "*Stigler Law*" [27] and the original Cartesius book "*Geometry*" [10] has no such coordinates, or even such an explicit concept. The epoch-making book introduced a new and unique concept – how to unite two distinct scientific disciplines that had been separate until that time: Geometry and Algebra. In the following development of this approach appears the coordinate system which now become a *logo* of analytical geometry. Leaving these historic details we shall turn toward our goals. In parallel to Euclid geometry, the first step is to identify points, but now using the coordinate system in Fig. 1.6 there are two points: A (3, 4) and B (-5, 2). Their coordinates are defined by an ordered couple, which means that the succession of the coordinates **cannot** be interchanged. The number "3" of Point A denotes the coordinate along the "*x*", axis and the number "4" gives the "*y*" coordinate. Interchanging the coordinates leads to a different point in the plane unless the two coordinates are equal in value.

Two perpendicular lines *x* and *y* intersecting at point O determine a coordinate system *xOy*. The two axes must be *scaled* with a uniform scale. For other applications three dimensional coordinate systems are used, but for our applications we can narrow down considerations to analytic geometry in the plane.

Rene Descartes (1596-1650)

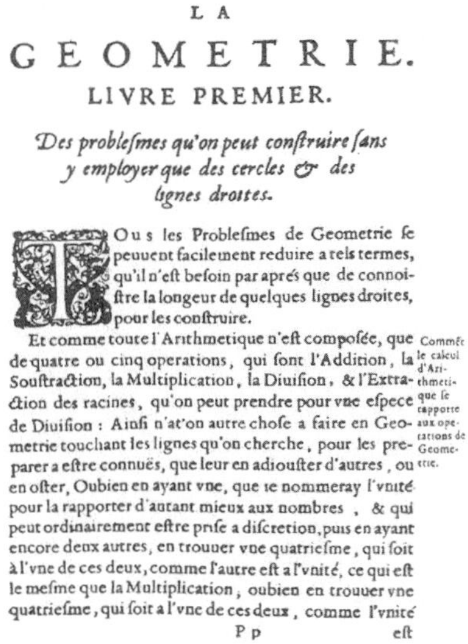

LA

GEOMETRIE.

LIVRE PREMIER.

Des problesmes qu'on peut construire sans y employer que des cercles & des lignes droites.

TOus les Problesmes de Geometrie se peuuent facilement reduire a tels termes, qu'il n'est besoin par aprés que de connoistre la longeur de quelques lignes droites, pour les construire.

Et comme toute l'Arithmetique n'est composée, que de quatre ou cinq operations, qui sont l'Addition, la Soustraction, la Multiplication, la Diuision, & l'Extraction des racines, qu'on peut prendre pour vne espece de Diuision : Ainsi n'at'on autre chose a faire en Geometrie touchant les lignes qu'on cherche, pour les preparer a estre connuës, que leur en adiouster d'autres, ou en oster, Oubien en ayant vne, que ie nommeray l'vnité pour la rapporter d'autant mieux aux nombres, & qui peut ordinairement estre prise a discretion, puis en ayant encore deux autres, en trouuer vne quatriesme, qui soit à l'vne de ces deux, comme l'autre est a l'vnité, ce qui est le mesme que la Multiplication, oubien en trouuer vne quatriesme, qui soit a l'vne de ces deux, comme l'vnité

Commẽt le calcul d'Arithmetique se rapporte aux operations de Geometrie.

Pp est

Fig. 1.5 Cartesius and the first page of his Geometry - first edition 1637

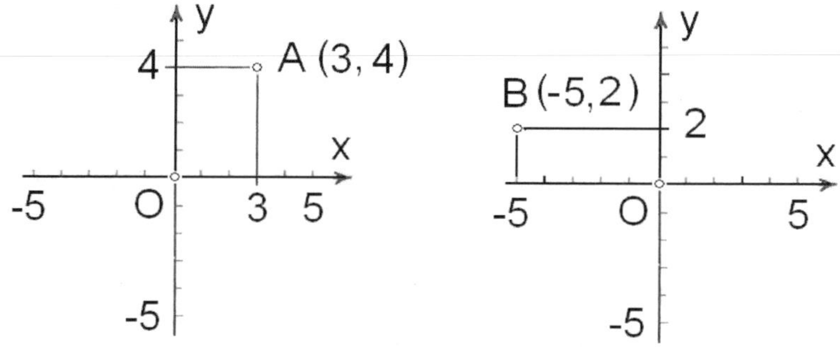

Fig. 1.6 Points in analytic geometry – ordered couple of numbers

The last topic in this introduction to analytic geometry is shown in Fig. 1.7. This figure shows line ℓ in the plane of orthogonal coordinates *xOy*. The bottom line of Fig. 1.7 features an analytic expression describing line ℓ.

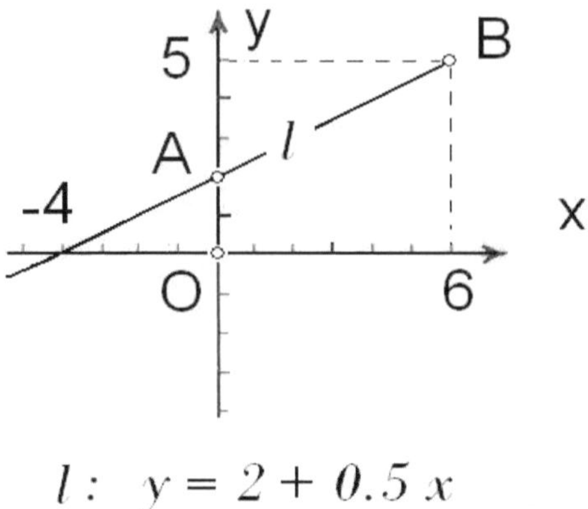

Fig. 1.7 A line in the coordinate plane xOy

It is evident in Fig. 1.7 that the line contains **two** points: A (0, 2) and B(6, 5). Analytic geometry confirms this fact in a suitable fashion. Substituting the coordinates of point "A" into the equation:

$$y = 2 + 0.5\,x \tag{1.7}$$

will satisfy this equation. Let us check this statement:

$$2 = 2 + 0.5 \cdot 0$$

Also substituting the coordinates of point B into (1.7) we should get similar a result:

$$5 = 2 + 0.5 \cdot 6$$

Closer inspection of Fig. 1.7 shows that also the point with coordinates (-4, 0), not labelled in the figure lies on line AB. The student is asked to provide a check similar to those given above. Moreover, the student after a further inspection of Fig. 1.7 will also be able to find at least one more point whose coordinates are given by the natural numbers lying on line ℓ.

The example above presents an introduction to the concept of the *linear transformation.* This meaning of equation (1.7) may also illustrate an example of statistics y_i obtained by substituting into such a linear equation successive values such as those read from Table 1.2 (or Table 1.3). Such statistics y_i inherit contributions related to the statistics x_i and also some features that can be predicted from the general rules related to the *linear transformation*, now applied to the original statistics x_i. The knowledge of those rules saves efforts in the further evaluation of the statistics y_i. The knowledge of those rules saves efforts in the further evaluation of the statistics

$$y = a + b\,x \tag{1.8}$$

There are two particular cases corresponding to two simplifying assumptions obtained by substituting into (1.8) either $a = 0$ or $b = 1$. Let us consider the case $b = 1$, then (1.8) takes the following form:

$$y = a + x \tag{1.9}$$

Transformation (1.9) applied to any descriptive statistics x_i for $i = 1, 2, \dots, N$ leads to descriptive statistics y_i due to:

$$y_i = a + x_i \tag{1.9a}$$

For instance – applying (1.9a) to x_i - given by 14, 14, 15, 17, 20 - and assuming that $a = -10$ we shall get statistics y_i given by 4, 4, 5, 7, 10 - which can be interpreted as statistics showing the age of the same children ten years earlier. Now let us ask about the consequences following the use of the transformation (1.9) with respect to the *mean* and *variance* of the resultant statistics. Regarding the first question the answer is:

$$\overline{y} = a + \overline{x} \tag{1.10}$$

The formal proof requires us to substitute relation (1.9a) into (1.1) applied to statistics y_i. This will give the following results:

$$\overline{y} = \frac{1}{N}\sum_{i=1}^{N}(a + x_i) \;\rightarrow\; \frac{1}{N}\sum_{i=1}^{N}(a) + \frac{1}{N}\sum_{i=1}^{N}(x_i)$$

We also add quite obvious result showing that:

$$\frac{1}{N}\sum_{i=1}^{N}(a) = \frac{1}{N}N\,a$$

Thus the proof of the result given by (1.10) has been completed. The second question with respect to the consequences of the transformation (1.9) regarding the variance has the following answer:

$$\sigma_y^2 = \sigma_x^2 \tag{1.11}$$

In other words – the variance of the statistics x_i becomes *invariant* with respect to a particular linear transformation in (1.9). Discussing the possible ways to prove the result given by (1.11), we may commence once more with a suggestion that Fig. 1.2 offers in fact *God's Proof* of this property. It is enough to see that the *shifting* does not modify the relative positions of all entries. They remain unchanged. The formal proof acknowledges the invariance of the *deviations* from the mean, which is formalized by the following equation, which is a simple consequence of (1.9) and (1.10):

$$\left(y_i - \overline{y}\right) = \left(x_i - \overline{x}\right) \tag{1.12}$$

The second particular linear transformation mentioned above is obtained by substituting $a = 0$ into (1.8), giving the form:

$$y = b\,x \tag{1.13}$$

Now let us ask again about the consequences following the use of the transformation (1.13) with respect to the *mean* and *variance* of the resultant statistics. Again we commence with the *mean* of y_i:

$$\overline{y} = \frac{1}{N}\sum_{i=1}^{N} y_i \tag{1.14}$$

Substituting (1.13) into (1.14) and performing simple manipulations, we get:

$$\overline{y} = \frac{b}{N}\sum_{i=1}^{N} x_i \quad \rightarrow \quad \overline{y} = b\,\overline{x} \tag{1.15}$$

A similar approach can be applied with respect to the variances of the two statistics y_i and x_i . A purely formal procedure with (1.4) gives

$$\sigma_y^2 = \frac{1}{N}\sum\left(y_i - \overline{y}\right)^2 \tag{1.16}$$

The student is most likely to accept that substituting into (1.16) the transformation in (1.13) and the result in (1.15) gives, after simple manipulations,

$$\sigma_y^2 = \frac{b^2}{N}\sum\left(x_i - \overline{x}\right)^2 \rightarrow \sigma_y^2 = b^2\,\sigma_x^2 \quad \text{and} \quad \sigma_y = b\,\sigma_x \qquad (1.17)$$

Having the above in mind we suggest that you focus your attention on a linear transformation that applies in the general case, but has a specific nature, and also a special name.

1.3.4 *Z-score* Statistics

Below we are going to introduce the *z-score statistics* which have many applications in statistics but play an especially important role as a part of normal statistics. Let us assume that the opening step is given by statistics x_i with given mean $\overline{x}$ and variance σ_x, and so define

$$z_i = \frac{x_i - \overline{x}}{\sigma_x} \qquad (1.18)$$

Transformation (1.18) leads to new statistics z_i called *z-score* statistics for which $\overline{z} = 0$ and $\sigma_z = 1$. Let's prove the first property regarding the zero-mean. Here we give essential steps of the proof:

$$\overline{z} = \frac{1}{N}\sum z_i \rightarrow \frac{1}{\sigma_x}\left(\frac{1}{N}\sum x_i - \frac{1}{N}\sum \overline{x}\right) \quad \text{also} \quad \frac{1}{N}\sum_{i=1}^{N}\overline{x} = \frac{1}{N}N\,\overline{x}$$

so $\overline{z} = 0$

On the other hand to derive the variance we may proceed as follows:

$$\text{(i)}\ \sigma_z^2 = \frac{1}{N}\sum\left(z_i - \overline{z}\right)^2 \rightarrow \sigma_z^2 = \frac{1}{N}\sum z_i^2$$

$$\text{(ii)}\ \sigma_z^2 = \frac{1}{N}\sum z_i^2 \rightarrow \frac{1}{N}\sum z_i^2 = \frac{1}{\sigma_x^2}\frac{1}{N}\sum\left(x_i^2 - \overline{x}\right)^2 \quad \text{so} \quad \frac{1}{\sigma_x^2}\,\sigma_x^2 = 1$$

Anticipating some of the material that will be developed and presented in Chapter 5, it may be now pointed out that the idea prescribing importance of this transformation with respect to statistics possessing normal distribution lies in the fact, that this distribution is invariant with respect to linear transformations. Therefore any normally distributed statistics may in a unique way be converted into *z-scored statistics* via transformation (1.18). Therefore, such unique statistics can be understood as a common representation for all normally distributed statistics. Therefore such a normal distribution may serve as a standardized normal distribution. In the last instant it may serve as a base for preparing special universal tables for normally distributed statistics. Such a table is also enclosed in this book.

The closure of this Chapter contains some biographical data and, as in every other chapter in this book, it is supplemented with a list of references.

1.4 Famous and Admired

Islanders – *fin de scièćle* generation

Francis Galton (1822-1911): The quotation enclosed here is from Karl Pearson [28] – and may serve here as a *motto*. It has been taken from an *obituary* published in *Nature* very shortly after Galton's death:

He belonged to that small group of inquirers, who do not specialize, but by their wide sympathies and general knowledge demonstrate how science is a real unity, based on the application of a common logic and a common method to the observation and treatment of all phenomena.

Pioneering accomplishments of Galton cover a wide range of disciplines. Let's mention at least some of them with no certainty that it will be a complete list. Commencing with the concept of *correlation* with respect to which he was a tireless collector of illustrative examples, moving on to his anthropomorphic studies of genealogies, of diversity and specific genders, of intelligence, and to his contribution to genetics. He initiated *eugenics*, which was later infamous and abused by the Nazi in connection with the concept of *racism*, and coined the phrase "*natura versus nurtura*". From him begins *scientometrix* and in particular *psychometrix*. It was Galton who focused attention on *fingerprints* [15]. Galton also deserves to be named father of *meteorology* as an initiator of *weather maps*. In conclusion let us recall two of his inventions. The first – *Galton's whistle* – the original exhibit is on show at the British Museum (Fig.1.8). Is is used nowadays to call our dogs as the sound it emits has such high frequency that it cannot be heard by the human ear, but lies in the range of frequencies heard by dogs. The other invention is the famous *Galton's board* – an exceptionally simple implementation of the binomial distribution, which can be understood as a visualization of the second theorem of De Moivre-Laplace. By typing the keywords "*Plinko Probability*" in an Internet browser we get an animation of *Galton's board*, where we can watch the motion of small spheres filling pipes and read directly the statistics with the mean and the variance values, complemented by the frequency histogram.

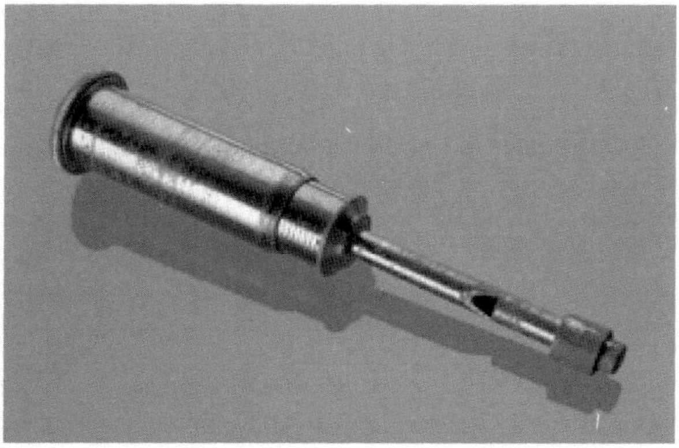

Fig. 1.8 *Galton's whistle – generating ultrasonic*

Galton's IQ was close to 200. From the earliest childhood he was a *wunderkind* – a *child prodigy*, speaking Latin and reading Shakespeare at the age of 6. He left about 340 published papers including "*Hereditary Genius*" – almost half of one thousand pages containing statistics of famous peoples throughout History–statistics collected in the course of his entire life. Gathering statistical data can be called his obsessive habit. In the end one more witty quotation – to add to the portrait of Francis Galton:

At the age of 30 years it was claimed that Galton's experience had been such that he knew more of mathematics and physics than nine biologists out of ten, more of biology than nineteen mathematicians out of twenty, and more of pathology and physiology than forty-nine out of fifty of the biologists and mathematicians of his day.

On the Internet there are more than 500 results if Galton's name is typed and some of them are extended pdf files. These document the interest that Francis Galton has convincingly retained until now. And our next biographical sketches may perhaps justify the somewhat melancholic judgemental saying that: *Galton was one of the last gentleman scientists.*

Karl Pearson was born in a typical *upper-middle class family* in mid 19th century (1857) London, and died at the age of 79 (1936) in Coldharbour, Surrey (sometimes indicated as London). Looking at Karl Pearson's biography and the biography of his major counterpart R. A. Fisher, it may be said that with utmost difficulty it would be possible to present for them both the account *sine ira et studio.* We refer to three biographical texts [29]-[31] in a particular effort to do justice. Biographers note that until the age of 23 he was Carl, and then changed his first name to Karl. Can it be considered as a symbol of his personal struggle in search of his own personal identity? Sometimes the name Karl is associated with Karl Marx due to the socialist ideas that Pearson cultivated in his youth. Born into the world in 1857, born into Cambridge University in 1875, he studied History and German Philology as well as philosophy of science publishing in 1892

"The Grammar of Science" [16] (printed and reprinted until now), – and finally concentrating on Mathematics, and founding Mathematical Statistics. About his Cambridge studies he said himself:

At Cambridge I studied mathematics under Routh, Stokes, Cayley, and Clerk Maxwell, but read papers on Spinoza. There was pleasure in the friendships, there was pleasure in the fights, there was pleasure in the coaches' teaching, there was pleasure in searching for new lights as well in mathematics as in philosophy and religion.

Then he studied for two years in Germany – mainly in Heidelberg and Berlin – such diverse fields as physics and metaphysics, Darwinism and Roman Law, physiology and German literature of the 16th century, also history of the Reformation, among many other fields. Again let's quote his own words regarding the progress of mankind (1905):

History shows me one way, and one way only, in which a high state of civilization has been produced, namely, the struggle of race with race, and the survival of the physically and mentally fitter race. If you want to know whether the lower races of man can evolve a higher type, I fear the only course is to leave them to fight it out among themselves, and even then the struggle for existence between individual and individual, between tribe and tribe, may not be supported by that physical selection due to a particular climate on which probably so much of the Aryan's success depended . . .

He was a doctoral student of Francis Galton (1879) at Cambridge (*The Grammar of Science*). He developed Galton's concept of correlation connecting it with regression lines. The popular– "*chi-squared*" distribution was not only invented by Pearson but he also found its application in *goodness of fit* tests. Cooperating with W. Weldon and F. Galton they founded in 1900 a scientific journal devoted to Statistics in Biology - "*Biometrica.*" This journal still exists today, and K. P. edited and published it until his death. There are different accounts about how many papers Pearson left behind (see [29]), but in every case they are counted in hundreds and show his *intellectual versatility* [30] – on the scale of his Master, Francis Galton. As has already been said *The Grammar of Science* (1892) remains in publication, as does *The Art of Travel* by Francis Galton. Udny Yule portrays Pearson in the following words: *a poet, essayist, historian, philosopher, and statistician.*

Starting from 1892 until his death K. P. lived in the same house in Hampstead. From 1911 he held the *Galton Chair of Eugenics* at University College, London until his retirement in 1933 at the age of 76. The following statement reflects a critical judgement of Karl Pearson (the source is left for the Student to discover):

… being the chairman of a first class academic department and the managing editor of a major journal, Pearson sometimes used his power to the detriment of other important scientists, such as R. A. Fisher and Jerzy Neyman…

Stephen Stigler consulted in this respect said: "I don't really think it correct to say he abused authority". It is also appropriate to add in the end that R. A. Fisher, the main target of his personal attacks, was able to retort to them in an equally severe manner. Moreover both of them had and still have supporters – so, opposition between them seems to have survived. Therefore the time for an unbiased account of their heritage is still ahead of us.

Walter Frank Raphael Weldon (1860-1906): This is not going to be a detailed biography of a man who was first of all a zoologist. Moreover God did not bless him with a lifetime of standard length. A student interested in this is advised to use the rich and accessible Internet resources. Nevertheless despite such a short life the biography of Weldon contains more than one episode significant for the development of Statistics. As a matter of fact we first quote a passage from Pearson's biography:

The importance for science of the intense friendship that sprang up between Pearson and Weldon, then both in their early thirties, can scarcely be exaggerated. Weldon asked the questions that drive Pearson to some of his most significant contributions.

Chapter 4 of this book presents and discusses Monte Carlo results of 26,306 rolls of a set of 12 dice obtained by Weldon in 1894. These data were utilized by Karl Pearson–confirming their close cooperation, which lasted until Weldon's death. It should also be recalled that both men contributed to the establishment of *Biometrica - new scientific journal* that soon became known for its high standards. It was at that time that Weldon wrote, quoting the authors of his biography [32], what follows:

... the questions raised by the Darwinian hypothesis are purely statistical, and the statistical method is the only one at present obvious by which that hypothesis can be experimentally checked ...

George Udny Yule (1871-1951) was born in 1871 in the infamous Paris Commune although this fact probably had no influence upon Yule's personality who was was the child of a Scottish family with impeccable reputation in the fields of scholarship, education and administration. He obtained his Bachelor's degree in civil engineering at University College, London in 1890 – then he started practising engineering "*working in engineering workshops. It was an experience which made him decide that engineering was not the subject for him, so, in 1892, he began to undertake research in physics.*"

So he moved to Germany for one year and found himself under the influence of Heinrich Hertz (1857-1894), the famous discoverer of electromagnetic waves. It was in Bonn, the town of Ludwig van Beethoven (1770-1827). On his return back to London, he came into contact with Karl Pearson, and this influence lasted throughout his entire life. Quoting here one of the opening lines of "*Eugene Onegin*" by A. S. Pushkin - "*и лучше выдумать не мог*" – "He couldn't have done better if he tried." From 1912 he was at Cambridge University where he worked for the rest of his life. The early years of close cooperation with Pearson resulted in the publication of papers related to problems of correlation and regression. He publicized his restrained attitude towards universality of the normal distribution. At the age 25 he became Fellow of the Royal Statistical Society and was very active in this respect. One of his extremely successful adventures was a book entitled "*An Introduction to the Theory of Statistics*" [20], published in 1911. "*The text was intended for those who possessed only a limited knowledge of mathematics and proved a great success*". During his lifetime it had 14 editions (the last three with M. G. Kendall as co-Author). Once more a quotation – this time from Jerzy Neyman – about this book "*In my opinion, this is the best book on statistics that has ever been written"*. His excellent relations with Pearson moved from the stage of joint holidays to the stage of *hard feelings*. The reason was a mistaken judgement of some property of the chi-square statistics related to the sample volume about which Pearson was wrong. In the closing part of this biographical note are some lesser known facts regarding his humanistic profile. It was a kind of fate – while Galton and Pearson began with humanistic studies and ended with statistics, Yule did the other way round. Approaching his retirement he decided to deepen his knowledge of Latin. This resulted in a statistical analysis of Publius Vergilius Maro's (70-19 BC) verses. What deserves special attention is his desire to become a pilot and to fly air planes. In his late 50s *he bought his own plane and acquired a pilot's license*. Unfortunately, at that time he developed a serious heart disease which practically disabled him. During German air raids over England in 1942 he confessed to Kendall that although he could fly he was unable to control the air plane! Now about his fascinations at end of his life. Let's quote [24] once more:

Yule tried to answer questions such as the following: Did Thomas à Kempis really write that little volume which passes under the title of its first chapter, the "De Imitatione Christi"? Did Shakespeare write the plays that are generally attributed to him? Did St. Paul write the Epistle to the Ephesians? What is the probable chronological order of Plato's works? (Yule, 1944)

In the end a quotation [24] with an account which was given by Frank Yates in Yule's obituary: "*To summarize we may, I think, justly conclude that though Yule did not fully develop any completely new branches of statistical theory, he took the first steps in many directions which were later to prove fruitful lines for further progress... He can indeed rightly claim to be one of the pioneers of modern statistics*" (Yates, 1952, p. 320);

***Short Hereditary Passage*:** Frank Yates (1902-1994) was in one respect a completely exceptional figure not only among the Islanders described here. This uniqueness refers to the *genetic material* which he inherited from his parents: he was a child of a couple where his mother was also a daughter of his father. The Author of this book can add in the same subject an episode from an Arab country. While giving me a lift to an optician where I intended to buy glasses, one of my students was asked whether the owner of the shop was family to him. The student replied: yes, he is – but it is a very special relationship. He said that several decades earlier, life in his country was very difficult and in poor families when the father died, the eldest son would take over as head of the family and marry his mother. So, he ended - I am a son of such a couple – the man we are going to visit is my father and my brother. I was shaken.

William Sealy Gosset – "Student" (1876-1937): – The author of this book could consider himself as a sort of descendant of Gosset – taking into account the coincidence of Gosset's year of death and the author's year of birth, – although there is serious doubt concerning the merits of the two persons, which cannot be resolved in this place. Among the Islander statisticians – Weldon lived the shortest, but next in order came Gosset. Moreover, Gosset was not an academician. In fact, Gosset in his entire lifetime was connected with Guiness' breweries and was eventually appointed director of the one of them. His best known paper [18] although written under the supervision of Karl Pearson, was his original achievement and not to be shared with anyone. Moreover it was not acknowledged by Pearson who neglected small samples and related to them statistical problems. On the other hand, it was R. A. Fisher who highly praised this approach calling it a "*logical revolution*".

Although the Student's distribution was in fact (in its exact mathematical form) invented by Fisher - nevertheless "*Gosset's idea of adjusting estimated standard deviation*" became the key idea and such association with him is fully justified (in defiance of "Stigler's law"). Marginally, a strangely forgotten person should be noted here–Herbert Edward Soper (1865-1930), for his related contributions - see [23]. His vivid and full *Obituary* is on the covers of *Journal of the Royal Statistical Society, (*Vol. 94, No. 1 (1931), p. 135-41), written by his younger colleague M. Greenwood.

Florence Nightingale (1820-1910) may also be mentioned here as a person who knew almost all the Islanders presented here. Her vivid account of them is now accessible on the Internet, however her stories cannot be taken as unbiased and/or as the only account.

Egon Sharpe Pearson (1895-1980): Despite the fact that he was the son of K. P. - (as Karl Pearson was known among his contemporaries) he had a personality radically different from that of his admired father. Quiet, introvert and frail in his mature years, he went through a serious crisis in 1925, he said about himself:

I had to go through the painful stage of realizing that K. P. could be wrong ... and I was torn between conflicting emotions:

- *a. finding it difficult to understand R. A. F.*
- *b. hating him for his attacks on my paternal 'god'*
- *c. realizing that in some things at least he was right.*

When K. P. reached his retirement age, against his will, the authorities of London College decided to divide Galton Chair into two separate units – dissatisfying both new chair heads: R. A. Fisher and E. S. Pearson. The struggle between K. P. and R. A. Fisher did not cease with the death of K. P., and the activities of R. A. Fisher were now aimed at Egon Pearson. In statistics the merits of E. S. P are in the theory of testing statistical hypothesis, although this was shared with Jerzy Neyman and is called the Neyman-Pearson theory, dating from 1926. Perusing [19] will show 10 joint papers published between 1928 and 1938, in five journals with three publications in *Biometrica,* two publications in the Bulletin of the

Polish Academy of Sciences (*Biuletyn Polskiej Akademii Umiejętności*). The Author of this book was able to buy [19] in 1967 for a modest (at that time) price of PLN 189 as for a book edited in the West. Both its authors were still alive. Pearson not only held the mentioned chair but also edited *Biometrica*.

Ronald Aylmer Fisher (1890-1962) formally belonged to the same generation as E. S. Pearson and Jerzy Neyman, nevertheless his professional position was much higher than the slight age difference would suggest. It does not seem very risky to quote Erich Leo Lehmann (1917-2009), one of Neyman's first American Ph.D. students from Berkeley, California, from an extended biography of J. Neyman [25]. The quotation showed Fisher during an incident which took place at the *Royal Statistical Society* when Jerzy Neyman in his paper expressed a critical remark regarding the *Latin Square* by Fisher. Here it is:

Fisher, who opened the discussion of the paper, stated that "he had hoped that Dr. Neyman's paper would be on a subject with which the author was fully acquainted, and on which he could speak with authority, as in the case of his address to the Society last summer. Since seeing the paper, he had come to the conclusion that Dr. Neyman had been somewhat unwise in his choice of topics"

In order to maintain a balance, below there is a less critical quotation from [35] – the official and extended *Obituary* for the author of [16] which was written by Frank Yates, co-author of R. A. Fisher's *Statistical Tables,* and the renowned specialist of genetics Kenneth Mather (1911-1990). The earliest years of Fisher document his mathematical abilities and an uneasy professional future of a person who was not a pupil of any famous patron. In 1919 he had choice between the position of the General Statistician in Galton Laboratory, under Karl Pearson, but he chose a similar position in Rothamsted Experimental Station, which was under a lesser known Sir John Russell but gave him open access to biological research and more independence. It has been recorded that he described this period, in a quite merciless manner, as *raking over the muck heap.* Nevertheless he was able to apply this experience to write a book - *Statistical methods for research workers* (1925). Next ten years in this job gave him Fellowship of the *Royal Society.* As we already know, Karl Pearson's retirement and his quick death resulted in the division of *Galton Chair* – which benefited Fisher as he got the Eugenics Chair of London College in 1933, and also the editorship of *The Annals of Eugenics.* Two years later his most renowned monograph [17] *The Design of Experiments* was published. Almost on the opening pages of this book (on p.11) one can find the famous problem: "*A LADY declared that by testing a cup of tea made with milk she can discriminate whether the milk or the tea infusion was first added to the cup*". Jerzy Neyman in his textbook [22] examined the same problem, although their two approaches have very little if any in common. The author of this book has been also magnetized by the charm of this invention by Fisher and placed in [2] an essay with a more detailed re-examination of the procedure given by Neyman. Within five years of publishing [17], Fisher, together with F. Yates, (as was mentioned already) published *Statistical tables.* To wind up this biographical note let us amend the information given in [35] regarding Fisher's family. Fisher was one of twins in his mother's eighth pregnancy. He was born first, however, his brother was still-born.

Yates/Mather wrote: "*He and his twin brother, who died in infancy, were the youngest of eight children,*", which is rather confusing. Regarding the matter of disturbing conflicts between Fisher and Karl Pearson their biographers Yates/Mather wrote:

The originality of his [Fisher's] *work inevitably resulted in conflicts with accepted authority* [Karl Pearson] *and this led to many controversies, which he entered into with vigour, but often in that indignant frame of mind that leads to a partial view of the problem and leaves unanswered objections that are obvious to the impartial observer.*

Regarding the disconcerting nature of this subject, note the following short statement from the same source:

His pungent verbal comments were well known; though frequently made without malice, they were nevertheless disconcerting to those of less robust temperament.

In 1957 Fisher retired and left his native England. His last doctoral student, Anthony W. F. Edwards (Professor Gonvill & Casius College at Cambridge) explains: "Arrangements for him to stay in a post-retirement academic position in England were not as attractive as in Adelaide where Henry Bennett, Professor of Genetics, and E. A. Cornish, a statistician at Commonwealth Scientific and Industrial Research Organization - CSIRO, were established, both of them being friends, collaborators, and great admirers of Fisher." And there the Almighty decided that his life would end as reports [35]: *he died following an operation, on 28 July 1962. He was still, up to a week before his death, full of intellectual vigour, and actively engaged in statistical research.*

Jerzy Spława Neyman (1894-1981) [25] was the grandson of a participant of 1863 January Uprising (an uprising in the former Polish-Lithuanian Commonwealth i.e. present-day Poland, Lithuania, Belarus, Latvia, parts of Ukraine, western Russia; against the Russian Empire), a deportee together with his family to Siberia, born in Tsarist Russia, and lived with Moldavia after his father's family was allowed to resettle to Central Russia. His mother Kazimiera Lutosławska and father Czesław, a lawyer who died early (in 1906), took care of Jerzy's education from his earliest years (including French and German instruction in his boyhood). Neyman attended a school in Kamieniec Podolski (Юрий Чеславович Нейман – secondary school certificates were issued under such a name) then his mother moved with him to Charkow and there in 1912 he began his university studies in physics and mathematics under the best mathematician/probabilist Siergej Natanowicz Bernstein (Сергей Натанович Бернсштайн, 1880-1968). In 1916 he got the academic position of Chair of Mathematics. He studied Lebesgue's theory of measure and integral. It was also Bernstein who acquainted him with "*The Grammar of Science*" by Karl Pearson. It is also possible to deduce that the personality of his master – exceptionally independent and original among the Tzarist Russia mathematicians – possibly affected Neyman, who, despite his associations with such influential scientists as Pearsons and Fisher, was able to go his own way. His return to his native Poland, his fatherland, was quite interesting and remains widely unknown. The story goes that during the war of 1919-20 which

Norman Davies called the war between the White Eagle and the Red Star, Neyman was placed under preventive arrest as a potential ally of the enemy. Then in due course an exchange of prisoners of war brought Neyman (aged 27) to Poland. His first employer was the Agricultural Institute in Bydgoszcz, but after one year he moved to Warsaw settling at SGGW, which still exists under this acronym initials as the Agricultural University. His Ph.D. thesis was defended in 1924 at Warsaw University under the supervision of two distinguished mathematicians – Wacław Sierpiński and Stefan Mazurkiewicz. Shortly after this he received a one-year grant for post-doctoral studies in London at the Laboratory of Karl Pearson, who was well recognized by that time also abroad. It is interesting to know what Lehmann writes (evidently based on Neyman's direct report) about Pearson's embarrassing ignorance of the basic theory of probability: *Pearson did not understand the difference between independence and lack of correlation*. This almost brought to an end Neyman's further presence in his Laboratory. Note here that only for two-dimensional random variables of the normal distribution is it necessary that their one-dimensional components are independent if they are uncorrelated. In general, independence implies zero correlation but not the contrary. One year later, due to the common efforts and support from Pearson and Sierpinski, Neyman got a one-year grant from Rockefeller scholarship funds, for which he has chose Paris with Lesbegue's lectures and Hadamard's seminars. As we know, the decisive turn towards Statistics came due to the cooperation with Egon Pearson. Initiated by Egon's cycle of investigations regarding testing hypothesis, Neyman proceeded with enthusiasm and new energy. Due to Lehmann's development, Neyman's engagement in Statistics in the 1920s and 1930s was so fast that by the time he was ending his cooperation with Pearson, the leader of the *duo* was Neyman. In 1933, sharp progress in Egon Pearson's independence allowed him to employ Neyman. After a while it became a permanent position. But what a fate! In Spring 1937 Neyman received a proposal for a completely independent position in Berkeley, to which he moved in Autumn 1937 and this resulted in his stay in the United States for the remainder of his life. He died of a heart attack in Oakland. The fall of communism materialized and an extremely dangerous superpower ceased to exist. Summing up Neyman's material achievements, it should be said that his gradual effort systematically paid off inasmuch as his employment in Berkeley led to the rise of the Department of Statistics, which became and remains the central scientific centre developing statistics in the United States. Lehmann lists the names of 50 doctoral students of Jerzy Neyman, commencing in Poland, then England, and finally 35 in the USA. The third position from the top is occupied by Lehmann. Regarding personal features of Neyman, Lehmann emphasizes his great *generosity*. For Neyman future biographers there remains the task of describing his family life and his early life in the rapidly changing Tzarist Russia. Regarding religion he was seemingly indifferent, although from childhood his heritage was Christianity (in his childhood he even served as an altar boy). He visited Poland but the author of this book never met him. The following quotation from Stanisław Brzozowski (1879-1911), an exceptionally controversial person, may serve to dramatize this biography: *"A man deprived of his nation resembles an empty soul – neutral, but sometimes even harmful if not dangerous*". Was he right at all?

References

[1] The Oxford Companion to Philosophy. Edited by Ted Honderich, Oxford (1995)

[2] Laudański, L.M.: Statystyka nie tylko dla licencjatów. In Polish: Statistics Not Only for Undergraduates, part1, part2, 2nd edn. Publishing House of the Rzeszow TU, Rzeszów (2009)

[3] Pogorzelski, W.: Zarys Rachunku Prawdopodobieństwa i Teorii Błędów. In: An Outline of Probability and the Error Theory. Towarzystwo Bratniej Pomocy Studentów PW (edited by the students organization soon confiscated by the communist government), Warsaw, pp. 1–100 (1948) (in Polish)

[4] Makać, W., Urbanek-Krzysztofiak, D.: Metody Opisu Statystycznego. In Polish: General Statistics, Outline, Wydawnictwa Uniwersytetu Gdańskiego (1995, 2001).

[5] Lewis, C.S.: Out of the Silent Planet. AVON Book Division, p. 159. The Hearst Corporation, New York (1949); Polish translation by Andrzej Polkowski: Z Milczącej Planety – Wydawnictwo M, Kraków, p. 160 (1989)

[6] Smoluk, A.: Mathematics – a Universal Science. In: Didactics of Mathematics, vol. (6), pp. 5–9. Wroclaw University of Economics, Wrocław (2005)

[7] Galileo, G.: Dialog o dwu najważniejszych układach świata Ptolemeuszowym i Kopernikowym. PWN, Warszawa (1962); Przełożył z języka włoskiego Edward Ligocki, pp. 314–316 (Dialogo sopra i due massimi sistemi del mondo Tolemaico e Copernico, Firenze 1632; Leyden - 1638)

[8] Gauss, C.F.: Theoria motus corporum coelestium in sectionibus conicis Solem ambienitum, Hamburg (1809); English translation by Davis, C.H.: Theory of the Motion of the Heavenly Bodies Moving About the Sun in Conic Sections, p. 416, accessible via Internet. Little, Brown, and Company, Boston (1857)

[9] Jacobs, H.R.: Geometry, 2nd edn., pp. 1–668. W.H. Freeman & Co., New York (1987)

[10] Descartes, R.: La Geometrie 1637. Appendix to Discours de la méthode. Translated into English by Michael Mahoney (New York: Dover, 1979). Internet offers pdf French Edition 82 pages. Edited by R. Hermann, Paris (1886)

[11] Weinberg, G.H., Schumaker, J.A., Oltman, D.: Statistics – An Intuitive Approach, 4th edn., pp. 1–447. Brooks/Cole, Monterey (1981)

[12] Romanowski, M.: On the Normal Law of Errors. National Research Council of Canada. Report APH-1178, Ottawa, pp. 1–29 (February 1964)

[13] Brown, L. (ed.): The New Shorter Oxford English Dictionary on Historical Principles. A-M, vol. 1, p. 1620. Clarendon Press, Oxford (1993)

[14] Laudański, L.M.: Statystyka Ogólna z Elementami Statystyki Matematycznej (in Polish: General Statistics and Probability). Wydawnictwa PWSZ Jarosław (2000)

[15] Galton, F.: Finger Prints, p. 247. MacMillan and Co, London (1892); 1892 appeared Second Edition of Hereditary Genius, p. 423–I Edition. 1869. Both books pdf copies offers Internet

[16] Pearson, K.: The Grammar of Science. Dover Publications (1892/2004)

[17] Fisher, R.A.: The design of experiments. Oliver & Boyd, Edinburg (1935)

[18] Student: The probable error of a mean. Biometrika 6, 1–25 (1908); accessible in Internet

[19] Neyman, J., Pearson, E.S.: Joint Statistical Papers, p. 300. Cambridge University Press (1967)

[20] Yule, G.U.: An Introduction to the Theory of Statistics. Charles Griffin and Co, London 1911 – edited 14 times, the Second Edition was translated into Polish by Z. Limanowski: Wstęp do Teorji Statystyki, Gebethner i Wolff, Warszawa 1921; pp. 1–446. 14-th Edition with M.G. Kendallem appeared in 1950 and was also translated into Polish Wstęp do Teorii Statystyki", PWN, Warszawa (1966)
[21] Laudański, L.M.: Dylematy jakości nauczania w epistolografii św, Pawła (Quality Dillemmas in Letters of St. Paul, Conference Proceedings) Materiały Konferencji Naukowej nt. Dylematy jakości kształcenia w uczelniach wyższych, pp.111–121. Politechnika Rzeszowska, Rzeszów (2008)
[22] Neyman, J.: First Course in Probability and Statistics. HR&W, New York - (1950); Polish translation Zasady Rachunku Prawdopodobieństwa i Statystyki Matematycznej, PWN, Warszawa. (1969), Russian translation Nauka, Moskwa (1968)
[23] Soper, H.E.: On the probable error of the correlation coefficient to a second approximation. Biometrica 9, s.91– s.115 (1913); Tables of Poisson's exponential binomial limit. Biometrika 10, 25–35 (1914)
[24] Williams, R.H.: George Udny Yule: Statistical Scientist. Human Nature Review 4, 31–37 (2004)
[25] Lehman, E.L.: Jerzy Neyman 1894-1981. A Biographical Memoir. National Academy of Sciences, 28 (1994)
[26] O'Connor, J.J., Robertson, E.F.: Sergei Natanovitch Bernstein. Wikipedia, 5
[27] Stigler Law, see Internet
[28] Karl, P.: Francis Galton. Nature 85, 440–445 (1911)
[29] von Collani, C.: Biography of Karl Pearson, 26 pages, `http://encyclopedia.stochastikon.com`
[30] Williams, R.H., Zumbo, B.D., Roos, D., Zimmerman, D.W.: On the Intellectual Versatility of Karl Pearson. Human Nature Review 3, 296–301 (2003)
[31] Stigler, S.: Karl Pearson's Theoretical Errors, and the Advanced They Inspired. Statistical Science 23(2), 261–271 (2008)
[32] O'Connor, J.J., Robertson, E.F.: Walter Frank Raphael Weldon – Wikipedia
[33] Kendal, M.G.: Ronald Aylmer Fisher. Biometrika 50, 17 (1963), parts 1 and 2
[34] Yates, F., Mather, K.: Ronald Aylmer Fisher. Biographical Memoirs of Fellows of the Royal Society of London 9, 91–120 (1963)
[35] Stigler, S.: Karl Pearson and the Rule of Three. To appear in Biometrika, circa, p.14

Chapter 2
Grouped Data. Introduction to General Statistics

Udny Yule's concept of statistical entries – entirely quantitative data. How to proceed when grouping statistical data. Graphical tools. Making use of combinatorial rules while processing attributed data. Defective histogram. Number of bins versus volume of the raw statistical data. Frequency histogram. Direct and coded methods for evaluating grouped data to derive their average and variance. Making use of percentiles

With the statistics of the second order of dimension one commences the study of "real" Statistics and its practice. The material included in Chapter 1 is quite often completely ignored in many textbooks – being *implicitly* incorporated into the opening pages of general statistics. That practice seems to be impractical. To point out a reference where material similar to that presented here in Chapter 1 has been presented let us mention, for instance, Hawkins [4], whose opening chapter is entitled "*Descriptive Statistics,*" and covers about 40 pages. But the leading example of such a textbook is a book [6] by Weinberg (principal author). On the other hand, it is rather easy to see that statistics after the number of their elements increases can hardly be treated by such a direct approach as described in Chapter 1. The remedy is called *grouping*. Naturally, in the beginning comes the question of what is going to be grouped. And here valuable guidance is offered in a book by Yule [1] which seems to be now forgotten/neglected. He said: *by Statistics we mean quantitative data affected to a marked extend by a multicity of causes.* The particular subject of statistical data may be called an *attribute* with which we associate qualitative character, therefore, following Yule, we commence with the theory of attributes. Next we present the theory of variables – or theory of numbered attributes. Let us repeat that in many contemporary books this initial part regarding the theory of attributes is completely ignored.

Regarding statistical data there are two points to be mentioned (not only for the beginners): how to evaluate them and how to interpret them? In this respect it should be stated clearly that this book is devoted to the first question – while to satisfy students seeking guidance regarding the second one, we can only express our believe that it is a matter of practicing Statistics. In this respect, seeing evident abuses of Statistics, an old quarrel between Mediterranean towns regarding the origin of prostitution may be recalled – success has many fathers, but it is not so with failure. Some time ago a British statistician William John Reichmann

L.M. Laudański: Between Certainty and Uncertainty, ISRL 31, pp. 37–65.
springerlink.com

(1920 - ?) wrote a very good book [3] under the provoking title "*Use and Abuse of Statistics*" published/printed many times since 1961 and still available (this book was also translated into Polish in 1968). Unfortunately there is no mention of him on the Internet although in the 1960s he was director of a number of companies, Fellow of the Royal Statistical Society, a member of the Mathematical Association and other professional societies. The first chapter of [3] is entitled *The Age of Statistics* – and we quote its opening (p.11, [3]):

The Age of Statistics is upon us. Almost every aspect of natural phenomena and of human and other activity is now subjected to measurements in terms of statistics which are then interpreted, sometime wisely, sometimes unwisely. Not even the more intimate details of human relationships have escaped the candid survey of the more relentless researches and, as they probe ever more deeply and more widely into our affairs, it is perhaps not surprising that the layman should begin to wonder whether the statisticians are not getting a little beyond themselves.

2.1 Grouping Due to Attributes

<u>Towards the Theory of Attributes.</u> The approach given below is based on an example shown in Table2.1. It can be understood as a modest introduction to the subject.

Example 2.1. There were 10 000 children who underwent a health examination of three attributes – A – *development defects*, B – *nervous symptoms*, C – *poor nutrition.* The results of the survey are given in Table 2.1.

Table 2.1 Warner's investigations – full record

Aggregate 1	Frequency	Aggregate 2	Frequency
(ABC)	57	(αBC)	78
$(AB\gamma)$	281	$(\alpha B\gamma)$	670
$(A\beta C)$	86	$(\alpha\beta C)$	65
$(A\beta\gamma)$	453	$(\alpha\beta\gamma)$	8 310

Instead of a full biographical reference (see [1]) for the data shown in Table 2.1 we give the year of publication –1895, and the name of the investigator F. Warner. The description commences with the terminology. Investigations of the data shown in Table 2.1 examined three *attributes* – designated as *A*, *B*, and *C*. The collection of all observations possessing a single or more attributes is called a *class*. Classes grouping observations of a single attribute are called *order 1.* A class containing two attributes is considered to be *order 2*. And so on. The attributes to which we limit our attention in this book belong to a group undergoing *a division by dichotomy.* The objects or individuals may either possess a particular attribute or not. The object/individual possessing a chosen attribute will be designated by a capital Latin letter A, and those who does not posses this attribute will be designated by a Greek letter α. Moreover, classes designated by capital Latin letters are called *positive classes*, while classes designated by Greek letters are called *negative classes.* In this respect if the total number of the investigated individuals is N, then the following equation *always* applies:

$$A+\alpha=N \tag{2.1}$$

The adverb "always" means that (2.1) remains in force for each *class* of order 1. As we have seen, the term *order* specifies how many attributes compose a considered class. In Example 2.1 the maximum order is 3. Among such classes there may also be empty classes. Note: the practical importance of this remark cannot be overestimated. Perhaps this is the place where the student may guess the necessity of applying some combinatorial tools/rules for the evaluation of such statistical data. For the sake of cohesion we propose to place a short supplement related to combinatorial analysis at the end of this *subchapter.* The main importance is attached here to combinations (without repetitions). Turning back to Example 2.1 we shall repeat that we have six classes of order 1. Let us now investigate how many classes of order 2 are present in this example. A class combining two positive attributes – *for instance* A and B - is designated by the symbol (AB) - which helps to understand that such a class has two attributes appearing simultaneously. Therefore the symbol (BA) could be used equally well for such a class. Such a class in general should not be an empty class, but it may happen from Logistics of investigations. On the other hand a class of kind $(A\alpha)$ will be *always* empty. It is obvious that there is no such object or individual who could belong to a class and simultaneously not belong to the same class – because it would be a contradiction. Now comes the first remark of combinatorial nature. Let us assume that some statistical investigations have considered n attributes – (referring to Example 2.1 - $n=6$) then there come the following question: in how many *ways* is it possible to combine n symbols into a pair of symbols? Which may be put more simply as the question posed above – how many classes of order 2 there will be in this case. The answer uses one of the basic combinatorial rules – called – combinations – which formally states:

$$\binom{n}{2}=\frac{n!}{2!(n-2)!} \tag{2.2}$$

The number of empty classes of order 2 in general will be equal to $n/2$ - therefore for the example under consideration there will be 3 empty classes: $(A\alpha)$, $(B\beta)$ and $(C\gamma)$. On the other hand, substituting $n=6$ from (2.2) we shall get the amount of the all classes order 2:

$$\binom{6}{2} = \frac{6!}{2!(6-2)!} \rightarrow \frac{6 \cdot 5 \cdot 4!}{2! \cdot 4!} = 15 \tag{2.3}$$

Therefore the number of non-empty classes of order 2 will be 15 – 3 = 12. It is obvious that the highest order of classes in the Example 2.1 is $r = 3$. Let us try to find how many non-empty classes of order 3 belong to the case under consideration. As the first attempt there may appear a suggestion that their number will be greater than 12, but this is wrong . Even though

$$\binom{6}{3} = 20, \tag{2.3a}$$

one finds that the number of the empty classes is greater than in the previous case. Let us use this opportunity to analyze the pattern of managing non-empty classes (from a purely formal point of view). First it may be noted that every empty class has the form of a three-element permutation with two initial symbols: $A\alpha$, $B\beta$ and $C\gamma$.

Generalizing this result leads to $n/2$ such cases (as we already know). Proceeding further, the first group of symbols describing empty classes can be symbolized by $(A\alpha\#)$ - so, any of the remaining symbols, that is B, β, C, γ should be used to fill the third *dummy* position. Therefore, in a general case their number will be given by $(n-2)$. The remaining classes $(B\beta\#)$ and $(C\gamma\#)$ should respect the same pattern. Such a procedure leads to $n/2$ cases – therefore the total number of empty classes of order 3 is given by: $n(n-2)/2$. Finally, n = 6, leads to 12 empty classes. Now it is possible to determine the number of all the classes of order 3. First, due to (2.3) we get :

$$\binom{6}{3} = \frac{6!}{3!(6-3)!} \rightarrow \frac{6 \cdot 5 \cdot 4 \cdot 3!}{3! \cdot 3!} \rightarrow 20 \tag{2.4}$$

Therefore, non-empty classes of order 3 are composed of 20 – 12 = 8 cases. Returning now to Table 2.1 we can ask whether Table 2.1 shows all the possible results. And now the above given considerations justify the final conclusion – that Table 2.1, collecting all possible cases for such an investigation, is complete. We expect that the student shall appreciate this conclusion.

An inquiring student will be able to deduce from the content of Table 2.1 the definition of an *aggregate* as a group of classes in which a single italic letter has been replaced by a Greek letter. All classes of order 3 can also be collected in a way shown in Table 2.2. It is a particular *decomposition* of Table 2.1. The idea makes use of the concept of *contrary classes* – that is classes with a formal contraction of the Italic and Greek letters. Corresponding frequencies for the contrary classes are adequately called contrary frequencies.

Table 2.2 Pairs of contraries

Classes	Frequencies	Contrary-classes	Contrary-frequencies
(ABC)	57	$(\alpha\beta\gamma)$	8 310
(αBC)	78	$(A\beta\gamma)$	453
$(A\beta C)$	86	$(\alpha B\gamma)$	670
$(AB\gamma)$	281	$(\alpha\beta C)$	65

Following the idea from the book [1] by Udny Yule – let us consider the problem in which we use the data from Example 2.1 to determine frequencies of the *positive classes* of all orders.

The problem will be solved gradually, step by step. Let us first state that we know frequencies for all the classes of order 3, they are disconnected and the total frequency is the sum of the all partial frequencies and is equal to $N = 10\,000$.

In the first step we shall point out all the *positive classes* – as (A), (B), (C), (AB), (AC), (BC) and (ABC). It should be noted, that among those classes there is class (ABC) for which the frequency is known. All the other frequencies should be determined. How? The pattern of all the solutions is the same. To get the frequency of the class under consideration we have to add the frequencies of all the classes which contain outcomes within the class under consideration. This will give the results given below:

(i). Class frequency of class (A) is equal to the sum of class frequencies for (ABC), $(AB\gamma)$, $(A\beta C)$ and $(A\beta\gamma)$ - they are given in Table 2.1 or in Table 2.2 leading to the result: 57 + 281 + 86 + 453 = 877 . The same pattern will give class frequency for (B) as 1 086, and class frequency for (C) as 286.

(ii). Class frequency of class (AB) will be the sum of class frequencies for (ABC) and $(AB\gamma)$. To derive numerical results one has to add 57 to 281 - getting 338 . In the same way class frequency for (AC) is calculated as 143, and class frequency for (BC) as 135.

Further details regarding the theory of attributes are to be found in the First Part of book [1] – with such headings as "consistence" (Chapter II), "association" (Chapter III) – and some others. Continuing these introductory considerations we come to the concept of *ultimate class* and *ultimate frequency*, with their definitions and some practical remarks. The adverb *ultimate* serves as an indicator of the completeness of the statistical description. It is described by the following definition. *The classes specified by all attributes noted in any case – i.e., the classes of the n-th order in the case of* n *attributes, may be termed the* ***ultimate*** *classes and their frequencies the* ***ultimate*** *frequencies.* Together with this the following property is used: *it is never necessary to enumerate more than the ultimate frequencies.* Moreover, ultimate frequencies form a natural set of which the data are completely given, but any other set containing the same number of algebraically independent classes (i.e. containing the same number of classes, as will be justified below 2^n classes) may be chosen instead. It will be proved below that the other set of this kind forms *positive-class frequencies.* The way to present such a proof requires the determination of the number of the ultimate classes in any given statistics – a result given by 2^n. The procedure for obtaining the right answer to this question is based upon mathematical induction. We present three initial steps, and then generalize the procedure:

(i). for the case of a single positive attribute, denoted by A, there are two *ultimate classes*: (A) and (α);

(ii). for two positive attributes (say A and B) the number of the ultimate classes is doubled – as we have classes (AB), $(A\beta)$, (αB) and $(\alpha\beta)$;

(iii). for three positive attributes denoted A, B and C, the number of ultimate classes will once more double the previous number of ultimate classes – for purely combinatorial reasons – giving the result shown in Table 2.1 in the form of eight classes: (ABC), $(AB\gamma)$, $(A\beta C)$, $(A\beta\gamma)$, (αBC), $(\alpha B\gamma)$, $(\alpha\beta C)$, $(\alpha\beta\gamma)$. The above given procedure, by using mathematical induction leads to the final result:

$$2\times 2\times 2\times 2\times\cdots\times 2 = 2^n \tag{2.5}$$

As was said above, the other set of the all class frequencies is represented by all positive classes reflecting all n attributes with additional appearance of the class of order 0.

order 0	*The whole number of observations*	$\binom{n}{0}$
order 1	*The number of attributes noted*	$\binom{n}{1}$
order 2	*The number of combinations of 2 things chosen from n things*	$\binom{n}{2}$
		(2.6)
order 3	*The number of combinations of 3 things chosen from n things*	$\binom{n}{3}$
…….	…… …… …… ……	
order n	*The number of combinations of n things chosen from n things*	$\binom{n}{n}$

The series composed of all the members of the components given above also leads to the binomial expansion:

$$\binom{n}{0}+\binom{n}{1}+\binom{n}{2}+\ldots+\binom{n}{n}=(1+1)^n \quad \rightarrow \quad (1+1)^n = 2^n \tag{2.7}$$

Therefore the above result (2.7) becomes the final proof that the set of all the *positive class frequencies* also contains the same number of algebraically independent frequencies as the set of all the *ultimate class frequencies* – that is 2^n.

The set of *positive class frequencies* is the most convenient one for both theoretical and practical purposes. To be certain about the truth of this statement compare the procedures of using the ultimate class frequencies with the positive class frequencies by using data from Example 2.1. The latter gives directly the whole number of observations and the totals of A's, B's, and C's – while the former gives none of those fundamentally important figures without doing more or less lengthy additions.

The expression of any class-frequency in terms of the positive frequencies is illustrated below:

(1) $(\alpha\beta) = (\alpha)-(\alpha B)$ because $(\alpha)=N-(A)$

moreover $(\alpha B)=(B)-(AB)$, therefore:

$$(\alpha\beta)=N-(A)-(B)+(AB) \tag{2.8}$$

(2) $(\alpha\beta\gamma)=(\alpha\beta)-(\alpha\beta C)$ - the result for $(\alpha\beta)$ has already been obtained, then $(\alpha\beta C) = (\alpha C)-(\alpha B C)$

The result for (αC) is derived like the result form (αB) has been derived in (1), while the last class results will be derived as follows:

$(\alpha B C)= (B C)-(A B C)$; by ordering all the obtained results we get:

$$(\alpha\beta\gamma)=N-(A)-(B)-(C)+(AB)+(AC)+(BC)-(ABC) \tag{2.9}$$

In the end we propose to solve the following problem.

Example 2.2

Check how Example 2.1 works by finding the ultimate frequencies from the positive class frequencies.

The solution presents results for the three chosen classes.

$(AB\gamma)=(AB)-(ABC)\rightarrow 338-57=281$

$(A\beta\gamma)=(A\gamma)-(AB\gamma)$ but $(A\gamma)=(A)-(AC)$ while $(AB\gamma)=281$,

therefore class frequency of $(A\beta\gamma)$ is 877 – 143 – 281 = 453; then:
$(\alpha\beta\gamma) = (\beta\gamma) - (A\beta\gamma)$ class-frequency of $(\beta\gamma)$ can be derived via (2.8) which allows to obtain

$$(\beta\gamma) = N - (B) - (C) + (BC) \tag{2.10}$$

Substituting into (2.10) the result obtained above for $(A\beta\gamma)$ the class-frequency for $(\alpha\beta\gamma)$ will be 10 000 – 1086 – 286 + 135 – 453 = 8310.

The other ultimate frequencies can be determined in a similar way.

Two Examples of a Graphic Representation. Especially this kind of grouped data which we associate under the common heading of "attributes" may be represented by special kinds of graphs. Presenting two examples of this kind we commence with the pictorial representation that we would like to call *'defected frequency histogram'* – due to its similarity to *frequency histograms* which are considered below. In the literature they are called *bar graphs.*

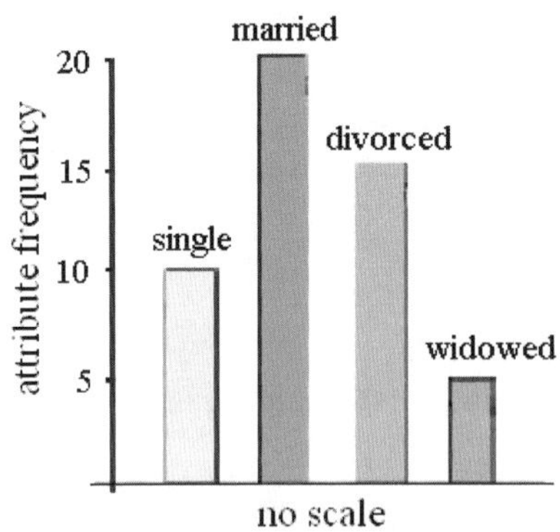

Fig. 2.1 Defected frequency histogram also known as a bar graph

Fig. 2.1 shows marital status of some office staff presented graphically with the vertical axis giving *case frequencies*, however the horizontal axis has no numerical values.

The second kind of graphic representation is shown in Fig. 2.2. These data depict a particular vote of the Security Council in a way which abuses the decision of this UN body. It is enough to say that in this particular case of voting, supporting votes came from four countries: the USA, Great Britain, Spain and Bulgaria, all the others voted against the resolution, despite this fact Fig. 2.2 suggests a false equilibrium. A proper diagram requires right visual proportions in this or other pie chart.

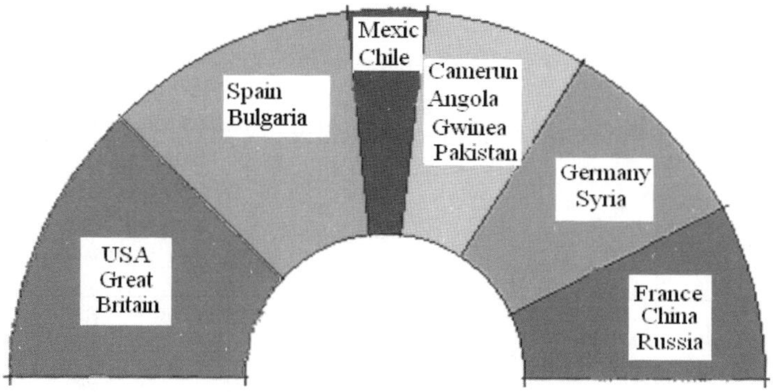

Fig. 2.2 Pie chart to represent a voting on the UN agenda

At the end of this subchapter an extract of combinatorial analysis is given. It is aimed at the actual requirements of Chapter 2 and Chapter 4. Some results exploring/extending this subject will also be provided later.

2.2 Inner Appendix

Factorial. Binomials. Pascal's Arithmetical Triangle.

An elementary definition of the factorial has the form of the following recurrence equation:

$$n! = n \cdot (n-1)! \tag{2.11}$$

The arguments n of the factorial (2.11) are natural numbers. This provokes the first question - How long is it allowed to go back? The value for $n = 2$ requires the meaning of factorial $1!$ We choose to say that

$$1! = 1 \tag{2.12}$$

We see now that allowing $n=1$ in (2.11) requires a value of $0!$ Therefore to get (2.12) we must define the *zero factorial* as

$$0! = 1 \tag{2.13}$$

All the above definitions lead numerically to what follows, which is frequently known from the secondary school:

$$n! = n \cdot (n-1) \cdot (n-2) \cdot \ldots \cdot 3 \cdot 2 \cdot 1 \cdot 1 \tag{2.14}$$

There is also a deeper understanding of the factorial. It recalls the so called - *Gamma Function* dating back to L. Euler and the integral known as Euler's integral of the second order:

$$\Gamma(p)=\int_0^{\infty} x^{p-1}\cdot e^{-x}\,dx \qquad \text{assuming} \quad p>0 \tag{2.15}$$

The Gamma Function is a continuous function for all the values of its argument p and has continuous derivatives of all orders. Regarding the demands of the foregoing considerations it has the property :

$$\Gamma(n+1)=n! \quad \text{here } n \text{ is a natural number (positive and integer)} \tag{2.16}$$

Therefore the factorial defined by (2.11) becomes a special case of the *Gamma Function* for arguments being natural (numbers). The *Gamma Function* has the property which generalizes (2.11) as

$$\Gamma(p+1)=p\cdot\Gamma(p) \qquad \text{for} \qquad p>0 \tag{2.17}$$

The biggest factorial obtained by using many *scientific calculators* is 69!. Nevertheless even market calculators allow one to calculate higher factorials than 69! by making use of so called Stirling's formulae. In fact there are three approximations [5], that are sufficient for our purposes. The first of them is

$$n!\cong\sqrt{2\pi n}\cdot n^n\cdot\left(\frac{n}{e}\right)^n \tag{2.18}$$

Newton's Symbol is defined as follows - with *n, k* being natural numbers:

$$\binom{n}{k}=\frac{n!}{k!(n-k)!} \qquad \text{moreover} \quad 0\le k\le n \quad n\ge 0 \tag{2.19}$$

Newton's Binomial is expressed by using (2.19) in two forms – concise, and expanded - as follows:

$$(a+b)^n=\sum_{k=0}^{n}\binom{n}{k}a^{n-k}b^k \tag{2.20}$$

$$(a+b)^n=\binom{n}{0}\cdot a^n\cdot b^0+\binom{n}{1}\cdot a^{n-1}\cdot b^1+\binom{n}{2}\cdot a^{n-2}\cdot b^2+\ldots+\binom{n}{m}\cdot a^{n-m}\cdot b^m+\ldots+\binom{n}{n}\cdot a^0\cdot b^n \tag{2.21}$$

Substituting $a=1$ and $b=1$ into (2.21) we get the formula already used, see (2.7) :

$$2^n=\binom{n}{0}+\binom{n}{1}+\binom{n}{2}+\ldots+\binom{n}{n} \tag{2.22}$$

Pascal's Arithmetical Triangle becomes the object with surprisingly old roots (see [7]). Here is just the first of numerous possibilities in presenting it - numbered as (2.23):

001	001	001	001	001	001	001	001	001	001
001	002	003	004	005	006	007	008	009	
001	003	006	010	015	021	028	036		
001	004	010	020	035	056	084			
001	005	015	035	070	126				
001	006	021	056	126					
001	007	028	084						
001	008	036							
001	009								
001									

(2.23)

On the other hand the same may be presented in symbolic fashion in a position rotated by 45 degrees:

$$\begin{array}{ccccccccc} & & & & \binom{0}{0} & & & & \\ & & & \binom{1}{0} & & \binom{1}{1} & & & \\ & & \binom{2}{0} & & \binom{2}{1} & & \binom{2}{2} & & \\ & \binom{3}{0} & & \binom{3}{1} & & \binom{3}{2} & & \binom{3}{3} & \\ \binom{4}{0} & & \binom{4}{1} & & \binom{4}{2} & & \binom{4}{3} & & \binom{4}{4} \end{array} \tag{2.24}$$

Below are listed some properties of binomial numbers – i.e. the entities of the triangle:

$$\binom{0}{0} = 1 \tag{2.25}$$

$$\binom{n}{0} = 1 \tag{2.26}$$

$$\binom{n}{n} = 1 \tag{2.27}$$

$$\binom{n}{k} = \binom{n}{n-k} \tag{2.28}$$

Properties (2.25)-(2.27) follow immediately from (2.19) while the property (2.28) defines the symmetry of the binomial numbers and may be traced to the description given by (2.23). (2.29) gives an arithmetic scheme of determining all binomial numbers – although the procedure should commence from the top of the triangle (2.24).

$$\binom{n+1}{k+1} = \binom{n}{k} + \binom{n}{k+1} \tag{2.29}$$

Already here – well ahead of the material presented in Chapter 4 – an impatient student may be advised to make use of the monograph by Edwards [7] containing an unusual richness of facts on this subject.

Three Basic Combinatorial Schemes

Combinatorial rules based on (so to speak) common sense describe how reduced/expanded the possibilities are in ordering different given objects – going from one combinatorial pattern to another. It is hard to devise a common method of presenting the ideas. In general a student may follow either formal or informal studies (see: [8]-[11]). Formal approaches demand more mathematical prerequisites. Informal approaches quite often use the pattern:

permutation → variation → combination

In the below sequence we shall follow the latter approach. Moreover there are so to speak *regional habits* – they are preferred in particular text books, and even by particular publishers. It is especially true when going eastwards.

Permutations. They give the number of the possible arrangements of a given *set*. A quick remainder: by a *set* we understand a collection of elements in which no two elements are the same. For instance – the letters of the Latin alphabet: A, B, C, ... In such circumstances a *permutation* means *any arrangement of the entire set.* Let us list *all permutations* for the set of $n = 2$ Latin letters:

A B   B A

A natural question concerns the number of all permutations for *n-members* set. The answer may be derived by using mathematical induction. Let us check the result obtained when the set of $n = 2$ is expanded into a set of $n = 3$ by adding a new member C – we get:

C A B   A C B   A B C
C B A   B C A   B A C

New permutations - allocate the element C within each subgroup of symbols AB, first in front of them, then between them, and then behind them. The same pattern is used regarding subgroup B A.

Therefore, introducing a third element has tripled the number of permutations. To see clearly what is going on we collect the results below:

Set of 2 elements – number of permutations $2 = 2!$
Adding a 3-rd element – triple permutations more –
hence: $2 * 3 = 6 \rightarrow 3!$
Adding a 4-th element – four times more permutations than for the previous case, so $6 * 4 = 24 \rightarrow 4!$

Final conclusion: with n elements the number of all permutations is determined by:

$$n! \tag{2.30}$$

<u>*Variations*</u> – Permutation of k elements chosen from an n-element set. Their number is given by:

$$n_k = \frac{n!}{(n-k)!} \tag{2.31}$$

Obviously for $k = n$ but also for $k = n - 1$ the number of permutations gives $n!$ – while for a smaller k it is given by (2.31) [we ask the Student why this formula does not show $k!$]. The proof of (2.31) may go backwards. As the first step an arbitrary element out of all n elements may be chosen. But in the second step the choice has $n-1$ possibilities, and so on. Coming to the last step the choice has $n-k+1$ possibilities; therefore the final result has the form:

$$n_k = n \cdot (n-1) \cdots (n-k+1) \tag{2.32}$$

Formula (2.32) is equivalent to formula (2.31).

<u>*Combination*</u> – a variation ignoring/neglecting succession of the elements. Example: a pack of cards lists n cards – how many different results may be obtained while drawing k cards? A system of k electrons arranged in n orbits with a single electron in each – giving *Fermi-Dirac Statistics* in terms of *Statistical Mechanics*. The impossibility of differentiating particular electrons corresponds to ignoring the sequence of cards in the first example. Here we encounter the key formula in the above used combinatorial schemes:

$$C_n^k \equiv \binom{n}{k} = \frac{n!}{k!\,(n-k)!} \qquad k \le n \tag{2.33}$$

Instant proof may indicate that a combination is a variant with n_k arrangements, but, on the other hand, the arrangement of k *elements* leads to $k!$ *permutations*. Therefore disregarding the succession of k *elements* in the previous number of arrangements will reduce the number by $k!$ times. Therefore theorem (2.33) is valid.

2.3 Grouping of Variables

An Algorithm

1. Range. Row statistics are given in Table 2.3. The statistics describe the percentage of the unemployed based on the Polish Statistical Year Book 1993 for 49 provinces of the country.

Table 2.3 Unemployment in Poland in 1992

Percentage of unemployed in Poland 1992 – data for all 49 provinces													
5.9	11.1	11.4	7.9	17.4	10.2	19.8	13.9	21.4	14.7	18.7	17.4	14.4	8.6
14.1	15.5	24.1	8.9	14.9	17.0	12.3	11.4	15.4	16.4	13.5	23.6	12.1	15.3
19.7	17.3	18.5	7.9	15.1	14.1	15.4	12.4	13.3	11.5	23.2	23.7	11.6	12.2
13.5	18.2	21.0	19.7	10.1	11.4	14.1	---	---	---	---	---	---	---

R - the Range/Interval

R = x_{max} - $x_{\min}$ $x_{\max} = 24.1$ $x_{\min} = 5.9$ therefore   R = 18.2

2. Number of Classes. This important and decisive step determines the shape of the frequency histogram, which is the main outcome of the grouping procedure, and is otherwise not accessible. Despite the fact that this statistical method becomes the subject of all such statistical procedures containing unusual quantities of data, there is no single commonly accepted procedure of doing it. Early approaches, as symbolized by [1], suggest satisfying demands of practice preferring the comfort/convenience of logistics in organizing the statistical procedure for evaluation of collected data.

In this book we recommend a particular formal condition (see for instance [6]). It is described below. If the total frequency symbolizes N while the number of classes symbolizes n then the requirement suggested is as follows:

$$2^n \leq N \tag{2.34}$$

For convenience of applying (2.34), some data are given in Table 2.4.

Table 2.4 The Rule of theThumb

n	N
4	16 - 31
5	32 - 63
6	64 - 127
7	128 - 255
8	256 - 511
9	512 - 1023
10	1024 - 2047

Therefore, following (2.34) and data given in Table 2.4 determines the choice of the number of classes

Number of Classes: $n = 5$ for Table 2.3

3. <u>*Class Range*</u> – the stage resulting from the two above completed stages. Having the range R of the grouped data and the n number of classes, the *class range* follows from the obvious division:

Class Range/Interval: $\Delta_x = \dfrac{R}{n}$, therefore $\Delta_x = \dfrac{18.2}{5} \rightarrow \Delta_x = 3.64$

4. <u>*Limits and the Middle Value of Each Class*</u> – this is the last step before the purely mechanical procedure of grouping collected data. The left edge of the frequency histogram is closely related to the value of $x_{\min}$ and can always be accepted as this limiting value. The problem which may occur in this procedure concerns the right edge. The numbers placed as the limits have to satisfy a condition called left side continuity. The simplest way to approach this condition requires one to have a look at Fig. 2.3. The left edge limit is the value 5.9, while the right edge is 24.2. The smallest value in Table 2.3 is 5.9. Due to the above condition this value belongs to the first class. To the first class, due to the requirement of left side continuity, belong numbers satisfying the inequality:

$$5.9 \leq x_i < 9.56$$

The Student's attention should be drawn towards both inequalities shown above: numbers x_i may include the value of the left limit, but cannot include the value of the right limit. If we purely mechanically calculate and collect all limits not respecting the condition of left side continuity, we shall get all the limits given in Table 2.5 . Then, as explained below, we arrive at a contradiction that will cause serious confusion.

Table 2.5 First attempt to determine class limits

Class limits	Median values	Class frequencies
20.46 – 24.1	22.28	5 (48)
16.82 – 20.46	18.64	11 (43)
13.18 – 16.82	15.0	14 (32)
9.54 – 13.18	11.36	13 (18)
5.9 - 9.54	7.72	5

5. Grouping Procedure. – The difficulty mentioned above will be hidden as long as we do not approach the last class for which limits are given by the set (20.46, 24.1). Among the entries in Table 2.3 there is 24.1. We know that each number x_i of the utmost right class has to satisfy the following requirement:

$$20.46 \le x_i < 24.1$$

So, the number 24.1 cannot be put in this class! The number 24.1 should be placed in the next class to the right, but we do not anticipate the presence of such a class! We can recall the reason: it follows from condition (2.34). which states that $n = 6$ is advised for the data counting up to 64 entries. So, this is the confusing situation and the problem has to be solved. One simple remedy may offer a slight correction of the value accepted as x_{max} - for instance to the value 24.2. The corrected *range/interval* will be **18.3**, and *corrected class range/interval* will be **3.66**. Therefore, new *boundaries* and new *median values* are obtained as found in Table 2.6. Comparing Table 2.5 and Table 2.6, it is seen that there are quite insignificant differences. Nevertheless, despite this fact, the columns containing *class frequencies* are ***not identical*** - among the five classes only two show the same class frequencies in the two cases. Nevertheless the desire to have *five* classes, satisfying (2.34), remains. Students frequently ask what may happen if condition (2.34) is not satisfied? Replying to this question, it first of all should be pointed out that temptation always goes, so to speak, upwards, to chose the value of n well above what is required by (2.34). As (2.34) stems from practice, the same source will tell us, that choosing higher values of the number of classes leads to a deficiency in the shape of the resulting histogram, as some classes will not be sufficiently filled, the diagram will be *jagged.* The student may easily find numerous examples of this kind in published statistical data (even in this book).

Table 2.6 Second attempt to determine class limits

Class limits	Class midpoints	Class frequencies
20.54 – 24.2	22.37	6 (49)
16.88 – 20.54	18.71	10 (43)
13.22 – 16.88	15.05	15 (33)
9.56 – 13.22	11.39	13 (18)
5.9 - 9.56	7.73	5

Some comments have to be made. The last columns of Table 2.5 and Table 2.6 contain so called *accumulated class frequencies,* to be used later on. The grouping procedure always reflects also the other conditions which we just mentioned – by referring to [1].

The final remark may be considered as specific for this book– the frequency histogram has to be regarded as a major outcome of grouping statistical data. It uncovers some hidden, so to speak, secrets of grouped data – namely, how they are distributed with respect to their frequencies.

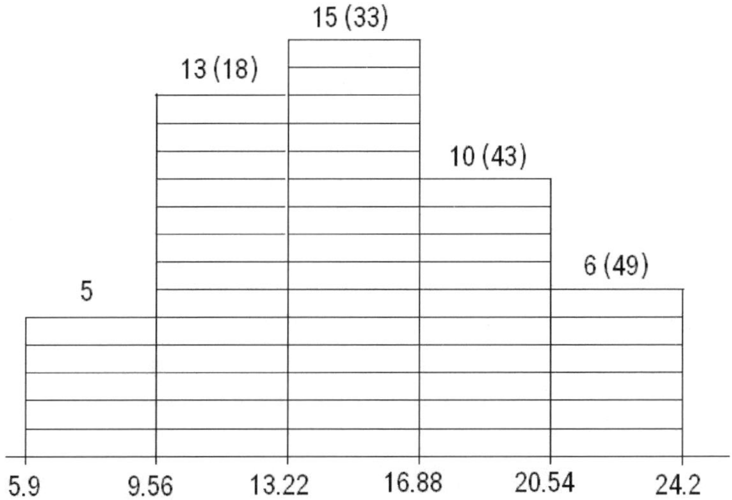

Fig. 2.3 Frequency histogram corresponding to Tab.2.6

Also, from the same point of view, one may emphasize the pioneering role of the book [1] . The content of Chapter VI *The Frequency Distributions* offers examples in a continuous search for the hidden shape of the frequency histogram associated with some kind of data. This is recommended for our Student as examples of the first category of inquiring explorations.

Data presented here rather have a purpose more closely related to satisfying the didactic task of Statistics. Therefore samples of grouped data are rather small, as is possible for presentation on a white board during classes. Also, their character hardly reflects the requirement which was also stated with respect to the statistical data as "*affected to a marked extend by a multiplicity of causes* " - the level of unemployment in Poland in 1992.

The next step should be the evaluation of *grouped statistical data*. First we are going to analyze two procedures to derive averages, and the second step will be to use cumulated histograms to demonstrate a practice of *percentiles.*

2.4 Direct Method to Derive Averages

The example of grouped data has been borrowed from [1]. It is a scanned copy – labeled here as Tab. 2.7 (data for England and Wales).

Table 2.7 Death-rates per Thousand of Population and per Annum, 1881-90

Mean Annual Death-rate	Number of Districts with Death-rate between Limits	Mean Annual Death-rate	Number of Districts with Death-rate between Limits
12·5—13·5	5	23·5—24·5	5
13·5—14·5	16	24·5—25·5	3
14·5—15·5	61	25·5—26·5	1
15·5—16·5	112	26·5—27·5	1
16·5—17·5	159	27·5—28·5	2
17·5—18·5	104	28·5—29·5	...
18·5—19·5	67	29·5—30·5	...
19·5—20·5	42	30·5—31·5	2
20·5—21·5	25	31·5—32·5	...
21·5—22·5	18	32.5—33·5	1
22·5—23·5	8	Total	632

Table 2.8 Direct method of evaluation – grouped data of Tab.2.7

x_{1-11}	x_{1-11}^2	f_{1-11}	$x_i \cdot f_i$	$x_i^2 \cdot f_i$	x_{12-21}	x_{12-21}^2	f_{12-21}	$x_i \cdot f_i$	$x_i^2 \cdot f_i$
13	169	5	65	845	24	576	5	120	2880
14	196	16	224	3136	25	625	3	75	1875
15	225	61	915	13725	26	676	1	26	676
16	256	112	1792	28672	27	729	1	27	729
17	289	159	2703	45951	28	784	2	56	1568
18	324	104	1872	33696	29	841	0	0	0
19	381	67	1273	25527	30	900	0	0	0
20	400	42	840	16800	31	961	2	62	1922
21	441	25	525	11025	32	1024	0	0	0
22	484	18	396	8716	33	1089	1	33	1089
23	528	8	184	4224			632	11188	203056

Basic mean: $\overline{x} = \dfrac{\sum x_i \cdot f_i}{\sum f_i} \rightarrow \overline{x} = \dfrac{11188}{632} \rightarrow \overline{x} = 17.70253165$

Mean square: $\overline{x^2} = \dfrac{\sum x_i^2 \cdot f_i}{\sum f_i} \rightarrow \overline{x^2} = \dfrac{203056}{632} \rightarrow \overline{x^2} = 321.2911392$

Variance: $\sigma_x^2 = \overline{x^2} - (\overline{x})^2 \rightarrow \sigma_x^2 = 7.911512422$

Standard deviation: $\sigma_x = 2.812741087$

As we see the above numerical results have been derived via a long arithmetic procedure, but the question – whether they are right – remains unanswered.

If we are not going to repeat in some way the above calculations, which may not lead to a reduction of uncertainty, the other way to find out if they are correct is to obtain the same results using another procedure. Before we resort to this concept there will be some intermediate stages. Examining the grouped data given in Tab. 2.7, it is seen that the condition (2.34) has not been met here. The apparent convenience of the collected data has been placed above the requirement expressed by (2.34). It is seen from Tab. 2.4 that for N = 632 the suggested class number is of order 8 – 9, while the data given in Tab. 2.7 show 21 classes – twice as many as advised. Therefore if we propose to double the interval of all the classes it automatically will reduce by two times the number of classes. Data grouped in this way have been inserted in Tab. 2.9.

Table 2.9 New grouped data of Tab.2.7

x	f	xf	xx	xxf
32.5	1	32.5	1056.25	1056.25
30.5	2	61.0	930.25	1860.5
28.5	2	57.0	812.25	1625.0
26.5	2	53.0	702.25	1404.0
24.5	8	196.0	600.25	4802.0
22.5	26	585.0	506.25	13162.5
20.5	67	1373.5	420.25	28156.75
18.5	171	3163.5	342.25	58524.75
16.5	271	4471.5	272.25	73779.75
14.5	77	1116.5	210.25	16189.25
12.5	5	62.5	156.25	781.25
---	632	11172	-----	201342

A short comment may help understand the procedure leading from Tab. 2.7 to Tab. 2.9. As the first step classes (32.5, 33.5) and (31.5, 32.5) were united giving a single class of (31.5, 33.5), for which the middle value is 32.5, and the corresponding class frequency is 1. Each of the following steps proceeds in the same fashion. Having in mind that the collection of 21 original classes was odd in number, the last class cannot be coupled with any other. Therefore the left side limit of the last class (12.5, 13.5) has been extended to preserve a constant interval, so the limits of this class show (11.5, 13.5), with the middle value of 12.5 and the class frequency remains 5.

Corresponding averages are listed/calculated below:

$$\overline{x} = \frac{\sum x_i \cdot f_i}{\sum f_i} \quad \rightarrow \quad \overline{x} = \frac{11172}{632} \quad \rightarrow \quad \overline{x} = 17.67721519$$

$$\overline{x}^2 = \frac{\sum x_i^2 \cdot f_i}{\sum f_i} \rightarrow \overline{x}^2 = \frac{201342}{632} \rightarrow \overline{x}^2 = 318.5791139$$

$$\sigma_x^2 = \overline{x}^2 - (\overline{x})^2 \rightarrow \sigma_x^2 = 6.095177026 \quad \sigma_x = 2.468841236$$

Conclusions: The aforementioned re-grouping does not have much effect on the basic average (it has been reduced by a negligible amount of 0.1%), but does significantly affect the variability, reducing the variance by 23%, and the corresponding value of the standard deviation by 12%. Hence, this re-grouping has far reaching consequences and raises a question of whether such a re-grouping does not constitute a *manipulation*. In a later part of this book the consequences of frequent practice resulting in changes to class ranges due to observed changes in the class frequency will be shown.

The student should be pleased that below, instead of using an elaborate solution based on statistics given in Tab. 2.7 (and Tab. 2.8), we rather use re-grouped statistics from Tab. 2.9 while presenting a new method of determining the required statistical averages. It is called the "coded method."

2.5 Coded Method

The essential evaluations of the new method are shown in Tab. 2.10.

Table 2.10 Coded method based on Tab. 2.9

x	f	U	Uf	UUf
32.5	1	7	7	49
30.5	2	6	12	72
28.5	2	5	10	50
26.5	2	4	8	32
24.5	8	3	24	72
22.5	26	2	52	104
20.5	67	1	67	67
18.5	171	0	0	0
16.5	271	-1	- 271	271
14.5	77	-2	-154	308
12.5	5	-3	- 15	45
---	632	---	- 260	1070

Averages obtained by the new method

$$\overline{U} = \frac{\sum U \cdot f}{\sum f} \rightarrow \overline{U} = \frac{-260}{632} \rightarrow \overline{U} = -0.411392405$$

$$\overline{U^2} = \frac{\sum U \cdot U \cdot f}{\sum f} \quad \rightarrow \quad \overline{U^2} = \frac{1070}{632} \quad \rightarrow \quad \overline{U^2} = 1.693037975$$

$$\sigma_U^2 = \overline{U^2} - \left(\overline{U}\right)^2 \quad \rightarrow \quad \sigma_U^2 = 1.523794264$$

<u>Conversion</u>

Linear transformation defines new entries:

$$U = \frac{x - R}{i} \qquad \text{or} \qquad x = i \cdot U + R \tag{2.35}$$

Basic mean:

$$\overline{x} = i \cdot \overline{U} + R \tag{2.36}$$

$$i = 2, \quad R = 18.5 \quad \rightarrow \quad \overline{x} = 17.67721519$$

Variance:

$$\sigma_x^2 = i^2 \cdot \sigma_U^2 \tag{2.37}$$

$$\sigma_x^2 = 6.095177056$$

<u>Short Description of the above Given Results</u>. It is seen that the basic mean and the variance calculated using this new method are almost identical with the results obtained using the direct method. Insignificant differences are due to the different rounding errors appearing in the course of both arithmetical procedures. The key idea lies in the appearance of a new variable U obtained by applying the linear transformation (2.35) regarding x the "old" variable. The choice of both parameters i and R of the linear transformation (2.35) is in part arbitrary, in part obligatory. The first simply marks *class range/interval* of the grouped data and *must be constant.* The choice of R is *arbitrary*. The choice of the row – in Tab. 2.9 – fixes the value of R. The structure of the data n Tab. 2.9 should obey the rule: *the lowest values are at the bottom of the table*. Then while filling the third column – it is done almost mechanically – by writing *integer numbers* – commencing from "zero" – upward for positive integers – and downward for negative values. Selecting the row containing "zero" – fixes the choice of parameter R – as it is seen in Tab. 2.9. Numerical practice advises to chose such a row for placing "zero" value of U which gives the smallest value for the total sum obtained for the next column. This value can be positive or negative. The smallest value of this sum simplifies further calculations.

2.6 Discussing Two Special Cases

1. The opening example examines some statistics from GUS, 1993. It is provided in Tab. 2.11 and has been chosen because of curious grouping applied by a professional statistician which we would like to comment.

Table 2.11 Statistical data – not recommended pattern

Family members	Family farm incomes per month and per person in thousands zlotys					Total
	600 & less	600 – 1 000	1 000 – 1 800	1 800 – 2 700	above 2 700	
1	–	–	27	75	64	166
2	1	35	243	257	170	706
3	13	116	557	292	106	1 084
4	41	384	734	175	17	1 351
5	47	229	247	20	5	548
6 & more	56	121	53	4	–	234
Total	158	885	1 861	823	362	4 089

Comments: grouped data shown in Table 2.11 contain 5 classes. Two of them do not specify the class limits – we can denote them as (?, 600) and (2700, ?). The first one due to its class frequency numbers about 4% of the entire statistics, the second one about 9%. Apart from that, the other three classes have three different intervals 400, 800 and 900 zł. In this way the rule advising to apply the same interval size for grouped data has been - without any visible reason - completely ignored. Therefore – the main purpose of the statistical data – their *quantitative nature* has been lost. The question: who it may concern – remains with no precise answer – those who do not care what they refer to?

Table 2.12 Age categories data, GUS 1993

Person age $(x_{i0} - x_{i1})$	Class interval (c_i)	Class frequency (n_i) thousands	%
0 – 2	3 (2)	1 581	4,1
3 – 6	4	2 343	6,1
7 – 14	8	5 353	13,9
15 – 19	5	3 068	8,0
20 – 24	5	2 601	6,8
25 – 29	5	2 498	6,5
30 – 44	15	9 297	24,2
45 – 64	20	7 645	19,9
65 – 90	(25)	4 032	10,5
	Total	38 418	100,0

2. The second example will require more of our attention – and it also comes from GUS sources. In fact more than in the data alone shown in Table 2.12 we are interested in some kind of a further re-evaluation.

Despite the critical character of this passage we quote the source of this re-evaluation as [12]. Grouped data presented in Table 2.12 reflect the status on 31 December 1992, *GUS Yearly Book, 1993*, p.46. It is possible to refer to the age categories shown in Table 2.12 by using some commonly understandable description: creche-age, nursery-school-age, school-age at primary and secondary stage, productive-age – in four sub-categories, and the past productive-age in a single last class. We should guess that these class intervals are of interest – at least for the people in charge of different districts of the country, although it may reflect an overestimation of such purposes. Let us have a look at the limits of the class intervals – to note, that we deal with *integers* – therefore to calculate class intervals we cannot use - so to speak - 'mechanical extraction' – as for instance - for the second row (6 – 3) does not result in 4 (and so on). The point concerns the first class: the left limit cannot be "0" – because "zero age" cannot be counted.

At least it is a "delicate point" which cannot be left without necessary comments. Also there is an open question of similar nature with respect to the last class – the discussion of which we leave for the Student. Here it may only be mentioned that the upper limit of this class defines the age of the eldest person in the country. To step one step further we enclose Table 2.13. It uses data given in Table 2.12.

Table 2.13 Statistics of population age in Poland 1992, GUS

x	f	xf	$x\,x$	$x\,x\,f$
77.5	4032	312480.0	6006.25	24217200.00
54.5	7645	416652.5	2970.25	22707561.25
37.0	9297	343989.0	1156.00	10747332.00
27.0	2498	67446.0	729.00	3079926.00
22.0	2601	57222.0	484.00	1258884.00
17.0	3068	52156.0	289.00	886652.00
10.5	5353	56206.5	110.25	590168.25
4.5	2343	10543.5	20.25	47445.75
1.5	1581	2371.5	2.25	3557.25
----	38418	1319067	-----	63538706.5

Results collected in Table 2.13 – have been used to derive the averages – mean, and variance. The Student should check the obtained below values by him/herself. They follow the so called *direct method.*

$$\overline{x} = \frac{1319067}{38418} \quad \rightarrow \quad \overline{x} = 34.33460878$$

$$\overline{x}^2 = \frac{63538706.5}{38418} \quad \rightarrow \quad \overline{x}^2 = 1653.87856$$

The last two values serve to derive both measures of variability:

$$\sigma_x^2 = \overline{x}^2 - \left(\overline{x}\right)^2 \quad \rightarrow \quad \sigma_x^2 = 475.0131995 \quad \rightarrow \quad \sigma_x = 21.79479753$$

It should be however noted that due to different class intervals the coded method simplifying calculations cannot be used. Nevertheless the above result does not close the case. As we already noted it is a book [12] where clever re-grouping of the data from Table 2.12 to data shown in Table 2.14 has been performed. To have material for commenting such an idea we once again obtained – also for this case - the major averages.

Table 2.14 Constant class intervals statistics

$x_i - x_{i+1}$	x	f	xf	xx	xxf
70-89(90)	80.0	3225.6	258048.00	6400.00	20643840.00
60 - 69	64.5	2717.7	175291.65	4160.25	11306311.43
50 - 59	54.5	3822.5	208326.25	2970.25	11353780.63
40 - 49	44.5	5010.2	222953.90	1980.25	9921448.55
30 - 39	34.5	6198.0	213831.00	1190.25	7377169.50
20 - 29	24.5	5099.0	124925.50	600.25	3060674.75
10 - 19	14.5	6413.7	92998.65	210.25	1348480.425
0 - 9	5.0	5931.4	29657.00	25.00	148285.00
------	-----	38418.1	1326031.95	----	65159990.29

$$\overline{x} = \frac{1326031.95}{38418.1} \quad \rightarrow \quad \overline{x} = 34.51581286$$

$$\overline{x}^2 = \frac{65159990.29}{38418.1} \quad \rightarrow \quad \overline{x}^2 = 1696.075295$$

$$\sigma_x^2 = \overline{x}^2 - \left(\overline{x}\right)^2 \quad \rightarrow \quad \sigma_x^2 = 504.7339576 \qquad \sigma_x = 22.46628491$$

Comments: The content of the numerical values given in Table 2.14 – together with the content which originates in book [12] will be examined below. It concerns the first three columns.

As we see, the intention was to get statistics, which preserves the constancy of all class intervals (except the highest class) by using non uniform intervals statistics given in Table 2.12. The prevailing class range counts 10 integers. A systematic check of the middle values given in the second column of Table 2.14 shows that they are right. It is not so easy to check the values in the third column.

Table 2.15 Density of the population birth/death processes

Person age $(x_{i0} - x_{i1})$	Class interval (c_i)	Class frequency (n_i)	Class density (g_i)
0 – 2	3	1 581	527,00
3 – 6	4	2 343	585,75
7 – 14	8	5 353	669,13
15 – 19	5	3 068	613,60
20 – 24	5	2 601	520,20
25 – 29	5	2 498	499,60
30 – 44	15	9 297	619,80
45 – 64	20	7 645	382,25
65 – 90	25	4 032	161,28
Total	-------	38 418	--------

The evaluation which we are going to examine was based on the assumption of uniform class intensity. An intriguing question comes up here: what are the reasons of such uniformity? There is – so to speak - an *interplay* between the intensity of the births and deaths. The final result is reflected by all class frequencies registered at the end of 1992. It was assumed, that for classes investigated in Table 2.12– each class intensity – given for convenience in Table 2.15 - certainly describes the entire period. Calculations provided to derive results given in Table 2.14 can be justified in this way. Let us have a look how *class frequency* for the *class range* (10 – 19) has been obtained and given in the second row from the bottom of Table 2.14 and equal to 6413.7. Assuming that *sub-class range* (10 – 14) – with 5 entries is characterized by the same *intensity* 669.13 as the intensity for the wider *class range* (7 - 14), then, further, taking *sub-class range* (15 – 19) containing also 5 entries having known the *class intensity* equal to 613.6 the unknown *class frequency* for the *class range* (10 – 19) can be derived in the following way:

$$5 \cdot 669.13 + 5 \cdot 613.6 = 3345.65 + 3068 \rightarrow 6413.65 \cdot$$

In a similar fashion - to derive the value in the first – the lowest row of Tab. 2.14 – appropriate calculations are: $3 \cdot 527 + 4 \cdot 585.75 + 3 \cdot 669.13 = 5931.39$. In the third step - to get class frequency for the highest row it is necessary to perform: $20 \cdot 161.28 = 3225.6$. This result corresponds to class range (70 – 89), which has 20 entries. To obtain a middle value equal to "20" - for this case (we do not say "class") it is advisable to replace class range (70 – 89) – by class range which includes the entry "90"- leading to (70 – 90). To check the remaining values we encourage the Student – our hint: they do not rise more doubts.

Final Conclusions - Results concerning averages – basic mean, and standard deviations - obtained following re-grouping given in Table 2.14 – are practically the same as results obtained following original data given in Table 2.13. Despite this important an annoying question may be posed here: does the above procedure not deserve to be called a *manipulation* of the statistical data? This question seems to be a difficult one – although the answer seen in the light of the obtained here results can be rather negative. Nevertheless in general – any intrusion into statistical data should be treated with justified suspicion. An accusation of any kind of manipulation of the statistical data is a serious one. The best way is to pay sufficient attention at the stage of Logistics while designing the project for a subsequent statistical investigation. It may eliminate temptations to abuse the already collected statistics – whether consciously or unconsciously.

2.7 Percentiles for the Grouped Data

The term "percentile" in fact denotes exactly the same as "percent" nevertheless – maybe due the fact that it sounds new, it appears in many text-books of Statistics (for instance [1] and [6]) being apparently originated by Francis Galton. The quotation from [1] may give a valuable definition:

>> If the values of the variable be ranged in order of magnitude, and a value P of the variable be determined such that a percentage p of the total frequency lies below it and $(100 - p)$ above, then P is termed *a percentile.* <<

Following [6] we also present *Percentile Ranks* – however, we moved the material to an appropriate place in Part II of this book. The concept of the *Percentile Rank* as we shall see it in Part II – presents an idea which for the given statistical data presents the same problem – but so to speak – from the other side. Looking at the arrows in Fig. 2.4 they have to be reversed.

Fig. 2.4 Determining the position of 60th Percentile – rough account

The numerical example shown in Fig. 2.4 follows data given in Tab. 2.3 and also shown in Fig. 2.3. Although the frequency histogram in Fig. 2.3 may serve as a final resort to express the percentile position, the cumulative histogram (or its appropriate fragment) – as shows Fig. 2.4 must be used for its numerical derivation. Data in brackets in the last column in Tab.2.6 have been used to draw Fig. 2.4. The problem requires: to determine position of 60-th Percentile – which we symbolize by P_{60}. Having in mind the value of the total frequency equal to 49 – we see that 60% of this value will be equal to 29.4. This quantity is identified by point A in both vertical scales of Fig. 2.4. Next we draw a horizontal line through this point – to intersect the line BD and derive the point C. Finally – projecting vertically this point C into horizontal scale line the position of P_{60} will be determined in this graphic way:

$$P_{60} \approx 15.9$$

Making use of a finer scale shown between numbers 13.22 and 16.88 - the numerical value shown above can be approximately read. This result is considered as a rough account of the required value. To get more accurate numerical result it is advised to proceed as indicated below:

1. The value to begin with serves the indicated above value obtained by $0.6 \cdot 49 = 29.4$ - the corresponding point A shows its position on both vertical scales;
2. The rectangular BGDH shown in Fig. 2.4 in measures 15 units of the "*physical*" scale vertical direction, so BG = 15; the *altitude of point* B determines "18" on the same scale; due to calculations shown above the vertical position of point F being the same as point A - equal to 29.4; therefore the length of the segment of BF results from (29.4 – 18) = 11.4 . With these results the ratio BF : BG is equal to the ratio 11.4 : 15. Applying Thales Theorem we find that ratios BK : BH = BF : BG = 0. 76 will be the same. From Fig. 2.4 it is clear that the segment BH has a length of

   (16.88 – 13.22) = 3.66,

   therefore distance BK will be equal to 0.76 BH which gives the value 0.76 * 3.66 = 2.7816 . Adding this distance to value 13.22 we shall get the value 16.0016 which gives the exact value of the desired percentile.

3. *Conclusion*: 60% of all unemployment figures is below 16% - while for the remaining 40% of the cases – it is the lowest rate of unemployment.

In the end of this passage it has to be pointed out that the geometrical procedure applied here is based on the *Theorem of Thales,* a Greek who opens the list of *seven sages* of the antiquity. In Greek alphabet his name is written so *Θαλῆς* - he is known to have lived from about 624 BC – to about 546 BC. He was a descendant of a rich Phoenician family from Miletus. The Theorem using his name is known round the world. But History of Mathematics does not ascribe this

Theorem to him – it is rather a symbolic tribute. In Poland it is usually taught at primary school.

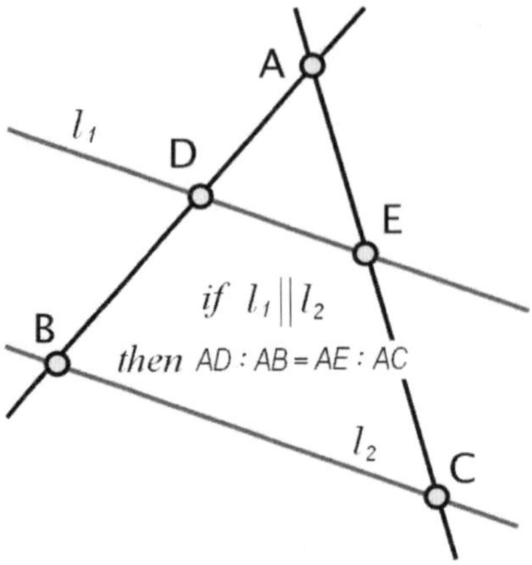

Fig. 2.5 Exposing the Thales's Theorem

As a matter of *curiosity* it may be said here that *Pergamon Museum* in *Berlin* – Germany has on display a re-construction of the original Miletus market square - although the Author of this book cannot say whether the monumental ruins found their way there thanks to Heinrich Schliemann (1822-90) – the famous explorer of Troy or thanks to some other German explorer?

References

[1] Yule, G.U.: An Introduction to the Theory of Statistics. Charles Griffin and Co., London (1911); 2-nd Edition translated into Polish by Z. Limanowski: Wstęp do Teorji Statystyki, Gebethner i Wolff, Warszawa 1921; pp. 1–446. VI-th Edition of 1922 accessible by Internet, pp. 1–415. 14-th edition, co-author M.G. Kendall, 1950 translated into Polish as Wstęp do Teorii Statystyki. PWN, Warszawa (1966)

[2] Laudański, L.M.: Dylematy jakości nauczania w epistolografii św. Pawła (Quality Dilemmas in Letters of St. Paul, Conference Proceedings) Materiały Konferencji Naukowej nt. Dylematy jakości kształcenia w uczelniach wyższych. Politechnika Rzeszowska, Rzeszów, pp. 111–121 (2008)

[3] Reichmann, W.J.: Use and Abuse of Statistics. First published by Methuen 1961, Published in Pelican Books since 1964, p. 345. Polish translation by Robert Bartoszyński (1933-1998): Drogi i bezdroża statystyki. PWN, Warszawa, pp. 1–395 (1968)

[4] Hawkins, C.A., Weber, J.E.: Statistical Analysis. Applications to Business and Economy, pp. 1–626. Harper & Row, New York (1980)

[5] Laudański, L.M.: Statystyka nie tylko dla licencjatów. In Polish: Statistics Not Only For Undergraduates, part1, part2, 2nd edn. Publishing House of the Rzeszow TU, Rzeszów (2009)

[6] Weinberg, G.H., Schumaker, J.A., Oltman, D.: Statistics – An Intuitive Approach, 4th edn., pp. 1–447. Brooks/Cole, Monterey (1981)

[7] Edwards, A.W.F.: Pascal's Arithmetical Triangle. The Story of a Mathematical Idea, p. 202. JHP, Baltimore (1987/2002)

[8] Gerstenkorn, T., Śródka, T.: Kombinatoryka i rachunek prawdopodobieństwa (In Polish: Combinatorics and Probability), 3rd edn. PWN, Warszawa (1976)

[9] Wilenkin, N.J.: Kombinatoryka, translated into Polish PWN, Warszawa (1972) (Original in Russian, Nauka, Moscow 1962)

[10] Flaschmeyer, J.: Kombinatoryka – podstawowy wykład w ujęciu mnogościowym, translated into Polish: Combinatorics, basic theory by using theory of set. PWN, Warszawa (1974) (original in German Kombinatorik, VEB, Berlin 1969)

[11] Diamond, S.: The World of Probability. Statistics in Science, p. 155. Basic Books, New York (1970) (Russian translation Statystyka, Moskwa)

[12] Makać, W., Urbanek-Krzysztofiak, D.: Metody Opisu Statystycznego (in Polish: General Statistics, Outline). Wydawnictwa Uniwersytetu Gdańskiego 1995 (2001)

Chapter 3
Regression versus Correlation

Idea of the linear regression. Auxiliary material - straight lines in Cartesian geometry. The concept of distance – geometrical distance and equivalent distances in two orthogonal directions – as a prerequisite to the concept of two regression lines. Frequent errors in interpreting two regression lines. The least square concept - Adrien Marie Legendre, Francis Galton, Karl Pearson [2]. *Formal derivation of the two regression lines. Correlation coefficient as a geometrical mean of the two directional coefficients of regression lines. Which in fact measures the correlation coefficient? Udny Yule* [1] *and his warnings regarding abuse of correlation analysis. Further reading* [3-4], [6-9].

3.1 Linear Regression – The Idea

The problem requires that we determine a straight line l *which best fits the relation between the two descriptive statistics* x_i *and* y_i *which have been combined into a single, two dimensional statistics made up of the couple* (x_i, y_i). At the turn of the 19th century formalism was developed (F.Galton, K. Pearson, U.Yule) which relates the concept of the best fitting line to the concept of the *least square* of A. M. Legendre. In this approach, the procedure meets the formal condition of the minimum value of the sum of the all squared distances between a set of the points (x_i, y_i) and the desired line l in quite a specific way. Below we present this approach describing its origins related to the concept of the geometric distance.

Distance between Two Points. In this place we have to recall a short reference to geometry from Chapter 1. After the problem of how to draw a line in coordinate geometry and of how its analytic equation is given are considered as resolved – we have to turn our attention to the problem of how the distance between a given point and given line is solved in this geometry. This problem can be reduced to the problem of determining the distance between two given points. Fig. 3.1 tells us to draw an auxiliary line perpendicular to the line l through given point P_1 - and

L.M. Laudański: Between Certainty and Uncertainty, ISRL 31, pp. 67–85.
springerlink.com

then – find the point P_2 - in which this auxiliary line (not named) intersects the line l. In this way the initial problem is equivalent to the problem of determining the length of a given segment – this time – segment $P_1\,P_2$.

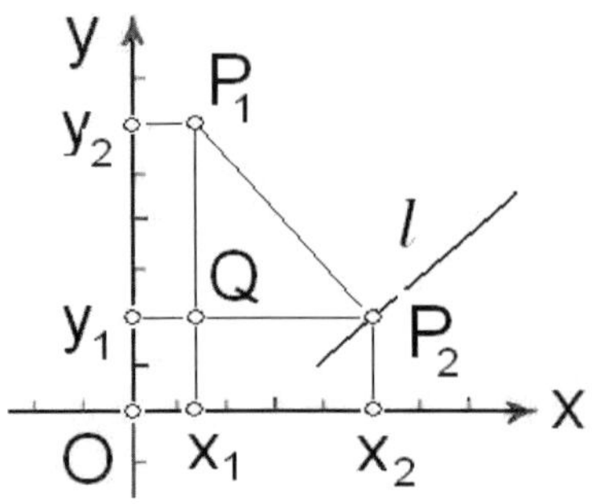

Fig. 3.1 Defining the distance between point P_1 and line l

Again with the help of Fig. 3.1 it becomes clear that the position of point P_1 is given by its coordinates (x_1, y_1), and similarly – the position of point P_2 - defines a pair of coordinates (x_2, y_2). Euclidean geometry defines the distance between two points as the shortest distance i.e. equal to the length of segment $P_1\,P_2$. To determine this value analytic geometry makes use of *Pythagoras Theorem* regarding the rectangular triangle $Q\,P_1\,P_2$ - by writing:

$$(P_1 P_2)^2 = (x_2 - x_1)^2 + (y_2 - y_1)^2 \quad \rightarrow \quad P_1 P_2 = \sqrt{(x_2 - x_1)^2 + (y_2 - y_1)^2} \tag{3.1}$$

Interestingly enough, the formula (3.1) regarding the considered problem was not applied in the regression analysis which eliminated the possibility of obtaining a single regression line. The procedure applied in statistical practice can be seen as the procedure which determines the distance between P_1 and P_2 taking advantage of the two definitions (3.2) and (3.3) given below (see Fig.3.1):

the distance defined by measuring it in x-direction, requiring that it is

$$x_2 - x_1 \tag{3.2}$$

the distance defined by measuring it in y-direction, requiring that it is

$$y_2 - y_1 \tag{3.3}$$

As a consequence of this approach – there are *two regression lines* - and neither of them can be considered superior with relation to the other. This is the first time where we take an opportunity to warn our Student not to be drawn into confusing explanations especially if using the Internet.

3.2 Regression Lines

In fact, the material shown in Fig. 3.2 precedes the topic currently discussed – nevertheless it may attract the Student's attention and push him/her to further studies. The statistical data in Tab. 3.was published by the US State Department to show community income for *Whites* and *Blacks* between 1980-96.

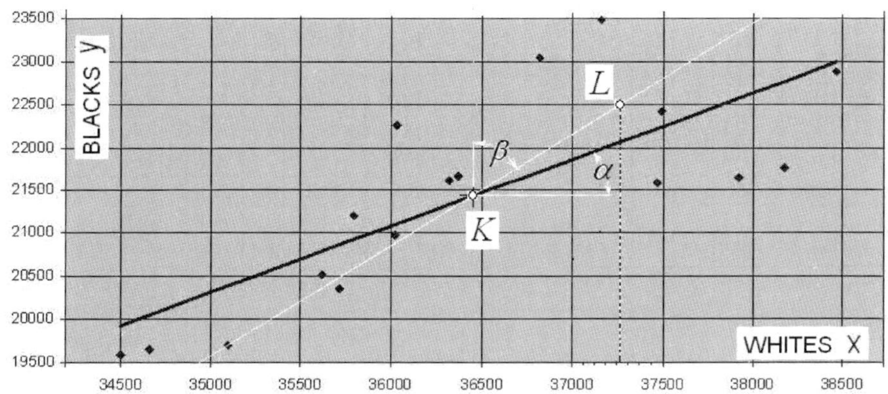

Fig. 3.2 Yearly income in the US – following Table 3.1

The first variable, and the second variable statistics can be understood as a descriptive statistics. But their association has to be considered as a two dimensional variable – which we propose to call statistics of the *first order dimension two*. The subject of this Chapter deals with such kind of statistical data to be given by the *ordered couples coordinates* (x_i, y_i).

Table 3.1 collects seventeen such *pairs,* while in Fig. 3.2 all such data is shown as *graphic data* – and due to conventions of *analytic geometry* – they present a set of points filling the plane **xy** . Moreover, between those points there are two lines – a *black* one, and a *white* one – which must be the already mentioned two *regression lines* – calling for an answer to the question of how they were determined. As we know "*there is no royal road to Mathematics*" and elucidation of the answer takes at least a few pages. Apart from that, in the beginning a purely symbolic character of these two groups of data may be noted – dividing them into "whites" and "blacks" due to their origin. To proceed further the Student has to be equipped with a pencil and a piece of paper – and like in every kind of poetry – he/she has to be a poet in his/her soul. Especially so keeping in mind an aspect of data in Table 3.1. – to which the usual *scientific calculator* with its 10 digits will be insufficient as the numerical results require at times 11 digits. The calculations included here were made using a calculator from more advanced Word versions.

Table 3.1 Yearly income in the United States

No	Year	x	y	x x	y y	x y
4	1980	35620	20521	1268 784 400	421 111 441	730 958 020
3	1981	35094	19693	1231 588 836	387 814 249	691 106 142
2	1982	34657	19642	1201 107 649	385 808 164	680 732 794
1	1983	34502	19579	1190 388 004	383 337 241	675 514 658
5	1984	35709	20343	1275 132 681	413 837 649	726 428 187
9	1985	36320	21609	1319 142 400	466 948 881	784 838 880
13	1986	37471	21588	1404 075 841	466 041 744	808 923 948
15	1987	37924	21646	1438 229 776	468 549 316	820 902 904
16	1988	38172	21760	1457 101 584	473 497 600	830 622 720
17	1989	38473	23000	1480 171 729	529 000 000	884 879 000
14	1990	37492	22420	1405 650 064	502 656 400	840 570 640
10	1991	36367	21665	1322 558 689	469 372 225	787 891 055
7	1992	36020	20974	1297 440 400	439 908 676	755 483 480
6	1993	35788	21209	1280 780 944	449 821 681	759 027 692
8	1994	36026	22261	1297 872 676	495 552 121	801 974 786
11	1995	36822	23054	1355 859 684	531 486 916	848 894 388
12	1996	37161	23482	1380 939 921	551 404 324	872 614 602
		619618	364446	22606 825 278	7836 148 628	13 301 363 896

We commence the calculations by determining the basic averages for single variables x and y.

Mean value for *whites*

$$\overline{x} = \frac{619618}{17} \quad \rightarrow \quad \overline{x} = 36448.1176470588235294\ 11764705882 \tag{3.4}$$

It is interesting to note the presence of a cycle of 16 digits in the above decimal extension.

Mean value for *blacks*

$$\overline{y} = \frac{364446}{17} \quad \rightarrow \quad \overline{y} = 21438 \tag{3.5}$$

is surprisingly an integer number! It may also be noted that Johan von Neumann (1903-57) was likely to consider number "17" as a first *true* prime number.[1]

[1] According to an American mathematician of Polish origin Mark Kac (1914-84) in a personal communication.

Inserting the above determined point ($\overline{x}$, $\overline{y}$) - in the plane (x, y) as seen in Fig.3.2 we conclude that the couple (36448, 21438) belongs to *both regression lines* – at their intersection – although for the time being there is no theoretical evidence to prove it. The point is called the *arbitrary point.*

As the first step towards determining *regression lines* an auxiliary geometric result from Chapter 1 has to be recalled. Due to it the first regression line will be given by:

$$y = A^{*} + B^{*} x \tag{3.6}$$

This line will be recognized as the ***black*** *regression line* - determined by the coefficients A^{*} and B^{*}.

To satisfy the intuition – it must be the closest line with respect to all 17 points of the statistics under consideration. From the formal point of view *the closest line* satisfies the *least squares condition* minimized the total squared distances of all the points referring them to the regression line. To explain all the required details we selected a single point $\left(x_1, y_1\right)$ - determined in Table 3.1 (see the first column) as point "1" with coordinates $(34502, 19579)$. We also selected the distance definition (3.3). Due to (3.3) the distance is measured along direction "*y*" – in other words - vertically . Therefore, it has to be recognized that coordinate y_2 points out the *black regression line* and according to (3.3) the x coordinate of this point belonging to the regression line – and denoted by x_2 has to be equal to x_1 so $x_2 = x_1$. Consequently, the unknown coordinates y_2 following (3.6) are determined by:

$$y_2 = A^{*} + B^{*} x_2 \tag{3.7}$$

Similar considerations are valid for the remaining sixteen points. This procedure leads to the distribution well known in Statistics – but not the subject of this course - called - χ^2 "*chi-square -distribution*". For our purposes it is enough to know that the final step in this procedure takes the form of a formal condition requiring to determine:

$$\min \chi_y^2 = \min \sum_{i=1}^{N} \frac{\left(y_i - A^{*} - B^{*} x_i \right)^2}{\sigma_y^2} \tag{3.8}$$

Condition (3.8) formalizes the above expressed *least square* requirement of finding the minimum of all the squared distances (using all the considered statistical points) between all the points and the (black) regression line. Let us look at how the purely mathematical way is developing further. Taking into account that the variance σ_y^2 is a constant value, condition (3.8) is equivalent to the simpler condition, below given by:

$$\min \chi_y^2 = \min \sum_{i=1}^{N} \left(y_i - A^* - B^* x_i \right)^2 \tag{3.9}$$

In order to satisfy (3.9) one must solve the problem of determining the minimum of an unknown linear function of two variables A^*, B^*. In the course of such a formal procedure two equations are derived which assure (3.9):

$$\frac{\partial \chi_y^2}{\partial A^*} \equiv -2 \sum \left(y_i - A^* - B^* x_i \right) = 0 \tag{3.10}$$

$$\frac{\partial \chi_y^2}{\partial B^*} \equiv -2 \sum x_i \left(y_i - A^* - B^* x_i \right) = 0 \tag{3.11}$$

In the last step, both equations (3.10) and (3.11) have to be solved simultaneously to give the values of the two coefficients A^*, B^*. Equations (3.10) and (3.11) arc a system of two linearly independent algebraic equations of the first order with respect to the two unknown coefficients of the regression line occurring in (3.6). Prior to the final formulae it is important to note that we expect to see those formulae as functions of the following quantities whose numerical values have been already determined in the last row of Table 3.1:

$$\sum x,\ \sum y,\ \sum x^2,\ \sum x \cdot y \quad \text{and} \quad N \tag{3.12}$$

The appearance in this list of symbol N - expressing *total frequency* - is the result of the following:

$$\sum_{i=1}^{N} A^* = N \cdot A^* \tag{3.13}$$

Let us comment briefly on how to solve equations (3.10) and (3.11) . By expanding the sums in both of these equations and then reordering results in a suitable way, we can derive as an intermediate stage the following two equations:

$$N \cdot A^* + \sum x_i \cdot B^* = \sum y_i \tag{3.10a}$$

$$\sum x_i \cdot A^* + \sum x_i^2 \cdot B^* = \sum x_i \cdot y_i \tag{3.11a}$$

The Student should note that the unknown coefficients in both equations remain outside the sums. This system of linear algebraic equations of the first order with respect to the unknown coefficients A^*, B^* can be solved by elimination and the final results given below can be derived:

$$A^* = \frac{\sum x^2 \cdot \sum y - \sum x \cdot \sum x \cdot y}{N \sum x^2 - \left(\sum x\right)^2} \tag{3.14}$$

$$B^* = \frac{N \cdot \sum x \cdot y - \sum x \cdot \sum y}{N \sum x^2 - \left(\sum x\right)^2} \tag{3.15}$$

Substituting the numerical values at the bottom of Table 3.1 into (3.14) and (3.15) gives after simple evaluations the following values for the coefficients:

$$A^* = -7180.9783952668169102631357930 94 \tag{3.16}$$

$$B^* = 0.78519770684443622921618369460312 \tag{3.17}$$

The obtained results (which for practical purposes can be rounded up to four digits i.e. to -7181 and 0.7852) call for a check with respect to the *black regression line* in Fig. 3.2. It can be done in the following way. Let us substitute into (3.6) the above rounded up values of both coefficients, together with the value $x = 34500$. Then the rounded up value for the corresponding y is:

$$y = 19910 \tag{3.18}$$

By inspecting Fig. 3.2 the coordinates (34500, 19910) evidently belong to the *black regression line* at its utmost left edge visible in the picture. The above check confirms satisfactorily that the below given approximate equation:

$$y = -\ 7181 + 0.7852\ x \tag{3.19}$$

can be understood as an analytical expression of the *black regression line* shown in Fig. 3.2. The above described procedure should also be considered as a guideline for the Student to follow independently to solve the problem of deriving the *white regression line*. The departing point of such a procedure would be the definition (3.2) giving the distance measured along the "*x*" direction. The final outcome should lead to the following equation:

$$x = A_* + B_*\ y \tag{3.20}$$

The initial step requires us to replace the condition (3.8) by the following one:

$$\min \chi_x^2 = \min \sum_{i=1}^{N} \frac{\left(x_i - A_* - B_*\ y_i\right)^2}{\sigma_x^2} \tag{3.21}$$

Accordingly, conditions (3.10) and (3.11) should be replaced by:

$$\frac{\partial \chi_x^2}{\partial A_*} \equiv -2 \sum \left(x_i - A_* - B_* \, y_i \right) = 0 \tag{3.22}$$

$$\frac{\partial \chi_x^2}{\partial B_*} \equiv -2 \sum y_i \left(x_i - A_* - B_* \, y_i \right) = 0 \tag{3.23}$$

While the replacement of the equations (3.10a) and (3.11a) by suitable ones is here left for the Student who should be able to check that the below given formulae expressing the desired coefficients of the *white regression line* are to be considered as proved:

$$A_* = \frac{\sum y^2 \cdot \sum x - \sum y \cdot \sum x \cdot y}{N \sum y^2 - \left(\sum y\right)^2} \tag{3.24}$$

$$B_* = \frac{N \cdot \sum x \cdot y - \sum x \cdot \sum y}{N \sum y^2 - \left(\sum y\right)^2} \tag{3.25}$$

Here we recommend that the Student compare formula (3.24) with (3.14) to recognize the possibility of deriving (3.24) by direct replacement of the symbols appearing in (3.14) in this way: every coordinate "x" has to be replaced by "y" and *vice versa.* A similar procedure is also true regarding the couple composed by formulae (3.15) and (3.25). Apart from this *rule of the thumb*, we consider this place as an opportunity to repeat our initial warning to the Student – to avoid a serious mistake in which such a *mirroring* may lead to misinterpretation of both regression lines. We shall explore this remark in detail somewhat later. Here we present numerical values for these new coefficients:

$$A_* = 19789.34786081568589514433196484 \tag{3.26}$$

$$B_* = 0.77706734705864062105921414036021 \tag{3.27}$$

At this point it has to be first stated that the graph of the *black regression line* was taken for granted when copying Fig. 3.2 from a student's work. Therefore to check the quality of the obtained coefficients A^*, *and* B^* we proceeded in the way presented above. The *white regression line* was drawn using the derived numerical values of the coefficients A_* and B_*. It is seen in Fig. 3.2 that in order to do that, an initial point L has been chosen. Its y coordinate has been arbitrarily fixed as y = 22500,- while its x coordinate has been derived by placing $A_* = 19790$ and $B_* = 0.7771$ into (3.20) together with the above chosen y coordinate - obtaining:

$$x = 37270 \tag{3.28}$$

In this way point L with coordinates (37270, 22500) has been determined as the second point belonging to the *white regression line* (3.20) . Therefore the *white regression line* has been finally drawn throughout points K and L - as seen in Fig. 3.2.

And now we recall the *initial warning* – and replace it by a *substantial* warning. To begin with it must be stated that it is quite common to come across remarks saying that the equation (3.6) determines the regression of y on x, while the equation (3.20) describes the regression of x on y. One of the possible consequences of this terminology would be a suggestion to solve equation (3.6) for variable x . Let us demonstrate what will happen in this case, it will lead to the equation:

$$x = 9145 + 1.273\,y \tag{3.29}$$

Though we already derived the equation of the *white regression line* which assigned (3.20) the following particular form:

$$x = 19789 + 0.7771\,y \tag{3.30}$$

Regarding equation (3.29) it must be stated that it is in fact another description of equation (3.19) determining the *black regression line.*

To close these considerations we will present proof that the point $(\overline{x}, \overline{y})$ denoted as "K" in Fig. 3.2 and called the *arbitrary point* - is the point of intersection of the two regression lines. We commence by recalling these two equations in their general form:

$$y = A^* + B^* x \tag{3.6}$$

$$x = A_* + B_* \, y \tag{3.20}$$

Applying to each of them the operation determining the mean value leads to:

$$\overline{y} = A^* + B^* \, \overline{x} \tag{3.31}$$

$$\overline{x} = A_* + B_* \, \overline{y} \tag{3.32}$$

The results (3.31) – (3.32) prove that the point $K(\overline{x}, \overline{y})$ belongs to both of them.

3.3 Arithmetical Appendix without Comments

8 238 967 045 265 988 – 8 241 764 494 511 728 = - 2 797 449 245 740

384 316 029 726 - 383 926 465 924 = 389 563 802

A^* = - 7180. 97839526681691026313579309 4

226 123 186 232 – 225 817 301 628 = 305 884 604

$$B^* = 0.78519770684443622921618369460312$$

$$x_2 = 34500$$

$$27089.32088613304990795833746380 8 - 7180.9783952668169102631357930 94$$

$$y_2 = 19908.342490866232997695201670714$$

$$4855418740584104 - 4847628866441616 = 7789874142488$$

$$133214526676 - 132820886916 = 393639760$$

$$A_* = 19789.347860815685895144331964 84$$

$$226\ 123\ 186\ 232 - 225\ \ 817\ 301\ 628 = 305\ 884\ 604$$

$$B_* = 0.77706734705864062105921414036021$$

$$17484.015308819413973832318158105 + 19789.347860815685895144331964 84$$

$$y = 22500 \qquad x = 37273.363169635099868976650122945$$

3.4 Correlation – Descriptive Statistics

Introductory Remark. The interplay between Statistics and Probability manifests its strength in different ways at different places. This remarks seems to be very much appropriate in this part of Chapter 3. Considering two dimensional random variables and defining the first mixed moment, leads, very naturally, to the correlation coefficient. Formally it has an analog in the basic mean for one dimensional random variable. Such a remark may throw some light on General Statistics which conventionally disregards Probability completely. What may be also noted is a hidden contradiction: authors of Statistics books are fully conscious of this situation, but the readers of their books – not at all! Such a situation frequently causes erroneous interpretations on one side and unjustified expectations on the other side. Let us try to be more specific: though on the basis of Probability, a two dimensional random variable appears as a well defined study object, on the basis of Statistics we commence with two one-dimensional statistics as given in Table 3.1 – without any formal reason to consider descriptive statistics of blacks y and whites x as the two dimensional entity (x, y). We know nothing at all about the possible two dimensional probability density of such a variable. Therefore, from the opening of Chapter 3, we are subject to a serious accusation which may be reduced to one word: WHY?! In fact this kind of objection also applies to considering the nature of variability of one dimensional statistics. Here we may recall the definition of Statistics given by Udny Yule who talks about "*the data affected to a marked extent by a multiplicity of causes*," which is responsible for the application of probabilistic tools. In order to arrive at some positive explanation, attention may be turned toward two-dimensional normal distribution: to determine it in a unique way it is required to know the mean, the variance, and the *correlation coefficient*. The above property can serve to justify, to some extent, the blind search for the entity from the title of Chapter 3. The objections expressed above were the same in case of Udny Yule and Ronald Fisher. The latter gave a famous example showing high correlation between imported apples and increasing divorce factor – and concluded that correlation does not imply causality.

Covariance is defined as follows:

$$\sigma_{xy} = \frac{1}{N}\sum_{i=1}^{N}(x_i - \bar{x})\cdot(y_i - \bar{y}) \tag{3.33}$$

The expression in (3.33) can be understood as a definition following the same chain of definitions as initiated in Chapter 1. To illustrate the determination of the numerical value of covariance due to (3.33), one can use the data given in Table 3.1. The Student can extend the content of Table 3.1, adding two more columns directly with the data required by (3.33). From covariance, determination of the correlation is straightforward:

$$r = \frac{\sigma_{xy}}{\sigma_x \cdot \sigma_y} \tag{3.34}$$

These two definitions can be complemented by known formulae giving variances of single statistics:

$$\sigma_x^2 = \frac{1}{N}\sum(x_i - \bar{x})^2 \tag{3.35}$$

$$\sigma_y^2 = \frac{1}{N}\sum(y_i - \bar{y})^2 \tag{3.36}$$

The most useful way to derive numerically the *correlation* is based on the definition attributed to Karl Pearson (1857-1936) which uses (3.33) and a set of inputs collected above as (3.12):

$$r = \frac{N\cdot\sum(x\cdot y) - \sum x \cdot \sum y}{\sqrt{N\cdot\sum x^2 - \left(\sum x\right)^2}\cdot\sqrt{N\cdot\sum y^2 - \left(\sum y\right)^2}} \tag{3.37}$$

Also from an efficient computing point of view – to determine variances instead of (3.36) and (3.37), the below formulae, also already known, are used:

$$\sigma_x^2 = \frac{\sum x^2}{N} - \left(\frac{\sum x}{N}\right)^2 \tag{3.38}$$

$$\sigma_y^2 = \frac{\sum y^2}{N} - \left(\frac{\sum y}{N}\right)^2 \tag{3.39}$$

Below particular steps leading to correlation by using (3.38) are provided. The denominator is given by:

$$226\ 123\ 186\ 232 - 225\ 817\ 301\ 628 = 305884604 \tag{3.40}$$

while the numerator is obtained as:

$$19737.370696219899492127156535353 * 19840.356851629458532076374041938 =$$
$$= 391596477.925896979076646269390 14 \tag{3.41}$$

Taking the ratio of (3.40) and (3.41) gives the desired correlation:

$$r = 0.78112194884930373689231902982605 \tag{3.42}$$

For the sake of completeness we also determine the values of both variances, and the corresponding standard variations:

$$\sigma_x^2 = \frac{\sum x^2}{N} - \left(\frac{\sum x}{N}\right)^2 = 132981325.64705882352941176470 59 +$$

$$- 1328465280.0138408304498269896194 = 1347971.6332179930795847750865052$$
$$\sigma_x = 1161.0218056599940877721856785502 \tag{3.43}$$

$$\sigma_y^2 = \frac{\sum y^2}{N} - \left(\frac{\sum y}{N}\right)^2 = 460949919.29411764705882352941176 - 459587844 =$$
$$= 1362075.2941176470588235294117647$$

$$\sigma_y = 1167.0798148017328548280220024669 \tag{3.44}$$

There is an interesting and important relationship between the values of the standard deviations for the two single statistics (i.e. σ_x and σ_y) on the one hand and the values of directional coefficients of the regression lines (i.e. B^* and B_*) on the other hand – both combined by the correlation r. They are given below:

$$B^* = r \cdot \frac{\sigma_y}{\sigma_x} \qquad B_* = r \cdot \frac{\sigma_x}{\sigma_y} \tag{3.45}$$

altogether these expressions lead to the apparently most interesting result :

$$r = \sqrt{B^* \cdot B_*} \tag{3.46}$$

Substitution into (3.46) of the already obtained numerical values for B^* and B_* gives the value:

$$r = 0.78112194884930373689231902982605 \tag{3.47}$$

which is exactly the same as (3.42) obtained via (3.37).

<u>Note</u> that the directional coefficients B^* and B_* of the two regression lines describe the tangential direction of those lines with respect to *two different coordinates*. The first coefficient B^* is the slope relative to the x direction and denoted by the angle α; the second coefficient B_* is the slope relative to the y direction and is denoted by the angle β - as shown in Fig. 3.2. Therefore the *geometric average* determined by (3.46), and equal to correlation r does not possess such a suggestive interpretation as it might at the first seem.

Demonstration that the second formula of (3.45), for instance, is right is provided by the check given below. Substitution of the values given by (3.43), (3.44) and (3.47) into (3.45), as shown below:

$$r\frac{\sigma_x}{\sigma_y} = 0.78112194884930373689231902982605\frac{1161.0218056599940877721856785502}{1167.0798148017328548280220024669}$$

leads to the final result:

$$B_* = 0.77706734705864062105921414036023 \tag{3.48}$$

which is identical with (3.27). A check of the remaining case has been left for the Student.

Closing this paragraph we would like to present the third way of defining *correlation*, which according to some biographical references was given for the first time also by Karl Pearson. This approach makes use of a definition introduced in Chapter 1 – namely, the linear transformation that leads to ***z-score*** statistics. Therefore, advising the Student to recall the above definition, it is also recommended to apply it to the two single statistics, which will be denoted by the symbols $z_x^{(i)}$ and $z_y^{(i)}$. Correlation becomes the mean value of their product:

$$r = \frac{\sum z_x^{(i)} \cdot z_y^{(i)}}{N} \tag{3.49}$$

Summarizing all the presented definitions of correlation, it may be repeated once more, that commencing evaluation of two single statistics in the way shown in Table 3.1, straightforward calculations leading to its numerical value are given by formula (3.37) despite its apparently discouraging appearance.

3.5 Correlation – Grouped Data

The opening example of grouped data presents the results published by John Houbolt ([5]), which it is proposed to name as *Houbolt's Cloud*. The specific character of this two dimensional statistic set is the absence of numbers, except the numbers provided for the two coordinate scales.

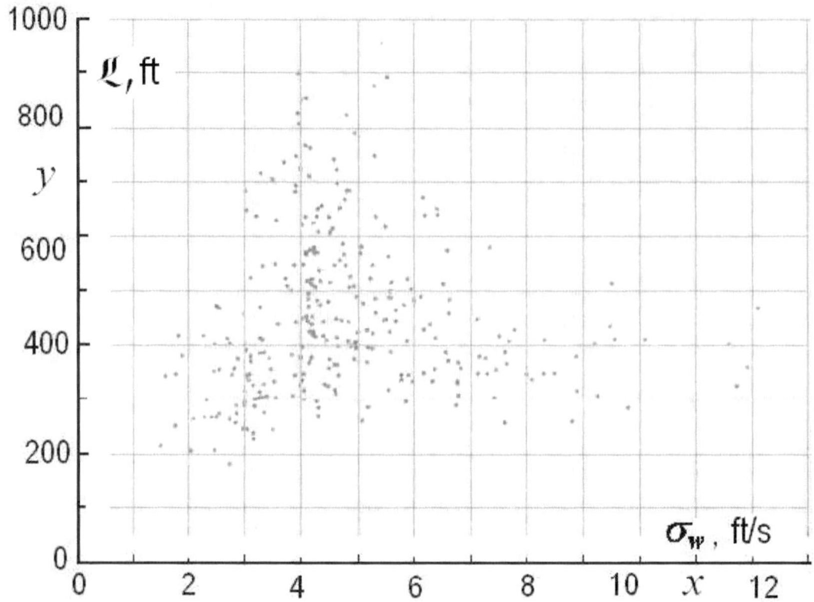

Fig. 3.3 Houbolt's Cloud – intensity versus scale of atmospheric turbulence

The problem before us is the same as that presented so far – to determine regression lines and correlation, but in these new circumstances. An apparent difficulty mentioned above and connected with a lack of numbers will be resolved in a way that is not seen in the literature on the subject. The entire solution is obtained with the help of Table 3.2 – which is proposed to be called

3.6 The Great Table of Correlation

The foregoing material is presented in order to instruct the Student how to perform such a new kind of calculations as collected in Table 3.2. The opening part of the evaluation is presented in the top left part of Table 3.2, a part which is also distinguished graphically by double-line frames. These data have been obtained by an appropriate use of the data given in Fig. 3.3, which deserve to be considered as raw statistical data. In this place we can briefly discuss the hypothetical case that would follow the situation known and experienced in Chapter 2

Table 3.2 The great table of correlation

	w_j	-5	-4	-3	-2	-1	0	1	2	3	4	5	6	$n_{i\mu}$	$u_i \cdot n_{i\mu}$	$u_i^2 \cdot n_{i\mu}$	$\sum_j w_j \cdot n_{ij}$	$u_i \cdot \sum_j w_j \cdot n_{ij}$
u_i	y_i \ x_j	1.5	2.5	3.5	4.5	5.5	6.5	7.5	8.5	9.5	10.5	11.5	12.5					
-3	150	-	1	-	-	-	-	-	-	-	-	-	-	1	**3**	9	**4**	12
-2	250	2	13	7	3	3	1	1	1	1	-	-	-	32	**388**	776	**86**	172
-1	350	3	5	19	17	9	10	8	6	1	-	2	-	80	**80**	80	**102**	102
0	450	1	6	5	28	12	7	4	1	3	1	1	1	70	**0**	0	**82**	0
1	550	-	-	7	21	9	2	1	-	1	-	-	-	41	41	41	**68**	**68**
2	650	-	-	6	15	2	4	-	-	-	-	-	-	27	54	108	**50**	**100**
3	750	-	-	4	7	1	-	-	-	-	-	-	-	12	36	108	**27**	**81**
4	850	-	-	2	2	2	-	-	-	-	-	-	-	6	24	96	**12**	**48**
5	950	-	-	-	-	1	-	-	-	-	-	-	-	1	5	25	**1**	**5**
$n_{\mu j}$		6	25	50	93	39	24	14	8	6	1	3	1	270	**311**	1243		**16**
$w_j \cdot n_{\mu j}$		**30**	**100**	**150**	**186**	**39**	0	14	16	18	4	15	6	**432**	$\overline{w}$ \ $\overline{u}$	$\overline{u^2}$		
$w_j^2 \cdot n_{\mu j}$		150	400	450	372	39	0	14	32	54	16	75	36	1638	$\overline{w^2}$			
$\sum_i u_i \cdot n_{ij}$		**7**	**34**	6	57	14	**2**	**9**	**8**	**2**	–	**2**	–			**Bold** fonts denote negative numbers		
$w_j \cdot \sum_i u_i \cdot n_{ij}$		35	136	**18**	**114**	**14**	–	**9**	**16**	**6**	–	**10**	–	**16**				

as given, for instance, in Table 2.7, showing grouped data of one-dimensional statistic. Here the procedure should be simply doubled – adding the grouped data to some other one-dimensional variable. And now, examining the suggested part of Table 3.2, we see the two statistics represented by the grouped data are shown simultaneously, due to the fact that they are interrelated. The point now is how the values have been obtained. As they are grouped data, there are routine questions regarding the two variables – first about their class limits and their number of classes, and after that about their class frequencies. Here attention must be turned towards Fig. 3.3 which suggests to use class intervals equal to $\Delta\mathfrak{L}=100\,ft$ for the scale of turbulence, and to use class intervals equal to $\Delta\sigma_w = 1\,ft/s$ for the intensity of turbulence. These intervals are also identified in the content of Table 3.2. In the table it is seen that the columns y_i define the middle values for all the classes with respect to the scale of turbulence, while the rows x_j define the middle values for all the classes with respect to the gust intensity (which is identified with the variance of the gust velocity). The Student should also recognize the customary system of units with respect to length, and velocity. Regarding the class frequencies – which take a substantial part of the table - they are determined simply by the number of dots lying inside each particular square shown in Fig. 3.3. So the effort of finding the numbers boils down to counting to determine the number of dots belonging to each particular "square" appearing in Fig. 3.3. The remaining part of Table 3.2 results from a suitable manipulation of the data determined in the above way, which we are going to describe in more detail.

The following comments will describe each result presented in Table 3.2 – step by step. Columns y_i and $n_{i\mu}$ present middle values and class frequencies for the scale of turbulence, in two appropriate columns. Regarding the symbol $n_{i\mu}$, the first subscript i is the index which changes its values along the column – indicating the number of the appropriate row, commencing from the top. The subscript μ remains a "*dummy index*" occupying the place which in the active case refers to the column number (column numbers increase from left to right). Similarly, rows x_j and $n_{\mu j}$ give the middle values and corresponding class frequencies of the intensity of turbulence. It should be noted that the total sum for the column $n_{\mu j}$ and the row $n_{i\mu}$ give the same numerical value, which is equal to $n = 270$ – where n denotes the total frequency of raw statistics data (appearing as dots in Fig. 3.2). This property can be written in a symbolic form as:

$$n = \sum_{i=1}^{N} n_{i\mu} = \sum_{j=1}^{M} n_{\mu j} \qquad N = 9,\ M = 12 \tag{3.50}$$

The next two steps lead to averages: the means and variances for each of the single statistics. As can be seen in Table 3.2 the procedure follows the coded method. For this reason *coded variables* for the scale of turbulence are given in the column u_i, while for the turbulence intensity they are given in the row w_j. To determine the coded mean of the scale of turbulence, one must use the total sum of the column $u_i\, n_{i\mu}$,- then the necessary evaluation is completed as follows:

$$\sum_{i=1}^{N} u_i\, n_{i\mu} = -311 \qquad \bar{u} = -\frac{311}{270} \Rightarrow -1.151851(851) \tag{3.51}$$

For the turbulence intensity the total sum of the row $w_j\, n_{\mu j}$ is required, leading to:

$$\sum_{j=1}^{M} w_j\, n_{\mu j} = -432 \qquad \bar{w} = -\frac{432}{270} \Rightarrow -1.6 \tag{3.52}$$

To determine the physical value of the basic average for the scale of turbulence $\bar{\mathfrak{L}}$, the Student should check further details presented below, given in three distinct steps:

$$\bar{\mathfrak{L}} = \bar{u}\, i_u + R_u \tag{3.53}$$

$$i_u = 50 \qquad R_u = 450 \tag{3.54}$$

$$\bar{\mathfrak{L}} = \bar{u}\, i_u + R_u = -1.151851(851) \cdot 50 + 450 \rightarrow \bar{\mathfrak{L}} \cong 392\, ft \cong 120\, m \tag{3.55}$$

A very similar procedure will lead to the value of the mean intensity which in customary units shows 4.9 ft/s, and when converted into SI units gives almost exactly a velocity of 1.5 m/s.

Another step towards determining numerical values for the coded variances of both variables will be described roughly in order to encourage the Student to make an effort to verify the other entries. The efficient numerical procedure requires us to extract from the mean square value the squared mean value. The Student has to decipher how the first two denoted by $\overline{w}^2$ and by $\overline{u}^2$ are calculated in Table 3.2. It will help him/her to check the following numerical results regarding the two *coded* variances:

$$Var\,w = \overline{w}^2 - (\overline{w})^2 \qquad \overline{w}^2 = \frac{1638}{270} \qquad \rightarrow \qquad Var\,w = 3.5006(6) \tag{3.56}$$

$$Var\,u = \overline{u}^2 - (\overline{u})^2 \qquad \overline{u}^2 = \frac{1243}{270} \qquad \rightarrow \qquad Var\,u = 3.276941016 \tag{3.57}$$

As the last step, the procedure of deriving true physical values of variances (and standard deviations) of the turbulence scale and its intensity has to be conducted. This is shown below:

$$Var\,\mathfrak{L} = i_u^2\,Var\,u \qquad \rightarrow \qquad \sqrt{Var\,\mathfrak{L}} \approx 90.5116155\,ft \cong 27.6\,m \tag{3.58}$$

$$Var\,\sigma_w = i_w^2\,Var\,w \qquad \rightarrow \qquad \sqrt{Var\,\sigma_w} \approx 1.870990663\,ft/s \cong 0.571\,m/s \tag{3.59}$$

If we are going to retain the same order as in the previous paragraph, we must now turn our attention towards the regression lines. And here the Student faces another challenge. The procedure given above to determine directional coefficients has used formulae (3.15) and (3.25), while below we will find the formulae:

$$B_1 = \frac{E(uw) - \overline{u}\,\overline{w}}{Var\,u} \qquad\qquad B_2 = \frac{E(uw) - \overline{u}\,\overline{w}}{Var\,w} \tag{3.60}$$

So before making use of this procedure it is desired to examine the validity of this approach – proving the equivalence of formulae (3.60) and (3.15) - (3.25), and this is left to the Student. In order to trace all the steps of this rather long procedure, it is necessary to note that formulae (3.60) make use of the coded variables, while the previously examined (3.15) and (3.25) operate with physical variables of the scale of turbulence and atmospheric turbulence intensity. Assuming that the above procedure has been successfully completed, full attention can be turned to (3.60), commencing with the derivation of the average $E(uw)$ which may be done in both of the following ways:

$$E(uw) = \frac{1}{n}\sum_{j=1}^{M} w_j \sum_{i=1}^{N} u_i\,n_{ij} = \frac{1}{n}\sum_{i=1}^{N} u_i \sum_{j=1}^{M} w_j\,n_{ij} \tag{3.61}$$

The numerical results taken from Table 3.2 lead to the desired average:

$$E(uw) = -\frac{16}{270} = -0.059259(259) \tag{3.62}$$

Substituting (3.62) together with (3.51), (3.52), (3.56) and (3.57) into both formulae of (3.60) gives, after simple calculations, the following values for the directional coefficients of the two regression lines:

$$B_1 \cong -0.580487171 \qquad B_2 \cong -0.542458808 \tag{3.63}$$

Together with these values the correlation will give the following result:

$$r = \sqrt{B_1 \cdot B_2} \quad \rightarrow \quad r \cong -0.561150941 \tag{3.64}$$

Before we explain why the correlation in this example is *negative*, which does not follow in a unique way from formula (3.46), we first propose to determine both regression lines. The obtained analytical expressions for the two of them will justify the conclusion. This opportunity will be used to demonstrate a procedure that omits derivation of the *free terms* A_1 and A_2. The point is that the two regression lines have a common point with coordinates given by $\left(\bar{\sigma}_w, \bar{\mathfrak{L}}\right)$ - the so called *arbitrary point.* Therefore substituting the coordinates of this point into the general formula expressing the regression line - either (3.6) or (3.20) – together with the appropriate value of the directional coefficient given by (3.63) will result in deriving the free terms A_1 or A_2. Following the described procedure gives:

$$y = 395.2517946 - 0.580487171\, x \tag{3.65}$$

$$x = 217.7648545 - 0.542458808\, y \tag{3.66}$$

The Student is advised to continue these considerations by trying to locate the regression lines in Fig.3.3 – first the regression line given by (3.65) . In order to do that the *arbitrary point* $K\,(4.9,\,392)$ has to be marked in Fig. 3.3. Then one has to derive any other point belonging to this regression line, such as for $x = 10$. Approximating the y coordinate gives $N\,(10,\,389.5)$. Therefore, it is seen that the first regression line, given by (3.65), slightly declined downward remains almost parallel to the coordinate x. In other words it is almost a horizontal line. Then one has to determine the second regression line, given by (3.66). It should also contain the *arbitrary point K.* To determine another point belonging to this line, one can choose again the same $x = 10$, which leads to the point $M\,(10,\,383)$. Therefore, it is clear that the second regression line is declined slightly more than the first, but with the scales of Fig. 3.3 these two regression lines are almost identical. Moreover, these numerical results confirm that the correlation coefficient is negative, which means that with increasing values of turbulence intensity the corresponding values of the turbulence scale decrease. Due to physical circumstances this result seems to be justified. We may add, in closing, that the quoted paper of Houbolt [5] has no such information and even erroneously suggests a positive correlation among these variables. Ending this study, though, it must be added that the absence of the regression lines in Fig. 3.3 is justified by the fact that they would overshadow the original data.

Data shown in Table 3.3, copied from Udny Yule book [1], present an example published in 1903 by Karl Pearson in Vol. 2 of *Biometrica* [10], analyzing the stature of father and son, in connection with Galton's results which gave rise to the term "regression". This example is considered in a detail in Part II.

Table 3.3 Correlation between Stature of Father and Stature of Son [1], [10]

(2) Stature of Son.	(1) Stature of Father.																	Total.
	58·5–59·5.	59·5–60·5.	60·5–61·5.	61·5–62·5.	62·5–63·5.	63·5–64·5.	64·5–65·5.	65·5–66·5.	66·5–67·5.	67·5–68·5.	68·5–69·5.	69·5–70·5.	70·5–71·5.	71·5–72·5.	72·5–73·5.	73·5–74·5.	74·5–75·5.	
59·5–60·5	—	—	—	—	·5	·5	1	—	—	—	—	—	—	—	—	—	—	2
60·5–61·5	—	—	—	—	·5	—	—	—	1	—	—	—	—	—	—	—	—	1·5
61·5–62·5	—	·25	·25	—	·5	1	·25	·25	·5	·5	—	—	—	—	—	—	—	3·5
62·5–63·5	—	·25	·25	2·25	2·25	2	4	5	2·75	1·25	—	·25	·25	—	—	—	—	20·5
63·5–64·5	1	—	1·5	3·75	3	4·25	8	9·25	8	1·25	1·5	·75	1·25	—	—	—	—	38·5
64·5–65·5	2	1	·5	2	3·25	9·5	13·5	10·75	7·5	5·5	3·5	2·5	—	—	—	—	—	61·5
65·5–66·5	—	·5	1	2·25	5·25	9·5	10	16·75	17·5	18	5·25	2	2·5	1	—	—	—	89·5
66·5–67·5	—	1·5	2	4·75	3·5	13·75	19·75	26·5	25·75	19·5	12·5	13·75	3·25	·5	1	—	—	148
67·5–68·5	—	—	1·5	2	7·5	10	10·25	24·25	31·5	23·5	29·5	13·25	8·5	9·5	2·25	—	—	173·5
68·5–69·5	—	—	1	—	5·25	5	12·75	18·25	16	24	29	21·5	10	8·5	2·25	—	1	149·5
69·5–70·5	—	—	—	—	1	2·5	5·75	18·75	11·75	19·5	22·5	19·5	14·5	6·25	3·5	1·5	1	128
70·5–71·5	—	—	—	—	—	3·25	5	8·75	10·75	19	14·75	20·75	10·75	8	5	1	1	108
71·5–72·5	—	—	—	—	—	·25	3	1·25	7	7·75	10·75	11·25	10	8·5	2·75	·5	—	63
72·5–73·5	—	—	—	—	—	—	·75	·75	2·5	7·5	8·5	6	7·5	6·25	3·25	·5	·5	42
73·5–74·5	—	—	—	—	1	—	1·5	1·5	—	5·25	2·25	2·5	6·5	3·25	3·25	—	2	29
74·5–75·5	—	—	—	—	—	—	—	—	—	1	2	—	2·5	·75	1·75	·5	—	8·5
75·5–76·5	—	—	—	—	—	—	—	—	—	1·25	·25	—	·5	1	1	—	—	4
76·5–77·5	—	—	—	—	—	—	—	—	—	1·25	·25	1	—	—	1·5	—	—	4
77·5–78·5	—	—	—	—	—	—	—	—	—	—	1	1	—	·25	·75	—	—	3
78·5–79·5	—	—	—	—	—	—	—	—	—	—	—	—	—	·25	·25	—	—	·5
Total	3	3·5	8	17	33·5	61·5	95·5	142	137·5	154	141·5	116	78	49	28·5	4	5·5	1078

References

[1] Yule, G.U.: An Introduction to the Theory of Statistics. Charles Griffin and Co., London (1911); 2-nd Edition translated into Polish by Z. Limanowski: Wstęp do Teorji Statystyki, Gebethner i Wolff, Warszawa 1921; pp. 1–446. VI-th Edition of 1922 accessible by Internet, pp. 1–415. 14-th edition, co-author M.G. Kendall, 1950 translated into Polish as Wstęp do Teorii Statystyki. PWN, Warszawa (1966)

[2] Stigler, S.: The History of Statistics. Harvard University Press, Belknap (1986)

[3] Weinberg, G.H., Schumaker, J.A., Oltman, D.: Statistics – An Intuitive Approach, 4th edn., pp. 1–447. Brooks/Cole, Monterey (1981)

[4] Fisher, R.A.: The design of experiments. Oliver & Boyd, Edinburg (1935)

[5] Houbolt, J.C.: Atmospheric Turbulence. AIAA Journal 11(4), 421–437 (1973)

[6] Laudański, L.M.: Statystyka nie tylko dla licencjatów (in Polish: Statistics not only for the undergraduates), Part 1 and 2, 2nd edn. Publishing House of the Rzeszow TU (2009)

[7] Soper, H.E.: On the Probable Error of the Correlation Coefficient to a Second Approximation. Biometrica IX, 91–115 (1913)

[8] Fisher, R.A.: Frequency Distribution of the Values of the Correlation Coefficient in Samples from Indefinitely Large Population. Biometrica X, 507–521 (1915)

[9] Fisher, R.A.: The Goodness of Fit of Regression Formulae and the Distribution of Regression Coefficients. J. Royal Statistical Society 85, 597–612 (1922)

[10] Pearson, K., assisted by Lee, A.: On the Laws of Inheritance in Man. Biometrica 2(4), 357–462 (1903)

Chapter 4
Binomial Distribution

Mysterious origins of the binomial distribution: binomial theorem (binomial coefficients) and combinatorial rules. Pascal's Arithmetical Triangle, Bernoulli's Trials. [John Arbuthnott's contribution]. Acquaintance with the binomial distribution – numerical examples, drawings of the distributions [their properties]. Jacob Bernoulli's Weak Law of Large Numbers. How to derive Poisson-Bortkiewicz distribution. Famous example from the chronicles of the Prussian Cavalry [Ladislau von Bortkiewicz's contribution: the law of small numbers]. Negative binomial, a revival of old ideas.

4.1 Tracing the Origin

To Begin with: H. Poincare allegedly said that the normal distribution must have something mysterious in itself if mathematicians consider it as a law of Nature, while physicists consider it as a mathematical theorem. Yet, as it will be shown in Chapter 5 devoted to normal distribution, its origins, despite having a variety of components, seem transparent. However, the origins of the binomial distribution are not like that. The proof of such an opinion has to be presented gradually, requiring some formal elements. It is interesting to add that neither Stigler in his very erudite "*History of Statistics*" [1], nor Edwards in his witty monograph [2] devoted to "*Pascal's Arithmetical Triangle*" – gave the origins in a complete form. Therefore, concerning the *distribution* – we are left with what is given in [2] and in Chapter 9 – entitled *The binomial and multinomial distributions* (strongly inclining the balance towards *multinomial* distribution). In the opening p. 112: "Pascal evidently possessed this form for $p = ½$ in 1654, though we must allow for the fact that he thought in terms of *expectations* rather than *probabilities*". The Author of this book was not able to go back before 1654. Nevertheless the Author is unable to resists a temptation to correct an erroneous opinion brought by Majstrow [3] – which attributes the binomial distribution to Jacob Bernoulli. This remark – based on the publication of a complete English translation of "*Ars Conjectandi*" [4] is firmly justified, and there is no reason to trace the origins of this error. Therefore, with these remarks presented openly to the Student, it is now time to decide which way of introducing *binomial*

L.M. Laudański: Between Certainty and Uncertainty, ISRL 31, pp. 87–127.
springerlink.com

distribution is to be chosen. Having in mind a very impressive paper by Arbuthnot [5] which is now easily accessible from JSTOR resources we propose to follow his approach. The Student is advised to at least look into the original paper.

Arbuthnot: As we shall see the *dramatis persona* of this passage are not Arbuthnot, the personal physician of queen Anna, because it is *Newton binomial* which we insert in this place. Beside this imaginary character, there will be another one, called *probabilistic experiment,* involving the idea of tossing a coin. With time it started to be referred to as *Bernoulli's trials*. Turning to practicalities, let us call the head *success S* and the tail *failure F.* It is quite natural to give them equal chances – therefore the chance value ½ appears in this account. Tossing a coin we get either "S" or "F". These results can be described as:

$$\text{S} \qquad \text{F} \tag{4.1}$$

Now, what will happen if the coin is tossed twice? Also this time without recalling any combinatorial rules, the result can be easily described/listed as:

$$\text{S, S} \qquad \text{S, F} \qquad \text{F, S} \qquad \text{F, F} \tag{4.2}$$

The only "agreement" here concerns the order – in the first position comes the result of the first tossing, and in the second place, the other result. Moreover, it follows from equal *chances* that each of the four above described compound results has the same chance of occurring equal to ¼ due to the fact that we have *four.* If we would now like to follow Arbuthnot and his paper [5] we have to use a universal formalistic description proposed by him. The method he used is easier than the above given procedure, which with an increasing number of tosses will present increasing technical difficulties in listing all the possible cases and their chances, even though it retains the highest readability of the final outcome.

Following Arbuthnot we have to replace (4.1) and (4.2) by the following two descriptions: :

$$\left(S+F\right) \tag{4.3}$$

$$\left(S+F\right)^2 = S^2 + 2S\,F + F^2 \tag{4.4}$$

This was the first step towards *binomials* - an ambiguous, multi-faceted term whose meaning has to be carefully considered in the context where it appears. The hint to the Student is to direct his/her attention towards an appropriate interpretation of coefficients expressions (4.3) and (4.4) noting an isomorphism between them - i.e. the unique correspondence, between descriptions (4.1) and (4.2) on the one hand and (4.3)-(4.4) on the other. For a Student less capable of abstract thinking the correspondence between (4.1) and (4.3) should be easier to digest, but to acknowledge the same correspondence (rule) between (4.2) and (4.4) it is advisable to re-write (4.4) in the form:

$$\left(S+F\right)^2 = S\,S + 2S\,F + F\,F \tag{4.4a}$$

According to Edwards [2], the above presented *isomorphism* was familiar to the Hindu mathematician Bashkara in twelfth-century India. Edwards presents (4.3) and (4.4) in modern terminology as

$$(pz + q)^n \tag{4.5}$$

calling (4.5) the *generation function* of the binomial distribution – with z being a *dummy variable*. Following a long tradition, symbols p and q denote *chances* of S and F respectively. To proceed one step further in the desired direction the following problem must be stated:

Determine the chances that tossing a coin n times leads to exactly k successes (4.6)

First we have to examine how the approach based on formulae (4.1)-(4.2) allows us to solve problem (4.6). For the time we assume equal chances for success and for failure. Taking $n = 1$ - the results are as follows: for $k = 0$ - ½ and for $k = 1$ - also ½ . Taking $n = 2$, we derive, for $k = 0$ - result ¼ , "for $k = 1$" - result ½ and for $k = 2$ - result ¼ . The above listed solutions are obtained due to the procedure that can be called "*direct inspection*". Let us now examine the solution of the problem (4.6) by using isomorphic procedure (4.3)-(4.4). To begin with, it must be seen that the symbols "S" and "F" stand for the *chances* of success and failure – that is p and q - which in this case are equal and have the value ½ . Therefore S^2 determines $k = 2$ - "double success" – so its chance becomes $\left(\frac{1}{2}\right)^2 = \frac{1}{4}$. The remaining terms of the expanded binomial (4.4) – should follow the same rule.

Here is a proposal to solve the case $n = 4$, which describes the case of tossing a single coin four times or of tossing four coins once. Let us make use of the second procedure, i.e., one based on (4.3)-(4.4) – by expanding the binomial as below::

$$(S + F)^4 = S^4 + 4\,S^3F + 6\,S^2F^2 + 4\,S\,F^3 + F^4 \tag{4.7}$$

The Student is advised to justify the results given in Table 4.1.

Table 4.1 Binomial distribution n = 4

Successes k	0	1	2	3	4
Chances	$\left(\frac{1}{2}\right)^4$	$4\left(\frac{1}{2}\right)^4$	$6\left(\frac{1}{2}\right)^4$	$4\left(\frac{1}{2}\right)^4$	$\left(\frac{1}{2}\right)^4$

Gradually proceeding in the presented way should lead us to the conclusion that we are looking for a generalization that should supply us with a general pattern where for instance (4.7) becomes just a particular case. And here we can still profit from the following Arbuthnot. The desired general formula given by him is enclosed below:

$$S^n + \frac{n}{1}S^{n-1}F + \frac{n}{1}\frac{n-1}{2}S^{n-2}F^2 + \frac{n}{1}\frac{n-1}{2}\frac{n-2}{3}S^{n-3}F^3 + \cdots \quad (4.8)$$

Nevertheless we may still expect more with respect to the coefficients of the binomial, which from this place onward we shall call also the *Bernoulli numbers.* They can be given most concisely according to the following definition:

Newton' Symbol defined for natural n, k gives the formula:

$$\binom{n}{k} = \frac{n!}{k!(n-k)!} \quad \text{for} \quad 0 \le k \le n \qquad n \ge 0 \qquad (4.9)$$

Making use of the above formula the most concise expansion formula for the binomial gives:

$$(S+F)^n = \sum_{k=0}^{n} \binom{n}{k} S^{n-k} F^k \qquad (4.10)$$

Fully expanded (4.10) has the form (4.11):

$$(S+F)^n = \binom{n}{0} \cdot S^n \cdot F^0 + \binom{n}{1} \cdot S^{n-1} \cdot F^1 + \binom{n}{2} \cdot S^{n-2} \cdot F^2 + \ldots + \binom{n}{m} \cdot S^{n-m} \cdot F^m + \ldots + \binom{n}{n} \cdot S^0 \cdot F^n$$

For the sake of completeness one has to define the *factorial* which in the simplest form presents the recurrent expression:

$$n! = n \cdot (n-1)! \quad \text{supplemented by} \quad 1! = 1 \quad \text{and} \quad 0! = 1 \qquad (4.12)$$

With (4.12) in mind, (4.9) leads to the following particular results:

$$\binom{0}{0} = 1 \quad \binom{n}{0} = 1 \quad \text{and} \quad \binom{n}{n} = 1 \qquad (4.13)$$

Pascal : Below the Student will find other details from Edwards' [2] book with regard to the distance in time necessary to take trace the essentials of the title story. Edwards proposed three groups of ideas while examining *Pascal's Arithmetical Triangle* . The oldest Edwards called *figurate numbers*. This is also the title of Chapter 1 which he opens with the phrase (see p.1, [2] where we replaced the suitable number of bibliographical reference):

>> *The longest of the threads which Pascal wove together in his [6] concerns the figurate numbers, and stretches back to the Pythagorean preoccupation with number-patterns 540 years before Christ.* <<

To explain the concept of the *figurate numbers* let us make use of them in relation to the numbers associated *with the figure of a triangle.* Here is an initial list of them: 1, 1 + 2 = 3, 3 + 3 = 6, 6 + 4 = 10, 10 + 5 = 15, which also suggests the way of deriving them. But there is the fundamental relationship:

$$f_2^{\,l} = f_2^{\,l-1} + l \qquad l = 1, 2, 3, \cdots \tag{4.14}$$

According to their geometric origin we present Fig. 4.1 which uses a pictorial way of defining them:

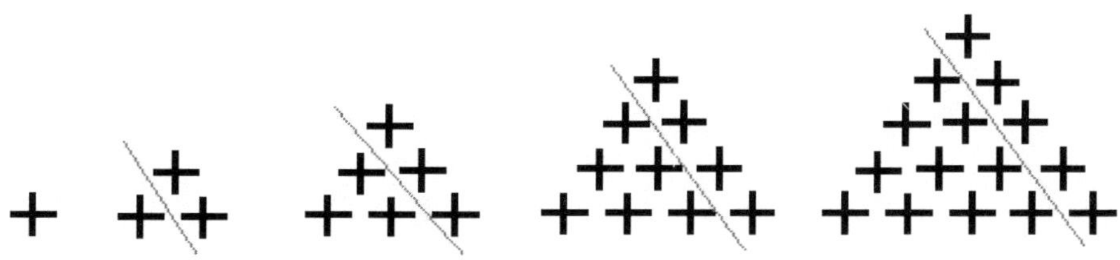

Fig. 4.1 Five initial triangle numbers

After making this first acquaintance which enables to list these numbers, our attention should turn to Pascal's arithmetical triangle in order to detect these numbers. Probably even for the Student who already knows Pascal's triangle this step will be surprising.

The exposition of the triangle shown in Fig. 4.2 allows us to see among the rows "*Rangs paralleles*" the *triangle numbers*, (displaying the initial 8 of them) in the third row. The rich content of this figure shows unusual mathematical objects, among them also *Bernoulli numbers*, but we suggest approaching them gradually. Keeping in mind that some of the Students may have heard of the renowned *Fibonacci numbers* (taking us back to the very beginning of the 13th century and to Italy) we recall their definition, to avoid confusion with the recurrence form of (4.14):

$$F_n = F_{n-1} + F_{n-2} \qquad F_0 = 0 \quad F_1 = 1 \tag{4.15}$$

Going back again to Fig. 4.2, its 4th row presents the so called *pyramidal* or *tetrahedral* numbers (Edwards gives Theon and Nicomachus as their inventors) denoted by $f_3^{\,l}$ - their fundamental definition is given by:

$$f_3^{\,l} = f_3^{\,l-1} + f_2^{\,l} \quad f_3^{\,1} = 1 \tag{4.16}$$

Finally we give the fundamental definition of the *figurate numbers* of all possible orders and note that it is possible to see them in the lower rows of the *triangle*::

$$f_k^{\,l} = f_k^{\,l-1} + f_{k-1}^{\,l} \quad f_k^{\,1} = f_0^{\,l} = f_0^{\,1} = 1 \quad \text{here } l = 2, 3, 4, \ldots \; k = 1, 2, 3, \ldots \tag{4.17}$$

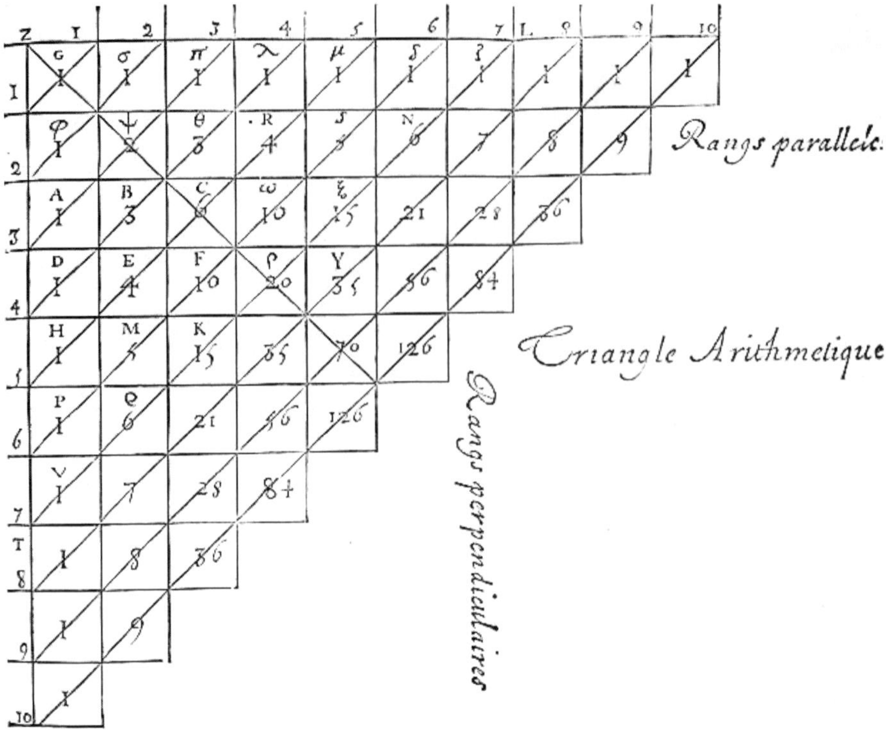

Fig. 4.2 Pascal's Triangle copied from his original Treatise [6]

Apparently the following step in this direction becomes the *solution* of the recurrence equation in (4.14) leaving aside its proof. The formula below may be assigned as the property of the *triangle numbers* (*figurate numbers* of dimension two):

$$f_2^l = \tfrac{1}{2} l \left(l + 1 \right) \tag{4.18}$$

Here the relation between *Pascal's triangle* and *Stifel's Triangle* (dated closely to AD 1544) may be mentioned as presented in Fig. 4.3.

It is seen in Fig. 4.2 that the *triangle numbers* (4.14) are given in the second *column*, then the numbers following (4.16) in the third column, and the subsequent columns give numbers that can be derived by using a general – fundamental relationship to generate *figurate numbers* of arbitrary order of (4.17). It is also seen that the missing initial numbers can be derived according to (4.17).

Reading chapters on the history of mathematics (see for instance [7]) regarding *Fibonacci numbers* one finds confirmation of the so called *Stigler's law*. They date back a few centuries before Leonardo from Pizza, known as Fibonacci, though the object called *Stifel's triangle* is attributed to Stifel – about whom Wikipedia tells us what is quoted below:

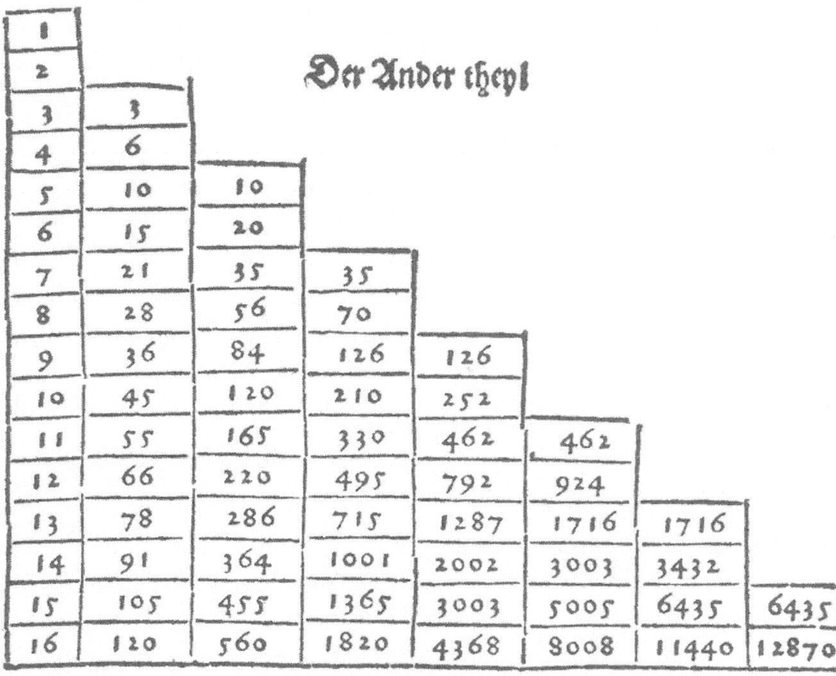

Der Ander theyl

1							
2							
3	3						
4	6						
5	10	10					
6	15	20					
7	21	35	35				
8	28	56	70				
9	36	84	126	126			
10	45	120	210	252			
11	55	165	330	462	462		
12	66	220	495	792	924		
13	78	286	715	1287	1716	1716	
14	91	364	1001	2002	3003	3432	
15	105	455	1365	3003	5005	6435	6435
16	120	560	1820	4368	8008	11440	12870

Fig. 4.3 Stifel's version of Figurate Triangle (corrected copy Fig. 9 from [2])

Michael Stifel *or* ***Styfel*** *(Esslingen 1486 or 1487 – April 19, 1567, Jena) was an Augustinian monk who became an early supporter of Martin Luther and was later appointed professor of mathematics at Jena University.*

At this point let us quote two results which Edwards [2] attributes to Stifel:

$$f_l^{\,l} = f_{l-1}^{\,l+1} \qquad {}^{n}C_r = f_r^{\,n-r+1} \qquad (4.19) \text{ and } (4.20)$$

but the explanation of their meaning will be given later on in the right context.

A new element of these investigations related to *Pascal's triangle* will be introduced with a story from Chapter 2 of Edwards' book [2], a chapter entitled "*Three combinatorial rules*". According to the story, commenting on Aristotle's logical *Categories* by writing *Isagoge* (comments), the author, a Greek philosopher Porphyry (234-c.305) included there the combinatorial problem - asking "*in how many ways can two things be chosen from* n *different things*"*?* Edwards describes thoroughly Aristotle's context, but we, trying to "*get to the point*" limit our report to the fact, that there was the particular case of $n = 5$. Therefore, combining the desired pairs, if in the first step we take the first "thing", we can choose 4 pairs; then if we take the next "thing", we can obtain only 3 new pairs; in the next step we can obtain 2 new pairs – and in the last step only 1 new pair; therefore: 4 + 3 + 2 + 1 = 10. And this correct answer was obtained by

Porphyry[1] in this way. In conclusion, taking into account that the correct result was obtained not by enumeration, but by following a combinatorial procedure, Edwards suggests that we may attribute to Porphyry the discovery that the number of unordered pairs which can be chosen from n things is given by the $(n-1)^{th}$ triangle number – or in our notation as f_2^{n-1} . Not so long after that, the work of Pappus (a Greek mathematician living in Alexandria ca. 300 AD) who was looking for a solution of the geometric problem concerning n intersecting lines with restrictions leading to the same problem as solved by Porphyry, left firm evidence that he not only knew that the result which can be written in the form

$$1 + 2 + 3 + 4 + \ldots (n - 1) = \tfrac{1}{2} n (n - 1) .$$

Pappus also says: "*it is unlikely that Euclid was ignorant of this [generalization]*". The Student who can sacrifice time and efforts to study the book by Edwards will find one more famous name in the context of the considered matter: it was - Anicius Manlius Severinus Boethius (ca. AD 480 – 23 October 524) - who translated *Isagoge* from Greek into Latin and who was familiar with the fact that $f_2^{n-1} = \tfrac{1}{2} n (n - 1)$, connecting this combinatorial rule with the triangular numbers. Therefore, this is the right place to connect the facts given in Chapter 2 regarding combinatorial rules with the triangular numbers discussed here by recalling formula (2.33) defining combinations ${}^{n}C_{r}$. It is interesting here, in particular, to consider the case with $r = 2$. And here is what we can get this way:

$$ {}^{n}C_{r} = \frac{n!}{r!(n-r)!} \equiv \binom{n}{r} \qquad {}^{n}C_{2} = \frac{n!}{2!(n-2)!} \rightarrow \tfrac{1}{2} n (n-1) \tag{4.21}$$

Also the above justifies the result (4.14) therefore, it can be written that:

$$ {}^{n}C_{2} = f_2^{n-1} \quad \rightarrow \quad \tfrac{1}{2} n (n-1) = f_2^{n-1} \tag{4.22}$$

In other words we have proved the *isomorphism* between *figurate numbers* and *combinatorial numbers* – specifically between *pairs* arranged by *n different things/objects* and the *triangular numbers* showing that (4.18) and (4.22) are identical. The second formula of (4.22) can be understood as the solution of the recurrence equation (4.14). But we cannot provide the details of such a procedure.

The final conclusion regarding the above is: the numbers seen in *Pascal's Arithmetical Triangle* apart from their *figurate* and *binomial* interpretation can be interpreted on *combinatorial grounds* .

And now we cannot resist the temptation to enclose two other combinatorial rules known in the ancient times despite the fact that they have no reference to the *arithmetical triangle* under discussion.

---

[1] On the Internet one can find an English translation of "*Isagoge*" – Porphyry in commenting Chapter 11 by Aristotle presents the above in its original form, curiously enough, he did not use the fonts describing numbers!

The second *ancient* combinatorial rule states that: $2^n - 1 - n$ = ${}^nC_2 + {}^nC_3 + \cdots + {}^nC_n$, therefore it describes *combinations* managed from n *things different* taking into account *permutations* of *two, three* … up to n things. Below is an illustrative example based on the initial letters of the Latin alphabet:

$n = 2$	1	ab	
$n = 3$	4	ab ac bc abc	(4.23)
$n = 4$	11	ab, ac, ad, cd, bd, cb, abc, abd, bcd, adc, abcd	

The third *ancient* combinatorial rule (and the oldest one) and defines the number of *arrangements* i.e. permutations of n things from a given *set* - in a form known from Chapter 2 as:

$$n! \qquad (2.30)$$

Tracing the possible origins, Edwards lists a number of sources taking us back to year 300 BC and ancient India using an example considering $n = 6$.

To comfort the Polish Student a few books are listed containing combinatorial analysis. [8]-[10]. The most accessible seems to be [9], but the broadest background is to be found in the Polish translation of a Russian book by Wilenkin. This book is entirely devoted to combinatorial analyses and presents a average level of mathematical treatment. To add a remark referring to the material listed here, it can be noted that the *second ancient rule* cannot be traced in any of these books (only implicitly appearing in [10], a book which was originally written in German). For English-speaking Students perhaps it is still Feller's book [25] which can be seen as a major reference. The book has also been translated into other languages (we know its Polish and Russian translations). We also list here [23], a small book in which its Authors especially devoted a lot of attention to *binomial numbers*. And this particular reference brings us also to the closing element regarding comments presented here relating to *Pascal's Arithmetical Triangle* – which we largely based on the book by Edwards [2], this closing element are *Bernoulli's numbers* which we also called *binomial numbers.*

Bernoulli's numbers – referring to Jacob Bernoulli and his "*Ars Conjectandi*" [4] are here considered as the third and closing interpretation of the content of the *arithmetical triangle* which is majestically shown in Fig. 4.2. Complementing this figure we present also its different orientation rotated *clockwise* by 45 degrees in Fig. 4.4 which allows a straight way of reading the binomial numbers.

Seemingly the configuration shown in Fig. 4 – giving *binomial coefficients* as numbers to be read horizontally and giving the initial binomials commencing with the power "zero" at the top of the figure - looks most natural. Especially with respect to the arithmetical rule which allows us to build the entire triangle in the unquestionably simplest way by using the property given below for which the formal description seriously overshadows the simplicity of the idea which it expresses:

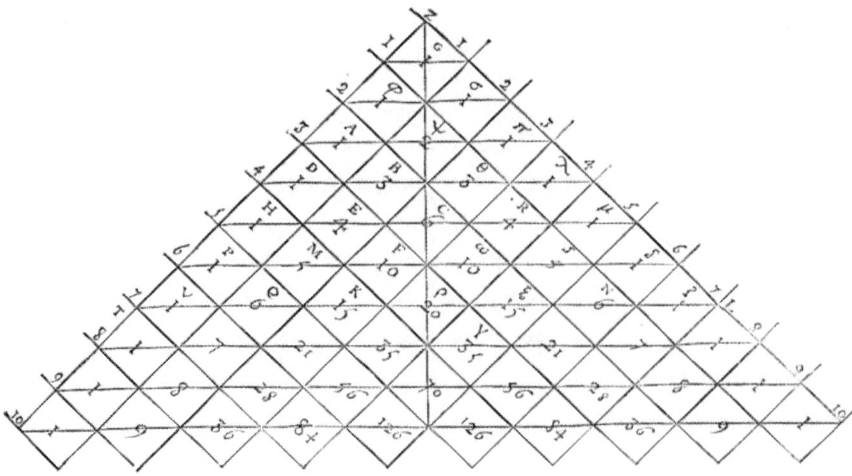

Fig. 4.4 Pascal's triangle exposing Bernoulli's numbers

$$\binom{n}{r} = \binom{n-1}{r-1} + \binom{n-1}{r} \quad \text{or} \quad {}^{n+1}C_{r+1} = {}^{n}C_{r} + {}^{n}C_{r+1} \tag{4.24}$$

To avoid misunderstanding, it must be mentioned that if intending to fulfill the entire triangle, from the top to the bottom, the simplest way follows the property (4.24); however, if intending to determine any particular *Bernoulli number* it is simpler to use the *combination formula* presented in this book first in Chapter 2 as (2.33), then recalled in the beginning of this chapter as (4.9) and later as (4.21). Considering that the auxiliary material focusing on historic origins of binomials has been exhausted, we come towards a modern approach regarding binomial distribution – an opportunity to start is offered by:

Bayes : Two years after the death of the Reverend Thomas Bayes (1702-1761) his paper [15] was published and it may up to now, in its most important part, be considered, so to say, the baking powder of further development of the concept of "*inverse probability*", and the origin of Monte Carlo method. Following Edwards' suggestive idea ([2], p.112-113) our attention will be directed to a passage from Bayes [15] - SECTION I, PROP. 7 - p.7-8. This passage will be the *faithfully quoted* original paper [15] apart from the symbols which will be replaced by our symbols used in this book and except an error which appears in the last line of the PROP.7:

If the probability of an event be "p", and that of its failure be "q" in each single trial, probability of its happening "k" times, and failing "n - k" times in "n" trials is $E\ p^k q^{n-k}$ *if E be the coefficient of the term in which occurs* $p^k q^{n-k}$ *when the binomial* $(p+q)^n$ *is expanded.*

In a comment to the above mentioned error of the original paper [15] - it seems to be most likely a kind of a printing error, as the correct symbol "*p*" was replaced by the wrong symbol "*b*". The enclosed text of *PROP. 7* presenting once again the binomial formula is now flawless.

With respect to the above definition of the *binomial,* it is seen that the symbol "*E*", the only symbol preserving the original Bayes' symbols, determines *Bernoulli's number* which in our notation is given by *Newtonian symbol* (2.19). Also with respect to the same, it may be seen that Bayes offers the solution to our problem (4.6). And changing the point of view once more it is justified to acknowledge that for Bayes the formula expressing the binomial distribution had the contemporary meaning commonly used now. The same conclusion can be found in Edwards' book [2]. Therefore we conclude that the binomial distribution is expressed in one of the two following ways:

$$ {}^{n}C_{k}\ p^{k}q^{n-k} \tag{4.25}$$

$$ \binom{n}{k}\ p^{k}q^{n-k} \tag{4.26}$$

Regarding the possible period of time when Bayes could have written [15], and keeping in mind that he spent the last 10 years of his life in complete isolation devoted entirely to religious contemplation, and so this period could not have been a time of possible mathematical activities, the approximate time of its genesis could have been 1750. The Author of this book cannot say anything about the possibilities of deriving formulae (4.25) – (4.26) by anyone else, rejecting as we already have the claim of Majstrow [3] who mistakenly attributes it to Jacob Bernoulli.

Terminology. Coming to the end of these considerations opening Chapter 4 let us focus our attention on the matter of terminology. Apparently this aspect of the matter is the last which can be explained in a satisfactory manner by referring to numerous books on the history of mathematics therefore, we provide just a few remarks which do not pretend to exhaust the subject. A careful Student of this course will be, as is the Author of this book, moved by the care of Bayes' [15] in this respect, so accurately described by his friend, Richard Price in his letter announcing [15] to John Canton, Fellow of the Royal Society. Bayes' paper opens Section 1 with a sequence of seven definitions. This may on the one hand remind one of Euclid's "Elements" and on the other of contemporary efforts in determining *axiomatic approaches* to different branches of mathematics, logics, and natural sciences in a view expressed by David Hilbert in his famous lecture addressed to mathematicians of the twentieth century. So, we shall point out a few remarks related to the above. It impossible for the Author of this book to say whether Bayes' was the first to use the term *event* now so frequently applied in mathematics and probability, designating *randomness* and now replaced by *random variable*. Bayes starts by defining events *inconsistent,* and events mutually *contrary*. He uses *determined* and *happening* instead of the word *occurred* which we would use today. Then similarly to the concept given later by Laplace he defines the *probability of the event* . And finally come his two last definitions. Def.6: "*By chance I mean the same as probability.*" Def.7 "*Events are independent when the happening of any one of them does neither increase nor abate the probability of the rest*".

In close relation with Bayes the above will be complemented by some biographical remarks evoked by papers and books appearing now in print. Let us start with a paper by Stigler [14] – whose suggestive title may cause wrong associations suggesting that the term "*chance*" has been used over the last 350 years! Contrary to this suggestion, most likely the author wrote the title with *marketing* in mind and it can be explained by his confession in a private letter that "*Titling is a dark art.*" Paper [14] becomes an evident tribute paid to Christian Huygens (1629-95) 350 years after his "*De Ratiociniis in Ludo Aleæ*" was published (exactly in 1657). In Latin of Huygens' paper there is no English term of "*chance*". As proof that the Author of this book is highly appreciative of historic publications by Professor Stephen M. Stigler – the Student shall find in the literature references two other papers by Stigler: [22] and [13]. Especially [13] presents little known facts regarding an English mathematician Thomas Strode (c.1620 - c.1690). Coming towards the end this section on terminology which is usually unfairly ignored, we propose one more term - "*distribution*" – short for "*probability distribution*". There are reasons to believe that this term appeared simultaneously with the term *random variable*. The latter originated in mid twentieth century. Looking for evidence we can briefly examine such a renowned book as [12] by James Victor Uspensky (1883-1947) who before leaving his native Russia became a doctoral student of A. A. Markov (1856-1922) at Sankt Petersburg University. The book [12] has 16 chapters, more than 400 pages and the concept of "*distribution*" is presented for the first time in Chapter XIII –"*The General Concept of Distribution*"(see p.260-282). This chapter is entirely devoted to considerations developing the concept which uniformly treats discontinuous and continuous variables. It will be reasonable to add a remark that such an idea was investigated by a Dutch mathematician Thomas Joannes Stieltjes (1856-1895) – who was quite an extraordinary person - a kind of a self-educated scientist who never obtained a university diploma (started Polytechnical School in Delft in 1873 – failed a few examinations and in 1876 definitely quit the university). The mentioned concept is widely known as Stieltjes' Integrals. Returning to the point, it has to be said that the term *random variable* – does not appear in this book – but instead Uspensky uses the term *stochastic variables* – which can be found earlier in Chapter IX "*Mathematical Expectation*". It persuades us to add a brief remark that the term *random variable* some later years could be found among the titles of books on probability, one of the first being a book by Athanasios Papoulis [20] – (a hard back copy in Poland of 1965 cost exactly 382.50 zlotys – or 20% of junior university lecturer's salary) – it was evidently a marketing sign indicating that: "*in my book this term is used*". For comparison let us take a book by a less known author who 20 years later followed the same idea (see [21]). Also for comparison and to finish this remark - Uspensky's paperback reprinted by the editor from his own publication of 1937 – at the same period of time – i.e. 1965 cost 88.5 zlotys. The black market exchange rate of the Polish zloty to USD was about 1 to 100.

4.2 Close Acquaintance

This is going to be the most practically useful part of Chapter 4. In its general view it is similar to Uspensky's [12] . But the real content of this sub chapter the Student is advised to reread after finishing this course. The very introductory pages will again present to the Student some historical episodes recalling S. Pepys, I. Newton, and F. Weldon. After making intensive use of calculators to collect particular examples of binomial distribution – the next passage will lead to the weak law of great numbers by Jacob Bernoulli. Then following De Moivre, an opportunity will be given to re-examine his method leading from binomial to the normal distribution. The material presented below –besides own calculations - follows Stigler [22], and is complemented by [24].

4.3 Three Problems by S. Pepys [22], p.400-401

A. Six fair dice are tossed independently and at least one "6" appears.

B. Twelve fair dice are tossed independently and at least two "6" s appear.

C. Eighteen fair dice are tossed independently and at least three "6" s appear.

A. $P(X \geqslant 1) = 31031/46656 = 0.665$
when $N = 6$ and $p = 1/6$.

B. $P(X \geqslant 2) = 1346704211/2176782336 = 0.619$
when $N = 12$ and $p = 1/6$.

C. Here Newton simply stated that, "In the third case the value will be found still less."

$P(X \geqslant 3) = 60666401980916/101559956668416 = 0.597$
result of George Tollet – contemporary of Pepys and Newton

At the top are problems which Samuel Pepys sent to Isaac Newton, asking him which one had the greatest chance. Then, below come numerical results quoted by Stigler [22]. Later on detailed solutions will be presented of all three problems obtained as well "*by direct enumeration of cases*" with the second one based on applying the binomial distribution. In fact Stigler did not say explicitly that the enclosed numerical results denoted as "*A*" and "*B*" were derived by Newton – instead he said: "*Newton worked from first principles assuming no knowledge of the binomial distribution*" what indirectly attributes them to Newton. Definitively it confirms [24]. Stigler does not comment the first approach *by direct enumeration of cases* while regarding the second one he states that "*solution as might be presented in an elementary class today*". In view of the Author's long teaching experience with Management students at Polish universities not studying for a degree in mathematics, his opinion is much more restrained.

Bernoulli trials Intending to solve the first problem posed by S. Pepys we start by defining the title entity understanding the term as such a *random trial* which has two outcomes – one denoted by "*S*" with probability "*p*" and the other denoted by "*F*" with probability "*q*" – where $q = 1 - p$. Following traditional terminology – the symbol "*S*" means "*success*", while "*F*" stands for "*failure*". This experiment may be repeated an arbitrary number of times – preserving independence of the all repetitions and the constancy of probabilities p and q.

To commence with the first problem by S. Pepys[2] let us briefly define the game of dice which according to the above defined *Bernoulli trial* - is formally described by the assumption that with the above symbols $n = 6$, $p = 1/6$ and $q = 5/6$ - and we are looking for the probability that the event "*S*" will appear at least once. This requirement defines the number of events "*S*" formally determined by the values of k equal to: 1, 2, 3, 4, 5, 6. Equivalently, the above probability is equal to the probability that the opposite event does not appear. The opposite event to the above describes a single event for which $k = 0$ - and its probability may be evaluated directly from (4.26) :

$$\binom{6}{0}(1/6)^0[5/6]^6 = [5/6]^6 = 0.334897976680384087791495198902 61$$

Therefore the desired probability will be 0.66510202331961591220850480109739.

Also by dividing 31031/46656 the same result 0.66510202331961591220850480109739 is obtained.

Let us examine more closely the solution obtained *"by direct enumeration of cases"*. The total number of all possibilities for throwing the die six times will be given by $6^6 = 46656$. On the other hand – the number of *non-six* appearances which corresponds to $k = 0$ - will be equal to $5^6 = 15625$ - therefore the number of events "*favorable*" for the appearance of the die-face showing "6" will be given by 46656 – 15625 = 31031. Their ratio describes the desired probability according to Laplace's definition of probability. These results can be commented by pointing out that although the second procedure seems to be less sophisticated, nevertheless it gives an *exact* final result – while the first solution gives only an *approximate* result.

Similar procedures are provided below with respect to the second problem defined by $n = 12$ and $p = 1/6$. At some point in the first solution – probabilities for $k =$ 0 and $k = 1$ have to be determined - with the results given below:

$$\binom{12}{0}(1/6)^0[5/6]^{12} = [5/6]^{12} = 0.11215665478461508427087861117227$$

[2] Samuel Pepys (1633-1703) an English naval administrator and Member of Parliament who is now most famous for his diary – adding for the benefit of the Polish Student - that it was translated into Polish by Maria Dąbrowska, further bibliographical details are to be found on the Internet.

$$\binom{12}{1}(1/6)^1[5/6]^{11} = 2\,[5/6]^{11} = 0.269175971483076202250108666681344$$

0.112156654784615084270878611117227 + 0.269175971483076202250108666681344 =

0.381332626267691286520987277979857

Probability of the opposite event: 0.618667373732308713479012722020143.

The second method results in:

1346704211 / 2176782336 = 0.618667373732308713479012722020143

As it was done above, let us examine the second method *"by direct enumeration of cases"*. And now the total number of the all possibilities for throwing the die 12 times will be given by $6^{12} = 2176782336$. Also the derivation of the all appearances corresponding to $k = 0$ will give $5^{12} = 244140625$. Apparently the new pattern will lead to derivation appearances for $k = 1$ in a view $\binom{12}{1}5^{11} = 585937500$. But in fact the previous case also corresponds to $\binom{12}{0}5^{12} = 244140625$ - having in mind that $\binom{12}{0} = 1$. Both considered cases give the total of 830078125 – therefore the *opposite events* –after subtracting them from overall total appearances determine *"favorable"* appearances as 1346704211. It may be noted by the way that the value of the coefficient $\binom{12}{1} = 12$ can be derived either from (4.9) or from Pascal's triangle. The last method was common in Newton's times (see [24]).

For Pepys' third problem it is defined by $n = 18$ and $p = 1/6$ - requiring to determine probabilities for $k = 0$, $k = 1$ and $k = 2$ - they are given by numerical results:

$$\binom{18}{0}(1/6)^0[5/6]^{18} = [5/6]^{18} = 0.037561036758607910916762652168352$$

$$\binom{18}{1}(1/6)^1[5/6]^{17} = 3\,[5/6]^{17} = 0.135219732330988479300345547806O7$$

$$\binom{18}{2}(1/6)^2[5/6]^{16} = 17/4\;[5/6]^{16} = 0.229873544962680414810587431270 32$$

The total probability of all of them will be 0.402654314052276805027695631244. Probability of the opposite event gives the desired result of 0.597345685947723194972304368756.

Interestingly the second approach leads to <u>almost</u> the same result 0.59734568594772319497230436875526 showing an intriguing difference in the last digits – which perhaps results from rounding errors of the longer chain of calculation of the first approach. Below are enclosed the details of obtaining the above sum:

0.037561036758607910916762652168352 + 0.13521973233098847930034554780607 +
+ 0.22987354496268041481058743127032 = 0.402654314052276805027695631244

Following the procedures already presented – also in this case we re-examine the procedure *"by direct enumeration of cases"*. The total number of all possibilities for throwing the die 18 times will be given by $6^{18} = 101559956668416$;

Appearances for $k = 0$ give $5^{18} = 3814697265625$

Appearances for $k = 1$ describe $\binom{18}{1} 5^{17} = 13732910156250$

And the third term for $k = 2$ shows $\binom{18}{2} 5^{16} = 23893554687500$

Taking all three results together gives 40893554687500 – and finally the amount of *"favorable"* appearances will be 60666401980916. Which confirms the impeccable character of the last check.

The Student who shall make use of [22] – supported by reading [24] will not be disappointed – as there are other interesting details – especially when perusing the original letters given in [24]. Regarding the Polish Student it may be added that the Polish translation [33] contains no trace of the correspondence between Pepys and Newton. Stigler [22] indicates which editions of Pepys' diaries include these letters.

4.4 Weldon's Dice Data

The following data are according to Professor W. F. R. Weldon, F.R.S.,[3] and give the observed frequency of dice with 5 or 6 points when a cast of twelve dice was made 26, 306 times:

The book by W.Feller [25] (see pp. 148-9) which frequently recommended as an easily accessible reference for an inquiring Student offers very little. Of course the reference to K. Pearson [34] is recommend from every point of view. For a serious study of the problem put forward by Weldon the paper [35] can be recommended. To satisfy the purposes of this Chapter Weldon's approach has significant meaning as the place where the problem of fitting theory and Monte Carlo practice was seriously examined. In this particular approach the problem was to consider whether the twelve dice were fair dice or not, and how to prove this kind of question.

[3] The sentence is literally quoted from K. Pearson's paper [34] – pp.167-9 – Illustration I & II.

Table 4.2 Weldon's Dice Data

k	$b(k;12,⅓)$	Monte Carlo.	$b(k;12,0.3377)$
0	203	185	187
1	1217	1149	1146
2	3345	3265	3215
3	5576	5475	5465
4	6273	6114	6269
5	5018	5194	5115
6	2927	3067	3043
7	1254	1331	1330
8	392	403	424
9	87	105	96
10	13	14	15
11	1	4	1
12	0	0	0
	26306	26306	26306

Commencing from Karl Pearson – to check the hypothesis about the fair dice a tool invented by him in a view of χ^2 testing was used. The method belongs to mathematical statistics exceeding the scope of this book. Nevertheless very primitive comparison results included in the second and the last columns of Tab. 4.2 may suggests that there were some *discrepancies* which allow to question this hypothesis. The authors of [35] included results justifying the shifting of the mean value – but even this simple outcome exceeds the scope of this lecture – as we are slowly approaching the values of the basic mean and the variance for the binomial distribution.

4.5 Two Shores – Two Tails

The main theoretical results regarding the binomial distribution will be discussed on the pages indicated below and in our own book [19] presenting here an expanded and corrected proposal. Regarding terminology and symbols we propose to follow Feller's book [25]. Therefore we commence once again by giving the definition of the binomial distribution – this time stemming from a desire to satisfy purely formal reasons:

Theorem let $b(k;n,p)$ denote the probability that in n *Bernoulli's trials* - determined by probability of a single success p and failure $q = 1 - p$ - appear exactly k successes and $n-k$ failures – preserving the condition $0 \le k \le n$ - then $b(k;n,p)$ describes the formula:

$$b(k;n,p) = \binom{n}{k} \cdot p^k \cdot q^{n-k} \quad \text{where} \quad 0 < p < 1 \tag{4.27}$$

Simultaneously the following shortened formula will be useful:

$$b(k;n,p) = P(S_n = k) \tag{4.28}$$

This is the right place to call S_n a *random variable* and (4.27) the *probability distribution* – in short the *distribution.* The symbol b can stand either for *binomial* – or for *Bernoulli.* For a long time the numerical values of these distributions were given in mathematical tables. Rapid development of numerical tools stopped this practice.

The basic mean and the variance of the binomial distribution are given by:

$$\overline{S_n} = n \cdot p \tag{4.29}$$

$$\sigma_S^2 = n \cdot p \cdot (1-p) \tag{4.30}$$

Binomial distribution belongs to the class of discrete distributions determined on the additive half of the axes only in the points given by natural integers. To derive its basic mean we propose to use the verbal definition given in Chapter 1 – it leads to the formal definition which generalizes the formal rule of determining the mean derived for the grouped data and is given below:

$$\overline{S_n} = \sum_{k=0}^{k=n} k \cdot b(k;n,p) \tag{4.31}$$

The symbol $b(k;n,p)$ can be read as *normalized class frequency.* The below proof showing that the formula (4.31) finally results in (4.29) offers successive steps of which possible mutual equivalence has been left for the Student and should not cause difficulties in doing so (also [26] can be advised as a possible reference).

1-mo $$\sum_{k=0}^{k=n} k \cdot \frac{n!}{k!(n-k)!} \cdot p^k \cdot q^{n-k} = \sum_{k=1}^{k=n} k \cdot \frac{n!}{k!(n-k)!} \cdot p^k \cdot q^{n-k}$$

2-do $$\sum_{k=1}^{k=n} k \cdot \frac{n!}{k!(n-k)!} \cdot p^k \cdot q^{n-k} = n\,p \sum_{k=1}^{k=n} \frac{(n-1)!}{(k-1)!(n-k)!} \cdot p^{k-1} \cdot q^{n-k}$$

3-tio $$n\,p \sum_{k=1}^{k=n} \frac{(n-1)!}{k!(n-k)!} \cdot p^{k-1} \cdot q^{n-k} = n\,p \sum_{m=0}^{m=n-1} \frac{(n-1)!}{m!(n-1-m)!} \cdot p^m \cdot q^{n-1-m}$$

4-to $$n\,p \sum_{m=0}^{m=n-1} \frac{(n-1)!}{m!(n-1-m)!} \cdot p^m \cdot q^{n-1-m} = n\,p\,(p+q)^{n-1}$$

5-to $$n\,p\,(p+q)^{n-1} = n\,p$$

Technically much more difficult is the proof of the result given by (4.30) - therefore the Student who is looking for it has to make use of books presenting higher levels of mathematical tools than this book (Polish Students may make use of [11]). Regarding the book by Feller [25] there are both proofs but the Author of this book thinks that they are aimed rather too high (one of the Polish novels written by Jerzy Andrzejewski (1903-1983) is entitled "*He Cometh Leaping upon the Mountain*" – which seems to indicate similar sentiments).

Before moving on to the behavior of both tails of binomial distribution (4.27) - attention will be paid for the central term of this distribution. With respect to it let us examine the ratio of the two successive values of (4.27) – proving what follows:

$$\frac{b(k;n,p)}{b(k-1;n,p)}=\frac{(n-k+1)\cdot p}{k\cdot(1-p)}=1+\frac{(n+1)\cdot p-k}{k\cdot(1-p)} \tag{4.32}$$

The last term proves that since $k<(n+1)\cdot p$ the distribution increases, and decreases when $k>(n+1)\cdot p$. Moreover there is exactly a single integer m_{in} satisfying the condition:

$$(n+1)\cdot p-1<m_{in}\leq(n+1)\cdot p \tag{4.33}$$

Therefore – in general terms of $b(k;n,p)$ - with increasing k commencing from zero – there is a monotonic increase reaching the maximum at $k=m_{in}$ and then follows a monotonic decrease. This property occurs for all m_{in} up to the lowest $m_{in}=1$. Specific behavior characterizes $m_{in}=0$ - because only its decreasing branch remains for these distributions. A specific situation is also related to all the cases when $(n+1)\cdot p$ - becomes an integer – let us denote it by m - because the distribution has double maximum i.e. $b(m;n,p)=b(m-1;n,p)$. Also once $m=1$ the distribution has no monotonic increasing branch. Several such cases are depicted by Fig. 4.5-4.7.

Numerical values corresponding to Fig. 4.5:

f1(0) = 0.2097152	f1(1) = 0.3670016
f1(2) = 0.2752512	f1(3) = 0.114688
f1(4) = 0.028672	f1(5) = 0.0043008
f1(6) = 0.0003584	f1(7) = 0.0000128

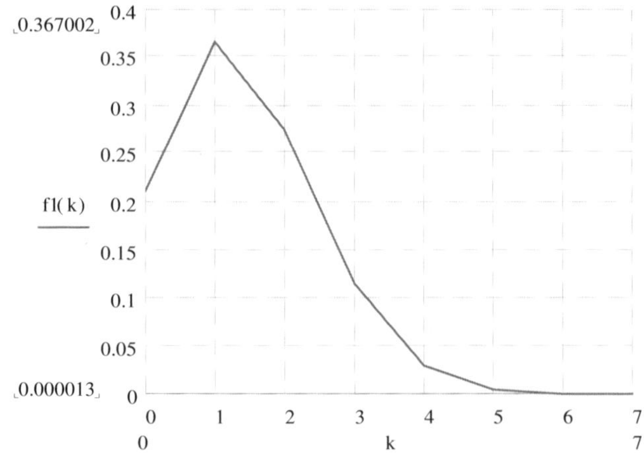

Fig. 4.5 Special case of $b(k;7,\frac{1}{5})$ $m_{in}=1$

Numerical values to Fig.4.6:

f1(0) = 0.392695903778076 f1(1) = 0.392695903778076

f1(2) = 0.168298244476318 f1(3) = 0.0400710105896

f1(4) = 0.005724430084229

f1(5) = 0.000490665435791

f1(6) = 0.000023365020752

f1(7) = 0.000000476837158

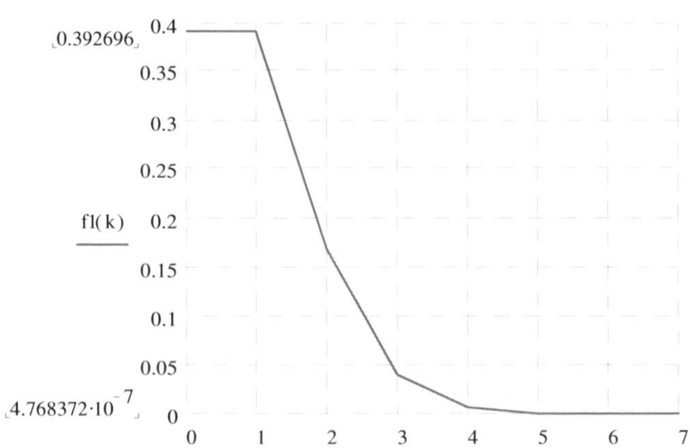

Fig. 4.6 Special case of $b(k;7,\frac{1}{8})$ m = 1

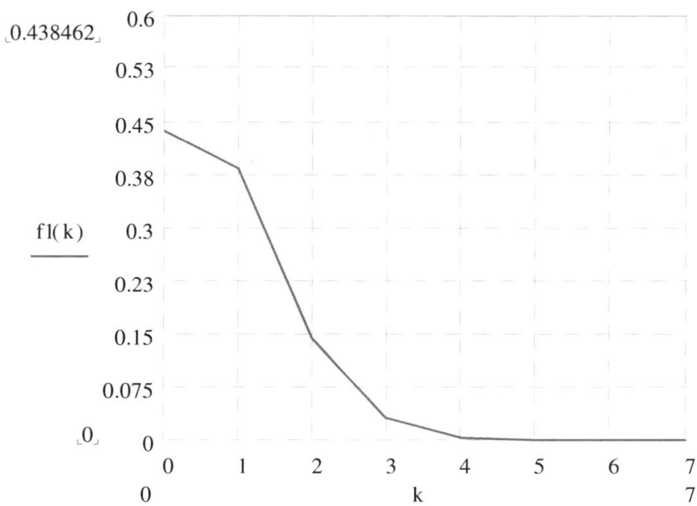

Fig 4.7 Special case of $b(k;7,1/9)\quad m_{in}=0$

Numerical values to Fig.4.7.

f1(0) = 0.438462386 f1(1) = 0.3836545878

f1(2) = 0.1438704704 f1(3) = 0.0299730147

f1(4) = 0.0037466268

f1(5) = 0.000280997

f1(6) = 0.0000117082

f1(7) = 0.0000002091

Intermediate formulae: to help the Student with difficulties in deriving the last term in (4.32) we show several intermediate steps towards it:

$$\frac{b(k;n,p)}{b(k-1;n,p)}=\frac{\dfrac{n!}{k!\cdot(n-k)!}p^k\cdot q^{n-k}}{\dfrac{n!}{(k-1)!\cdot(n-k+1)!}p^{k-1}\cdot q^{n-k+1}}=\frac{(k-1)!\cdot(n-k+1)!}{k!\cdot(n-k)!}\cdot\frac{p^k\cdot q^{n-k}}{p^{k-1}\cdot q^{n-k+1}}=\frac{(n-k+1)\cdot p}{k\cdot q}$$

Now we shall investigate the behavior of both tails of the binomial distribution. Commencing with the right tail i.e. deriving the probability of the appearance of *at least* r *successes*. The initial step requires:

$$P(S_n \geq r) = \sum_{v=0}^{\infty} b(r+v; n, p) \tag{4.34}$$

Important hint: the series (4.34) is only formally infinite – since the terms with $v > n - r$ vanish. The upper bound for probability denoted by (4.34) will be derived below. First to be considered is the case $r \geq n \cdot p$. Examining (4.32) it is clear that the terms of the series in (4.34) decrease faster than the terms of a geometric series with ratio $1-(r-n\cdot p)/r\cdot(1-p)$ - therefore

$$P(S_n \geq r) \;\leq\; b(r; n, p) \cdot \frac{r \cdot q}{r - n \cdot p} \tag{4.35}$$

Let us look closer at why (4.35) takes place. On the one hand (4.32) allows to determine the *reduction ratio* of the series (4.34) as:

$$1 - \frac{r - n \cdot p - p}{r \cdot q} \tag{4.36}$$

In turn (4.36) helps to see, that the geometric series - the *reduction ratio* of which is determined by:

$$1 - \frac{r - n \cdot p}{r \cdot q} \tag{4.37}$$

decreases slower than (4.36) – therefore (4.35) takes place.

It is possible to add some more details. The limit of the geometric series of the form:

$$1 + x + x^2 + x^3 + \ldots + x^n + \ldots \tag{4.38}$$

with the condition requiring $|x| < 1$ with respect to the series terms is equal to:

$$\frac{1}{1-x} \tag{4.39}$$

Regarding formulae (4.38) and (4.39), the symbol x denotes the *decrement ratio* of the terms of geometric series. Therefore the symbolic limit (4.39), by substituting the expression (4.37) instead of x – after a simple manipulation gives the ratio $r\,q/(r - n\,p)$ as indicated in (4.35). To complete the proof the formula (4.35) should be presented in view of an explicit algebraic form.

With respect to this requirement let us note that the values of k satisfying the following condition:

$$m_{in} \leq k \leq r \tag{4.40}$$

will be more than values determined by:

$$r - n \cdot p \tag{4.41}$$

Because the integer m_{in} - satisfying inequalities (4.40) is greater than the real $n \cdot p$ and moreover the condition (4.41) further involves two bounding values. It justifies the below given inequality:

$$b(r; n, p) < \frac{1}{(r - n \cdot p)} \tag{4.42}$$

Altogether the final upper bound value is given by:

$$P(S_n \geq r) \leq \frac{r \cdot (1-p)}{(r - n \cdot p)^2} \quad \text{where} \quad r > n \cdot p \tag{4.43}$$

The above procedure supplied us with the upper limiting value for the right tail of the binomial distributions. A similar procedure could be conducted to determine the lower limit for the left tail of these distributions. Instead we have decided to present only the final result which has the following form:

$$P(S_n \leq r) \geq \frac{(n-r) \cdot p}{(n \cdot p - r)^2} \quad \text{where} \quad r < n \cdot p \tag{4.44}$$

4.6 Jacob Bernoulli's Weak Law of Large Numbers

It is true that quite well-known *intuitive notion of probability* follows from the assumption, that if in n identical experiments the event A appears v times and moreover n is sufficiently great – then the ratio v/n should be close to the probability p of the appearance of the event A. Therefore – returning on the ground of *binomial distribution* – if S_n denotes the number of successes in n trials – then S_n / n is the average number of successes and it is natural to expect that this number is close to p - the probability of the appearance of success in a single *Bernoulli trial.* The above method leads to the title entity – the *Law of Large Numbers* as it first appeared in a work by Jacob Bernoulli entitled "*Ars Conjectandi*" [4], posthumously edited in 1713.

Here it can only be noted that the properties of the binomial distribution allow to prove what follows (see for instance Chapter VI of [12], "*Bernoulli's Theorem*", pp. 96-118):

$$P\left\{\left|\frac{S_n}{n} - p\right| < \varepsilon\right\} > 1 - \eta \quad \text{for small } \varepsilon, \eta \quad \text{and accordingly big } n \tag{4.45}$$

We may read (4.45) in the following manner: with increasing n the probability that the mean number of successes will differ from p more than by the arbitrarily chosen $\varepsilon > 0$ - tends to be zero. Comments and proofs of this theorem frequently called the *Weak Law of Large Numbers* – fill the pages of many advanced books on Probability. Among them Uspensky's [12] deserves special attention. He starts by giving the proof originally provided by the inventor of the theorem – Jacob Bernoulli – it consists of 5 pages and has Uspensky's recommendation who insisted that:

... *Several proofs of this important theorem are known which are shorter and simpler but less natural than Bernoulli's original approach* ...

It is interesting to know that the theorem (4.45) can be used to determine such a value of n with respect to assumed ε, η that the theorem will be satisfied:

$$n \geq \frac{1+\varepsilon}{\varepsilon^2} \ln \frac{1}{\eta} + \frac{1}{\varepsilon} \tag{4.46}$$

Let us derive, following (4.46), such an n which corresponds to $\varepsilon = 0.01$ and $\eta = 0.001$. By using a scientific calculator we shall get n = 69868.32832. Having in mind that n must be a natural number, we get 69 869 - which was also derived by Uspensky (see [12] p. 101). Further points closely related to what is being discussed here will be presented in Chapter 5 – keeping in mind a generalization of the Law of Large Numbers.

Keeping in mind the subject of the next paragraph – and intending to announce the problem, this is the right place to draw the Student's attention, still with respect to the consequences of increasing values of n, firstly to the fact of increasing numerical difficulties in operating with binomial distribution and secondly to the fact that apart from the troubling consequences, increasing values of n enable to make use of the normal distribution in approximating the values of the binomial distribution in a very simple way. Here we come close to the subject which was investigated successfully by Abraham de Moivre.

4.7 Following Abraham de Moivre

The approach presented below, leading from the binomial distribution to the normal distribution, may be found in a modest book by W. Pogorzelski [11] (known only to those who read in Polish). Out of two such proofs, the less rigorous one has been chosen which commences from the place already examined above. To begin with, the binomial distribution is expressed once more in a slightly different notation which proves useful:

$$P(k) = \frac{n!}{k!(n-k)!} \; p^k q^{n-k} \tag{4.47}$$

The presented approach examines the increment of the binomial distribution first as:

$$P(k+1) = \frac{n!}{(k+1)!(n-k-1)!} \; p^{k+1} q^{n-k-1} \tag{4.48}$$

This serves to derive the expression related to (4.32):

$$\frac{P(k+1)}{P(k)} = \frac{n-k}{k+1} \frac{p}{q} \tag{4.49}$$

With the help of (4.49) we can determine the relative increment of the function $P(k)$:

$$\frac{P(k+1) - P(k)}{P(k)} = \frac{n-k}{k+1} \frac{p}{q} - 1 \tag{4.50}$$

It will be reasonable to recall the essential means of the binomial distribution:

$$\overline{k} = n\,p \tag{4.29}$$

$$\sigma_k^2 = n\,p\,q \tag{4.30}$$

For the purpose of further considerations a new variable x has to be defined which belongs to the domain of real numbers, formally determined by:

$$x = k - n\,p \tag{4.51}$$

The numerator of (4.50) will be further denoted as:

$$\Delta P(k) = P(k+1) - P(k) \tag{4.52}$$

Formulae (4.50)-(4.52) allow to get the following approximate result:

$$\frac{1}{P} \frac{\Delta P}{\Delta x} = - \frac{x}{n\,p\,q} \tag{4.53}$$

The ratio (4.53), assuming that the increment Δx is sufficiently small leads to the functional equation:

$$\frac{1}{\tilde{P}}\frac{d\tilde{P}}{dx} = -\frac{x}{n\,p\,q} \tag{4.54}$$

Its solution is presented in three steps:

$$\frac{d}{dx}\left(\ln\tilde{P}\right) = -\frac{x}{n\,p\,q} \quad \text{then} \quad \ln\tilde{P} = -\frac{x^2}{2n\,p\,q} + \ln C \qquad (4.55)\ \&\ (4.56)$$

Finally (4.56) determines the continuous function $\tilde{P}(x)$ of the exponential form:

$$\tilde{P}(x) = C\exp\left(-\frac{x^2}{2n\,p\,q}\right) \tag{4.57}$$

So, in this way we obtained the result (4.57) which for the large numbers n well approximates the discrete function (4.47) in the neighborhood of the mean $n\,p$. Moreover it is seen that (4.57) – is an even function possessing the property $\tilde{P}(x) = \tilde{P}(-x)$; $\tilde{P}(x)$ presents a fast decreasing function due to its exponential form. Normalized requirements allow to determine the constant C:

$$C = \frac{1}{\sqrt{2\,\pi\,n\,p\,q}} \tag{4.58}$$

$\tilde{P}(x)$ in a view of (4.57) and (4.58) defines new probability distribution known as the *normal distribution* – called also the *Gaussian distribution.* Normal distribution plays a leading role in Statistics but it is sometimes overshadowed by a habit to present numerous distributions in Statistics courses. This particular course pays attention to this distribution was devoted to it the entire Chapter 5, which closed the theoretical part of the book. In the end of this paragraph we repeat – the first mathematician who derived this distribution was a French emigrant to England – Abraham de Moivre. The closing part of this chapter is devoted to a few concepts closely related to the binomial distribution – the opening paragraph presents the Poisson distribution.

4.8 Beyond the Binomial Distribution

4.8.1 Montmort, de Moivre, Poisson, Bortkiewicz

Empirical Background preceded by some Biographical Data

Distribution by Poisson-Bortkiewicz – commonly known as the Poisson distribution should be placed at the top of Statistical distributions – combining the binomial distribution, the normal distribution and the Poisson distribution. The reason why this distribution has such a high rank can probably be deduced from the content of this paragraph. But the Student who wants to know about its application is directed to go to Chapter 5 – Part two, Exercises. Jack Good examines the credit which goes to Denis Poisson (1781-1840) in a paper [27] consisting of 10 pages but devoting to Poisson distribution a single page (quoting about 100 papers of which only 25% are of his own). Jack Good himself deserves for a brief note. Good was born Isadore Jacob Gudak to a Polish-Jewish family in London 1916. Isadore later changed his name to Irving John Good – as was later known as Jack Good. His range of scientific interests is reminiscent of Francis Galton (presented in Chapter 1), he also died at approximately the same age. He left a good record of his war time activities working closely with Allan Turing (1912-54) at Bletchley Park on German Enigma. Later on he moved to the United States and got professorship at the Virginia Polytechnic Institute.

During this time he wrote numerous books on probability, and is known as the creditor of modern Bayesian methods. Writing a paper on Poisson close when he was practically in his seventies he demonstrated his deep knowledge of Probability, profoundness of his historical references, his general erudition and his good sense of humor. He died of natural causes in 2009. Let us now return to his paper [27] and its third part which is in the end due to the fact that "Poisson was scarcely responsible for introducing this distribution, nor for its application". Good refers to [36] which states that the discoverer of the distribution was Abraham de Moivre who derived it as a limiting form of the negative binomial (1718), and the same result was then obtained by Poisson (1837) in the same way, however neither of them knew the formula $e^{-\lambda} \cdot \lambda^k / k!$ used today (below we return to this analytical details). In an interesting excerpt Good [27] says:

Perhaps the Poisson distribution should have been named after von Bortkiewicz (1898) because he was the first to write extensively about rare events …

So, continuing our biographical data we finally arrive at two names which will be shortly mentioned: Poisson and Bortkiewicz. Although the life of Poisson is described in numerous sources, in the Author's opinion it is rare that we can find such an account as the one by Boyer [7] therefore we quote just a few lines from [7]:

Simeon-Denis Poisson (1781-1840) was the son of a small-town (Pithiviers) administrator who took charge of local affairs when the Revolution broke out, and the child was reared under republican principles; but he later became a staunch Legitimist and in 1825 was rewarded with the title of baron. In 1837, under Louis

Philippe, he became a peer of France. Relatives at first hoped that the young man would become a physician, but strong mathematical interest led him in 1798 to enter the École Polytechnique, where on graduation he became successively lecturer, professor, and examiner. He is said to have once remarked that the life is good for only two things: to do mathematics and to teach it. Consequently he published almost 400 works, and he enjoyed a reputation as an excellent instructor.

Unable to indicate another source which would supply us with such a brief biography of Bortkiewicz, we include here our own few lines regarding him. When crediting Bortkiewicz there is no better source than a short paper by J. E. Gumbel [28] which forces us to also include some remarks about the latter. Ladislaus von Bortkiewicz (1868-1931) written in Russian in the following way *Владислав Иосифович Бортжевич* and in Polish as Władysław Bortkiewicz – is called today a Russian economists and statistician of Polish descent (his mother was Helena Rokicka, father – a colonel of Russian cavalry). Born in Saint Petersburg, he graduated with a degree in law in 1890. Influenced by the lectures of mathematics given there by A. A. Markov he discovered in himself a talent for this subject. He went to Goettingen where under Wilhelm Lexis (1837-1914) he obtained his doctoral degree in 1893, then in Strasbourg he got his Habilitation in 1895. Shortly after publishing a book about the Poisson distribution "*Das Gesetz der Kleinen Zahlen*" (The Law of Small Numbers) – Gumbel calls it "*a brochure of sixty pages*" – which presented his significant contribution to the Poisson distribution theory and applications - he obtained a position at Berlin University where he worked there until his death. We shall describe this contribution below. He is moreover considered an important contributor to economy – especially with respect to a critical account of *Capital* (Das Kapital, vol.3) by Karl Marx in which he undermined the latter's claim to have provided a consistent account of capitalist economics. Bortkiewicz also contributed to the system of price index numbers used for a few decades before him. Gumbel claims that "*He was a true scholar of the old school and his life was passed in enviable quietness*" (perhaps having in mind his own life). And here are a few remarks about Emil Julius Gumbel (born on 18 July 1891, died on 10 Sept. 1966). We have to say in the beginning that his was a split personality – he possessed two faces (or better to say two faiths) – he was a mathematician and a pacifistic politician. Born in Munich, he graduated from University of Munich in 1913. Until 1932 he taught Statistics at the University of Heidelberg. Expelled from this post for his pacifism and leftist views he settled in the United States in 1941. Despite quite a long period of time which he spent there, he only enjoyed a part time job at the Columbia University at one point. His biographer (see [37]) says that until this death of cancer he was permanently struggling with financial troubles. His major scientific contribution was the theory of extreme values and "*Statistics of Extremes*" [39] was his crown achievement. Following the Russian translation of his work, Gumbel became known in Poland. The exact date of his birth was given here as it coincides with the Author's (July 18, however the Author was born almost half a century later) who became familiar with the book on extreme values at the time it was published by Mir. This was not incidental as the Author's main field of specialization is related to the gust-loads problems in Aeronautics. In the literature of his book Gumbel included a report by H. Press [40] documenting the applicability of the statistics of extremes also in the field of Aeronautical Engineering. And this is where we end the biographical passage and return to the empirical background of the Poisson distribution - commencing from Bortkiewicz.

The work which made Bortkiewicz's name widely known was a brochure [41]. It gives an example of the number of soldiers killed by horse kicks (per year in the Prussian army corps) documenting that the overall distribution is remarkably well fitted by the Poisson distribution. Bortkiewicz's results are provided here first as an extract from [26] – pp. 155 – 156 and given below:

$$P(0) = 0.545;\ \ P(1) = 0.325;\ \ P(2) = 0.110;\ \ P(3) = 0.015;\ \ P(4) = 0.005 \qquad (4.59)$$

The mean value of the empirical distribution given by (4.59) is determined below:

$$\hat{\lambda}_1 = 0 * 0.545 + 1 * 0.325 + 2 * 0.110 + 3 * 0.015 + 4 * 0.005 = 0.610 \qquad (4.60)$$

The variance estimate, obtained in a similar way as the above mean estimate, is:

$$\hat{\lambda}_2 = \sigma^2 = 0.6079 \quad \rightarrow \quad \sigma \approx 0.7797 \qquad (4.61), (4.62)$$

Original data by L. Bortkiewicz copied from the Internet are given as Table 4.3 (see also [29], p.20).

Table 4.3

(18)	75	76	77	78	79	80	81	82	83	84	85	86	87	88	89	90	91	92	93	94
G	–	2	2	1	–	–	1	1	–	3	–	2	1	–	–	1	–	1	–	1
I	–	–	–	2	–	3	–	2	–	–	–	1	1	1	–	2	–	3	1	–
II	–	–	–	2	–	2	–	–	1	1	–	–	2	1	1	–	–	2	–	–
III	–	–	–	1	1	1	2	–	2	–	–	–	1	–	1	2	1	–	–	–
IV	–	1	–	1	1	1	1	–	–	–	–	1	–	–	–	–	1	1	–	–
V	–	–	–	–	2	1	–	–	1	–	–	1	–	1	1	1	1	1	1	–
VI	–	–	1	–	2	–	–	1	2	–	1	1	3	1	1	1	–	3	–	–
VII	1	–	1	–	–	–	1	–	1	1	–	–	2	–	–	2	1	–	2	–
VIII	1	–	–	–	1	–	–	1	–	–	–	–	1	–	–	–	1	1	–	1
IX	–	–	–	–	–	2	1	1	1	–	2	1	1	–	1	2	–	1	–	–
X	–	–	1	1	–	1	–	2	–	2	–	–	–	–	2	1	3	–	1	1
XI	–	–	–	–	2	4	–	1	3	–	1	1	1	1	2	1	3	1	3	1
XIV	1	1	2	1	1	3	–	4	–	1	–	3	2	1	–	2	1	1	–	–
XV	–	1	–	–	–	–	–	1	–	1	1	–	–	–	2	2	–	–	–	–

To accompany this famous example two brief remarks will be added. Soldiers who died in the way investigated by Bortkiewicz – were referred to by Good [27] expressing his sense of humor as the victims of "Bortkiewicz's disease" which "is always fatal by definition". Secondly – although we can frequently read about the remarkably good fit of this empirical data with the Poisson distribution – we were able to find only one book [26] which contains the proof of this claim. Instead of providing other examples of empirical data fitting the Poisson distribution, we rather make use of a remark made by Gumbel in [28] which we consider very much up to the point and very much true (see pp. 24-25):

In this [Poisson] *distribution the variance (…) is equal to the expectation. The corresponding observed quotient should therefore be near unity. This is called "normal dispersion" in the Lexis theory. The law of small numbers says that rare events usually show normal dispersion. For a mathematical explanation of this fact consistent with Lexis theory, see Gosset* [42].*(…) Bortkiewicz created an important instrument for mathematical statistics and probability theory. However, the name he gave it was unfortunate because it implied a nonexistent contrast to the law of large numbers and led to much confusion and unnecessary arguments (…) It would have been better to speak of "rare events".*

Before we close such a brilliant reference as Good's paper [27], we would like to quote also a short passage (see p.166):

It is reasonable to maintain that even de Moivre was anticipated by de Montmort (1708) who discussed the matching problem (or "treize"). If two packs, each of "n" cards, the cards being labeled 1, 2, …, n – in each pack, are shuffled and laid out in two rows, the probability of exactly "r" matches, when $n \rightarrow \infty$*, tends to* $e^{-1}/r!$ *that is, to the Poisson distribution with mean 1.*

And in the end let us quote what Winston Churchill said according to Good's [27] and which remains true in many countries: "*Everybody has a right to pronounce foreign names as he chooses*".

4.9 Derivation of the Poisson Distribution

The proof presented below is the one which can be seen in contemporary books on Statistics and Probability. As it has been already noted the original proof was so specific that it has not been included (see Stigler [38]). The formal procedure below regarding limiting property the formal exposition should overcome inefficiency of the Word Equations Editor. The Student familiar with this Editor looking at what is shown below will recognize the point. Otherwise it would be difficult to explain. Having in mind our purposes, also the binomial formula has to be rewritten again in a slightly different form:

$$b(k;n,p_n) = \frac{n!}{k!(n-k)!}\, p_n^k\,(1-p_n)^{n-k} \tag{4.63}$$

The mentioned difficulty is found in a limiting procedure in which simultaneously with $n \rightarrow \infty$ there are gradually decreasing probabilities $p_n \rightarrow 0$

preserving the constancy of the multiplier $n\ p_n = \lambda$ which we have tried to express below:

$$\lim n\, p_n \xrightarrow{n\to\infty} \lambda \tag{4.64}$$

Such a double limiting procedure leads to the Poisson distribution:

$$\lim b\,(k; n, p_n) \xrightarrow{n\to\infty} \frac{\lambda^k}{k!}\, e^{-\lambda} \tag{4.65}$$

The theoretical problem under consideration can be stated as determining the limit:

$$\lim_{n\to\infty} b\,(k) = \lim_{n\to\infty} \frac{n!}{k!(n-k)!}\, p_n^k\, (1-p_n)^{n-k} \tag{4.66}$$

The next stage may have the shape:

$$\lim_{n\to\infty} \frac{n!}{k!(n-k)!}\, p_n^k\, (1-p_n)^{n-k} = \lim_{n\to\infty} \frac{n(n-1)(n-2)\cdots(n-(k-1))}{k!} \left(\frac{\lambda}{n}\right)^k \left(1-\frac{\lambda}{n}\right)^{n-k} \tag{4.67}$$

The right side of (4.67) may be transformed into what follows:

$$\frac{\lambda^k}{k!} \lim_{n\to\infty} \left(1-\frac{1}{n}\right)\cdot\left(1-\frac{2}{n}\right)\cdots\left(1-\frac{k-1}{n}\right)\cdot\left(1-\frac{\lambda}{n}\right)^n\cdot\left(1-\frac{\lambda}{n}\right)^{-k} \tag{4.68}$$

To develop the sequence of the particular limits we get the following one:

$$\lim_{n\to\infty} \frac{n}{n}\cdot\frac{n-1}{n}\cdot\ \cdots\ \cdot\frac{n-k+1}{n} = 1 \tag{4.69}$$

Afterwards (4.68) takes the shape:

$$\frac{\lambda^k}{k!} \lim_{n\to\infty} \left(1-\frac{\lambda}{n}\right)^n \cdot \lim_{n\to\infty} \left(1-\frac{\lambda}{n}\right)^{-k} \tag{4.70}$$

The two limits given below secure the final outcome :

$$\lim_{n\to\infty} \left(1-\frac{\lambda}{n}\right)^{-k} = 1 \tag{4.71}$$

$$\lim_{n\to\infty}\left(1-\frac{\lambda}{n}\right)^n = e^{-\lambda} \tag{4.72}$$

And the formula (4.68) becomes identical with (4.65). Which ends this proof.

The Student who can for instance consult [31] (pp.87-88), may find there proofs for the limits given by (4.70) and (4.71) – see also [32], pp.125-127 (the reference indicates a Polish translation).

Two Numerical Examples

The examples below show applicability of the Poisson distribution instead of binomial distribution in such cases when *'n'* is comparatively high while the probability of a single success p_n is comparatively low – and their multiplier is close to one– then the Poisson distribution proves to fitting such cases well.

Example 4.1. To determine probability that among 500 people chosen randomly exactly *'k'* people born in the New Year's Day can be found.

Solution. Binomial distribution under the assumption that children are born uniformly throughout the entire year seems to be well documented by statistical data. Therefore the above example falls under the binomial distribution with the following parameters:

$$p_n = 1/365 \quad \text{and} \quad n = 500 \tag{4.73}$$

The below calculations done with a help of *MathCad* package – provide the answers to the problem stated in the Example 4.1 – for both distributions – first taking binomial, then Poisson distributions. All these results for both distributions calculated for $k = 0$, 1, 2 and 3 are denoted as (4.74):

$$\left(\frac{364}{365}\right)^{500} = 0.253664443773374 \qquad \left(\frac{364}{365}\right)^{499}\cdot\frac{500}{365} = 0.348440170018371$$

$$500\cdot\frac{499\cdot\left(\frac{364}{365}\right)^{498}}{2\cdot 365^2} = 0.238834676976878 \qquad 500\cdot 499\cdot\frac{498\cdot\left(\frac{364}{365}\right)^{497}}{6\cdot 365^3} = 0.108919110929010$$

in the both cases $k =$ 0, 1, 2, and 3

$$\lambda := \frac{500}{365} \qquad \lambda = 1.36986301369863$$

$$p(k,\lambda) := \frac{\lambda^k\cdot e^{-\lambda}}{k!} \qquad p(0,\lambda) = 0.254141771109641 \qquad p(1,\lambda) = 0.34813941247896$$

$$p(2,\lambda) = 0.238451652382849 \qquad p(3,\lambda) = 0.108882033051529$$

The above obtained numerical results are convincingly close to each another. Moreover due to small p_n also we have $q_n = 1 - p_n \rightarrow q_n \approx 1$ therefore also $n \cdot p \cdot (1-p) \approx n \cdot p \rightarrow \sigma^2 \approx \lambda$. The Student is advised to check that coming up to $k = 5$ he/she should get $\sum_{k=1}^{5} k \cdot p(k, \lambda) \approx 1.353$ via the Poisson distribution taking an approximate value $\lambda \approx 1.369853$.

Example 4.2. Automatic production of nuts assesses their high quality, so the probability of getting a defective product is equal to 0.015. Determine probability that for hundred nuts random sample there is not a single defective one.

Solution requires to find a single numerical value. First, the exact result from *binomial*:

$$(1-0.015)^{100} = 0.22060891046938756292526835432722 \tag{4.75}$$

then for $\lambda = 100 \cdot 0.015 = 1.5$ the Poisson approximation is:

$$e^{-1.5} = 0.223130160148\ldots \tag{4.76}$$

Deriving the Basic Mean and the Variance

The simplest procedure leads to the formula giving the basic mean. Also this time to present the procedure we propose to write the Poisson distribution in a new form:

$$p(k;\lambda) = \frac{\lambda^k}{k!} \cdot e^{-\lambda} \tag{4.77}$$

Presenting the "Bortkiewicz's disease" data by (4.59) we already used the formula (4.60) determining the mean value of the Poisson distribution but in the general form it is given as below:

$$\sum_{k=0}^{\infty} k \cdot p(k;\lambda) = \lambda \tag{4.78}$$

As the essential stage of the evaluation of the above has to be written as follows:

$$\sum_{k=0}^{\infty} k \cdot \frac{\lambda^k}{k!} \cdot e^{-\lambda} = \lambda \cdot e^{-\lambda} \cdot \sum_{k=1}^{\infty} \frac{\lambda^{k-1}}{(k-1)!} = \lambda \cdot e^{-\lambda} \cdot \sum_{r=0}^{\infty} \frac{\lambda^r}{r!} = \lambda \cdot e^{-\lambda} \cdot e^{\lambda} \tag{4.79}$$

Commenting (4.79) it is visible that mostly simple formal steps have been taken there. First the exponential expression can be moved and placed in front of the sum. Together with this also the term λ has been extracted. Regarding the sum its first term can be increased to correspond to $k = 1$. Therefore the lower symbol will indicate the value $k = 1$ instead of $k = 0$. Then the ratio $k/k!$ can take the form of $1/(k-1)!$ and if we replace $(k-1)$ by r, it simplifies the term under the sign of summation (the change in the summation index needs to be noted). Finally it has to be noted that the sum presenting the infinite series converges with the exponential limit – being the expansion of the number $\exp(\lambda)$ - as shown below:

$$\sum_{k=1}^{\infty}\frac{\lambda^{k-1}}{(k-1)!}=\sum_{r=0}^{\infty}\frac{\lambda^{r}}{r!}=\frac{\lambda^{0}}{0!}+\frac{\lambda^{1}}{1!}+\frac{\lambda^{2}}{2!}+\frac{\lambda^{3}}{3!}+\cdots \rightarrow e^{\lambda} \tag{4.80}$$

The result (4.80) obviously concludes the procedure.

Below we also present the procedure leading to the variance of the Poisson distribution. In due course the well known property (used from Chapter 1) offers the departing point:

$$\sigma_k^2=\overline{k}^2-(\overline{k})^2 \tag{4.81}$$

The following result is an essential part of the proof:

$$\overline{k}^2=\lambda^2+\lambda \tag{4.82}$$

Substitution of (4.82) into (4.81) gives the final result:

$$\sigma_k^2=\lambda^2+\lambda-\lambda^2=\lambda \tag{4.83}$$

To prove a surprisingly simple property (4.82) we recall the formal definition of the mean square value:

$$\overline{k}^2=\sum_{k=0}^{\infty}k^2\cdot\frac{\lambda^k}{k!}\cdot e^{-\lambda} \tag{4.84}$$

The procedure leading from (4.84) to (4.82) includes an apparent digression presenting a new path leading to (4.78) . In this new procedure the opening step makes use of the result which states exactly the same as (4.80) although in a slightly rearranged formulation:

$$\sum_{k=0}^{\infty}\frac{\lambda^k}{k!}\cdot e^{-\lambda}=1 \tag{4.85}$$

The trick requires to differentiate (4.85) with respect to variable k . It leads to the following:

$$\sum_{k=0}^{\infty}k\cdot\frac{\lambda^{k-1}}{k!}\cdot e^{-\lambda}-\sum_{k=0}^{\infty}\frac{\lambda^k}{k!}e^{-\lambda}=0 \tag{4.86}$$

While substitution of (4.85) into (4.86) gives the new identity:

$$\sum_{k=0}^{\infty}k\cdot\frac{\lambda^{k-1}}{k!}\cdot e^{-\lambda}=1 \tag{4.87}$$

And now multiplying (4.87) by λ gives (4.78). This trick has to be conducted for a second time getting *per analogiam* the result (4.82) – as required. Again the first step requires to differentiate with respect to variable k the new identity (4.87) – and then proceed in the above way – which has been left for the Student.

Ending this analytical passage we add an interesting result which can be found in a small book by J. F. C. Kingman [30]. This time the Poisson distribution has been denoted by the symbol π_k then the following property (whose proof can be found in [30]) takes occurs:

$$\frac{d\pi_k}{d\lambda} = \pi_{k-1} - \pi_k \tag{4.88}$$

4.10 Notes on the Multinomial and Negative Binomial Distributions

Multinomial Distribution

Edwards in [2] p.113 writes:

The rule, that the number of arrangements of "a" things of one kind, "b" of another, "c" of another, and so on, is equal to:

$$\frac{n!}{a!\,b!\,c!\ldots} \tag{4.89}$$

first appears in the West in work of Mersenne in 1636 and was explained by Wallis in 1685; Bashkara had already given in the East.

The explanation of (4.89) is straightforward: n different things can be arranged in $n!$ ways, but if any a of them (which can be arranged among themselves in $a!$ ways) should be identical, the number of arrangements is thereby reduced to $n!/a!$, and so on for b, c, ...Therefore if we have only two different things – the probability distribution governing their arrangements will be given by the *binomial* distribution. To generalize formally this case the binomial distribution and its isomorphic counterpart can be written as follows:

$$\frac{n!}{k_1!\,k_2!}\,p_1^{\,k_1}\,p_2^{\,k_2} \qquad n = k_1 + k_2 \quad p_1 + p_2 = 1 \quad \text{and} \quad \left(p_1 + p_2\right)^n \tag{4.90}$$

Then every new case in which more than two components ("things") appear is called "multinomial". To illustrate the above we shall use the most popular example of throwing dice. Let us identify – following the formal rule expressed by (4.90) its formal items. Assuming fair dice it is obvious that $p_i = 1/6$ for $i = 1, 2, \ldots, 6$. But to present the applicability of (4.90) it has to be said how many times a die has been thrown, i.e. some particular value of n - for instance $n = 12$ and also – how many times each face of the die appeared, i.e. values $k_i = ?$ for $i = 1, 2, \ldots, 6$. Assume $k_i = 2$ - then:

$$\frac{12!}{(2!)^{12}}\left(1/6\right)^{12}=\frac{479001600}{4096\cdot 2176782336}$$

with scientific calculator accuracy the result 0.000 053 723 is obtained but using Word calculator we get 5.3723217092478280749885688157293 e-5 in any case - hardly a high probability!

Negative Binomial Distribution – Terminated Binomial

Suppose there is a sequence of Bernoulli trials with probability of success p and of failure $(1-p)$. The sequence is observed until predefined number r of successes has occurred. Then with the number of trials n the number $k = n - r$ failures will be associated having the negative binomial distribution given by:

$$f(k) = \binom{k+r-1}{r-1}\cdot p^r\cdot(1-p)^k \quad \text{for } n = r,\ r+1, r+2, \ldots \text{ or } \quad k = 0, 1, 2, \ldots \tag{4.91}$$

Let us include the following list:

$n =$ number of events
$r =$ number of successes terminating the game
$p =$ probability of success on a single trial
$q = (1-p)$ probability of failure

The mean, and the variance of the negative binomial defined as above are given by the two following formulae:

$$(1-p)r/p \quad \text{and} \quad (1-p)r/p^2 \qquad (4.92) \text{ and } (4.93)$$

Formulae (4.92)-(4.93) locate these values on the k scale – but to locate them on the n scale they have to be shifted by the value $r = 5$. This will be illustrated in a numerical example provided below. Before resorting to the example a formula equivalent to (4.91) will be given which is the consequence of the following equivalence which can be easily proved:

$$\binom{k+r-1}{r-1} = \binom{k+r-1}{k} \tag{4.94}$$

$$f(k) = \binom{k+r-1}{k}\cdot p^r\cdot(1-p)^k \qquad k \geq 0 \tag{4.95}$$

Let us examine the following example:

Example 4.3. Consider an unfair coin with probability of the appearance of heads *success* $p = 0.4$, and tails, *failure* $q = 0.6$. Assume the coin is tossed until $r = 5$ successes appears. Determine the probability of n tossing which secure the appearance of k *failures* with the fixed number of r *successes* .

Solution. To get the equivalent formulation of (4.91) suitable for this case, substitute $k = n - 5$ into the distribution function (4.91) to get the distribution of the trials (independent variable $n \geq 5$):

$$f(n) = \binom{n-1}{r-1} \cdot p^r \cdot (1-p)^{n-r} \qquad n \geq 5 \tag{4.96}$$

To derive the formula corresponding to the particular conditions stated in the above example, substitute numerical values $r = 5$, and $p = 0.4$ to get first $f(n) = \binom{n-1}{4} \cdot 0.4^5 \cdot (1-0.4)^{n-5}$ which can be transformed into what follows: $f(n) = \binom{n-1}{4} \cdot 2^5 \cdot \dfrac{3^{n-5}}{5^n}$ the formula most useful to perform calculations whose results are shown in Tab.4.4. By the way it can be noted that there is also the following equivalence: $\binom{n-1}{r-1} = \binom{n-1}{n-r}$. The proof goes directly from the definition of $\binom{a}{b}$.

Fig.4.8 presents the distribution of the negative binomial ($r = 5$, $p = 0.4$) even beyond the range values shown in Tab. 4.4.

Resorting to the history of mathematics, according to Gurland [43] this distribution was formulated by Pierre Montmort (1678-1719) in his Essay [44], which takes us back to 1713. For biographical data about this Frenchman the Student is advised to use the Internet. Now we briefly discuss the comparison of the averages for three distributions closely interconnected and discussed here: (positive) binomial, Poisson, and negative binomial. Regarding the (positive) binomial $n\,p > n\,p\,q$, its average is always above its variance. For the Poisson distribution – they are equal $\lambda_k = \sigma_k^2$. Then, in the negative binomial the variance is always greater than its mean $(1-p)r/p < (1-p)r/p^2$. These properties may help in statistical practice when trying to chose the distribution to fit some empirical results (see [42]).

And again in this context – we return to the above given Example 4.3. Applying the result (4.92), we shall get the mean value, which we denote by $\mu_k = 7.5$ (Greek letter corresponding to the Latin "*m*" – the initial of the word "*mean*"). In the n scale shifted by the value of $r = 5$, it corresponds to $\mu_n = 12.5$. The variance according to (4.93) is equal to $\sigma^2 = 18.75$ - disregarding the scale. The Student is here advised to recall an appropriate part of Chapter 1 discussing implications of linear transformation with respect to the main averages.

Table 4.4 Numerical results for Example 4.3

n	k	$\binom{n-1}{4}$	3^k	5^n	$f(n)$
5	0	1	1	3125	0.01024
6	1	5	3	15625	0.03072
7	2	15	9	78125	0.055256
8	3	35	27	390625	0.0774144
9	4	70	81	1953125	0.09289728
10	5	126	243	9765625	0.1003290624
11	6	210	729	48828125	0.1003290624
12	7	330	2187	244140625	0.09459597312
13	8	495	6561	1220703125	0.085136375808
14	9	715	19683	6103515625	0.0737848590336
15	10	1001	59049	30517578125	0.061979281588224
16	11	1365	177147	152587890625	0.050710321299456
17	12	1820	531441	762939453125	0.0405682570395648
18	13	2380	1594323	3814697265625	0.031830478600027392
19	14	3060	4782969	19073486328125	0.02455494063449 7024
20	15	3876	14348907	95367431640625	0.0186617548822177 3824
21	16	4845	43046721	476837158203125	0.01399631616166330368
22	17	5985	129140163	2384185791015625	0.010373740213938683904
23	18	7315	387420489	11920928955078125	0.0076074094902217015296
24	19	8855	1162261467	59604644775390625	0.0055253816297399726899 2
25	20	10626	3486784401	298023223876953125	0.003978274773412780336 7424

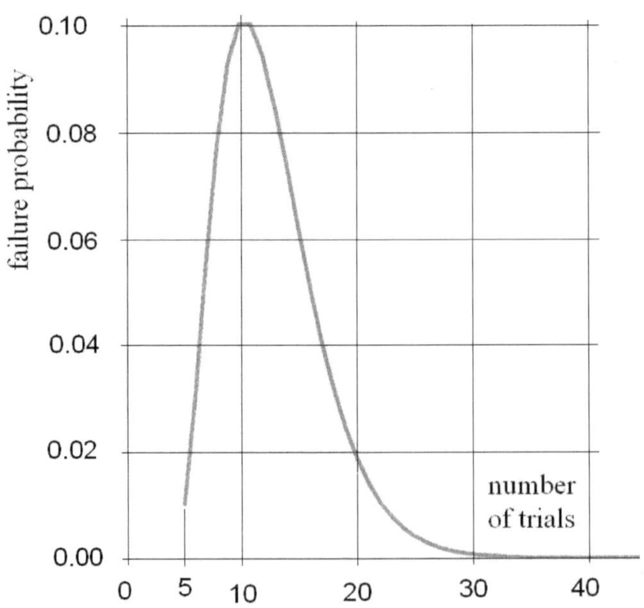

Fig. 4.8 Negative binomial $r = 5,\ p = 0.4$

Returning to the data in Tab. 4.4 – we can say that the cumulative probability for the last calculated result is 0.990529 - which means that less than 1% of all results remain outside the calculated statistics. The distribution is discrete and infinite.

References

[1] Stigler, S.M.: The History of Statistics – the Measurement of Uncertainty before 1900. The Balknap Press of Harvard UP, Cambridge (1986)

[2] Edwards, A.W.F.: Pascal's Arithmetical Triangle – The Story of a Mathematical Idea, pierwsze wydanie - w Anglii 1987 – Charles Griffin & Company Ltd., London, a od roku 2002 było drukowane jako Paperback, p. s.202. Johns Hopkins University Press, Baltimore

[3] Леонид Ефимович Майстров: Теория Вероятностей – Исторический Очерк, Наука, Москва (1967); also known as the English edition: Maistrov, L.E.: Probability Theory - A Historical Sketch. Academic Press, New York (1974)

[4] A complete translation of the Ars Conjectandi is available as Jacob Bernoulli: The Art of Conjecturing, together with Letter to a Friend on Sets in Court Tennis, trans. by E.D. Sylla, p. 580 $57.60. Johns Hopkins University Press, Baltimore (2006)

[5] Arbuthnott, J.: An Argument for Divine Providence, Taken from the Constant Regularity Observ'd in the Births of Both Sexes. Phil. Transactions (1683-1775) 27 (1710-1712), 186–190

[6] Pascal, B.: Traite du triangle arithmetique. Desprez, Paris (1665)

[7] Boyer, C.B.: A History of Mathematics. Princeton UP (1985)

[8] Wilenkin, N.J.: Kombinatoryka (in Polish). PWN, Warszawa (1972)(translated fron Russian, Nauka, Moscow 1962)

[9] Gerstenkorn, T., Śródka, T.: Kombinatoryka i rachunek prawdopodobieństwa (in Polish: Combinatorics and Probabilisty), 3rd edn. PWN, Warszawa (1976)

[10] Flaschmeyer, J.: Kombinatoryka – podstawowy wykład w ujęciu mnogościowym (in Polish: Combinatorics, basic theory by using theory of set). PWN, Warszawa (1974) (translated from German Kombinatorik VEB, Berlin 1969)

[11] Pogorzelski, W.: Zarys Rachunku Prawdopodobieństwa i Teorii Błędów (in Polish: An Outline of Probability and the Error Theory). Towarzystwo Bratniej Pomocy Studentów PW (edited by the students organization soon confiscated by the communist government), Warsaw, pp. 1–100 (1948)

[12] Uspensky, J.V.: Introduction to Mathematical Probability, p. s.411. McGraw-Hill, New York (1937)

[13] Stigler, S.M.: The Dark Ages of Probability in England: The Seventeenth Century Work of Richard Cumberland and Thomas Strode. International Statistical Review / Revue Internationale de Statistique 56(1), 75–88 (1988)

[14] Stigler, S.M.: Chance is 350 Years Old 20(4), s.33–s.36 (2007)

[15] Bayes, T.: An Essay towards solving a Problem in the Doctrine of Chance. Philosophical Transactions of the Royal Society of London 53, 370–418 (1763)

[16] de Moivre, A.: The Doctrine of Chances or A Method of Calculating the Probabilities of Events in Play. The Third Edition, Fuller, Clearer, and more Correct than the Former, London, A. Millar, pp. 1–378 (1756), Digitized by Google, Internet

[17] Wieleitner, H.: Geschichte der Mathematik. Part I from Descartes to about 1800. Leipzig 1911-21, 2 vols. (the reference based on the Russian translation: История Математики од Декарта до середины XIX столетия, Наука, Москва 1956)
[18] Simon, P., de Laplace, M.: A Philosophical Essay on Probabilities. Transl. Fr. French. Dover (1951)
[19] Laudański, L.M.: Statystyka nie tylko dla licencjatów (in Polish: Statistics not only for undergraduates) part1, part2, 2nd edn. Publishing House of the Rzeszow TU, Rzeszów (2009)
[20] Papoulis, A.: Probability, Random Variables, and Stochastic Processes, p. s.583. McGraw-Hill, Nework (1962)
[21] O'Flynn, M.: Probabilities, Random Variables, and Random Processes, p. s.523. Harper & Row, Cambridge (1982)
[22] Stigler, S.M.: Isaac Newton as a Probabilist. Statistical Science 21(3), s.400–s.403 (2006)
[23] Benjamin, A.T., Quinn, J.J.: Proofs That Really Count. The Art of Combinatorial Proof. Mathematical Association of America, p. s.194 (2003)
[24] Rubin, E., Schell, E.D.: Questions and Answers. The American Statistician 14(4), 27–30 (1960)
[25] Feller, W.: An Introduction to Probability Theory and Its Applications, 3rd edn. I, Posthumous Edition, p. 509. John Wiley and Sons, New York (1971)
[26] Fisz, M.: Rachunek Prawdopodobieństwa i Statystyka Matematyczna (in Polish: Probability and Mathematical Statistics). Posthumous 3rd edn., p. s.694. PWN, Warszawa (1967)
[27] Good, I.J.: Some Statistical Applications of Poisson's Work. Statistical Science 1(2), 157–170 (1986)
[28] Gumbel, E.J.: Bortkiewicz, Ladislaus von. International Encyclopedia of Statistics 1, 24–27 (1978); reprinted from the International Encyclopedia of Social Sciences (1968)
[29] Tennant-Smith, J.: BASIC Statistics. Butterworths, London (1985)
[30] Kingman, J.F.C.: Poisson Processes. Oxford UP (1993); translated into Polish. PWN, Warszawa (2002)
[31] Zubrzycki, S.: Wykłady z rachunku prawdopodobieństwa i Statystyki Matematycznej (in Polish: Lectures on Probability and Mathematical Statistics), p. 334. PWN, Warszawa (1966)
[32] Neyman, J.: Zasady rachunku prawdopodobieństwa i statystyki matematycznej, p. s.258. PWN, Warszawa (1969); English origin, First Course In Probability and Statistics. Henry Holt & Co., New York (1950)
[33] Dziennik Samuela Pepysa (The Diary of Samuel Pepys). Translated into Polish and selected Maria Dąbrowska, vol. 1(1660-65), vol.2 (1666-69). PIW, Warsaw (1952)
[34] Pearson, K.: On the Criterion that a Given System of Deviations from the Probable in the Case of the Correlated Systems of Variables Is Such That it Can be Reasonably Supposed to Have Arisen from Random Sampling. Philosophical Magazine 50, 157–175 (1900)
[35] Kemp, A.W., Kemp, C.D.: Weldon's Dice Data Revised. The American Statistician 45(3), 216–222 (1991)
[36] Haight, F.A.: Handbook of the Poisson Distribution. Wiley, New York (1978)

[37] A Guide to the Microfilm Edition of: The Emil J. Gumbel Collection contains 10 pages biography written by Arthur Brenner (Leo Baeck Institute, New York). University Publications of America. No year of publication (approximately 1990), p. 32
[38] Stigler, S.M.: Poisson on the Poisson Distribution. Statistics & Probability Lectures 1, 33–35 (1982)
[39] Gumbel, E.: Statistics of Extremes. Columbia University Press, New York (1958/1962); Russian translation of B. Ю.Тамарский (W.Yu. Tamarski), Mir, Moscow 1965, pp. 450, Lit.: 647 entries
[40] Press, H.: The Application of the Statistical Theory of Extreme Values to Gust-load Problems, NACA Report 991, Washington, p. 16 (1949)
[41] von Bortkiewicz, L.: Das Gesetz der kleinem Zahlen, p. 60. Teubner, Leipzig (1898)
[42] Gosset, W.S.: An Explanation of Deviation from Poisson's Law in Practice 12(3/4), 211–215; also: Paper in Student's, Collected Papers, London, Biometrica Office, pp. 65–69 (1942)
[43] Gurland, J.: Some Applications of the Negative Binomial and Other Contagious Distributions. Pdf file No.1388, Wikipedia, pp. 1–12; Printed in A.J.P.H. 49(10), 1388–1399 (October 1959)
[44] de Montmort, P.: Essay d'analyse sur les jeux de hazard, 2nd edn., p. 468. Jacque Quillau, Paris (1713) (pdf copy accessible by Internet)

Chapter 5
Normal Distribution Binomial Heritage

Acquaintance with the normal distribution, tables of the normal distribution. Probabilistic paper. Sample means distribution and Monte Carlo simulation. Two theorems of de Moivre-Laplace. When does normal approximation fit binomial distribution data? [Heritage of F.Gauss and Marquis de Laplace] –

5.1 Normal Statistics, Preliminaries

To understand how specific and how universal the normal distribution is, the point based on the Central Limit Theorems of the Theory of Probability should be taken. A routine course presenting this theory usually closes with the Central Limit Theorems. Therefore, the foregoing presentation may only present these results without supplying the Student with any rigorous proofs and leaving out the details. Thus, when indicating possible courses which present a similar approach, let us first mention a book by Weinberg [1] referred to earlier rather than that by Neyman [2], however, this remark is addressed more to the instructor than to the student. From an intuitional point of view a very important element of the limit theorems seems to be the fact that the normal distribution is the result of the sum of a number of random components (strictly speaking they are random variables) not necessarily of precisely defined nature (which stands for the knowledge of their distributions)[1]. How universal the normal distribution is has constituted a heated subject of discussions or even bitter quarrels among mathematicians and statisticians for more than a century. There is no doubt about its power, but there is also no doubt about its limitations. The human species displays a wide range of such applications, from the purely physical (stature or weight) to mental (such as IQ or grades). One of its special applications is mass products. From a theoretical point of view the normal distribution has a unique property: invariance regarding linear transformations. The first encounter with such a property was offered by Chapter 1.

---

[1] Just here we may recall a rule of the thumb well known in Statistics and using at least twelve uniformly distributed components to get a sample of the normal distribution.

L.M. Laudański: Between Certainty and Uncertainty, ISRL 31, pp. 129–157.
springerlink.com

Below we commence with presenting the normal distribution from scratch. It is given in general form by (5.1) showing a real function of the real variable with two parameters denoted by $\overline{x}, \sigma^2$ and these symbols suggest the special meaning of the parameters. Of course $\overline{x}$ denotes the basic mean, and σ^2 denotes the variance.

$$f(x) = \frac{1}{\sigma\sqrt{2\pi}} e^{-(x-\overline{x})^2/2\sigma^2} \quad \text{here: } -\infty < x < \infty \tag{5.1}$$

The above stated properties of the normal distribution will be proved (see p.5.4) – here we limit ourselves to presenting the defining steps. The basic mean is formally defined by:

$$\mu = \int_{-\infty}^{+\infty} x \cdot f(x)\, dx \tag{5.2}$$

To avoid a clash of symbols in (5.2) symbol μ also appears with respect to the first moment. The proof will justify that substitution of (5.1) by (5.2) will give:

$$\mu = \overline{x} \tag{5.3}$$

To make (5.2) more familiar, let us recall the formal rule to derive the basic mean on the ground of the grouped data $\mu = 1/N \sum_{i=1}^{N} f_i \cdot x_i$, assuming that f_i/N is replaced by $f(x)$ interprets (5.2) as the formula suitable for the continuous distribution such as normal distribution. Similar reasoning can be applied with respect to the formal rule deriving the variance:

$$\sigma^2 = \int_{-\infty}^{+\infty} (x-\overline{x})^2 \cdot f(x)\, dx \tag{5.4}$$

To be illustrated graphically the distribution given in (5.1) should be slightly modified. This modification means that the mean value has been assumed as zero. The result of such a transformation leads to the centered distribution (5.5):

$$f(x) = \frac{1}{\sigma\sqrt{2\pi}} e^{-(x)^2/2\sigma^2} \tag{5.5}$$

We shall comment the content of Fig. 5.1 depicting three well known bell-shaped curves. It gives the best opportunity to explain the role of the variance particularly regarding normal distribution due to the fact, that Fig.5.1 presents three diagrams preserving the real scales in both coordinates showing appropriate numerical values. Regarding these values we are also going to pay attention.

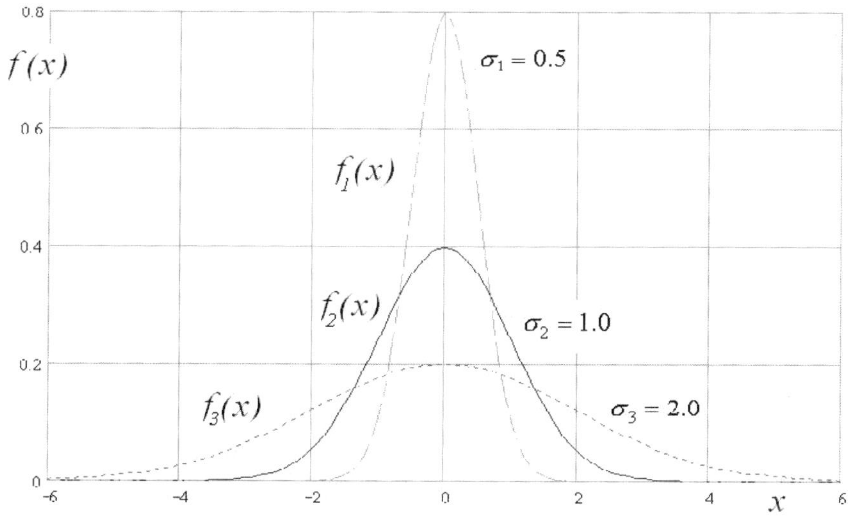

Fig. 5.1 Centered normal statistics with different variances

Fig. 5.1 gives the first opportunity to mark the ranges of the normal distribution with respect to changing variances. But first let us denote the maximum value for each diagram: they decrease in the following proportion: **0.79788456**, **0.39894228** and **0.19947114** and their relative values correspond to the ratios **4:2:1**. The above given maxima can be easily calculated by using a *scientific calculator* substituting into (5.6) the suitable values of σ :

$$f(x=0;\ \sigma) = \frac{1}{\sigma\sqrt{2\pi}} \tag{5.6}$$

This initial discussion directly leads to another simplification which this time plays a very important role in Statistics. Now we insert $\sigma = 1$ in (5.5). Such a value leads to the *standardized normal distribution* with $\mu = \overline{x} = 0$ and $\sigma = 1$. This important function is given by (5.7):

$$f(x) = \frac{1}{\sqrt{2\pi}} e^{-x^2/2} \tag{5.7}$$

Formula (5.7) will keep our attention for some time. However, earlier two intermediate definitions have to be given which due to the field of interest also play an important role. The first of these two is the so called *error function* defined by:

$$erf(t) = \frac{1}{\sqrt{2\pi}} \int_{-t}^{t} e^{-z^2/2} dz \tag{5.8}$$

Due to the importance of this function we depict it in Fig. 5.2.

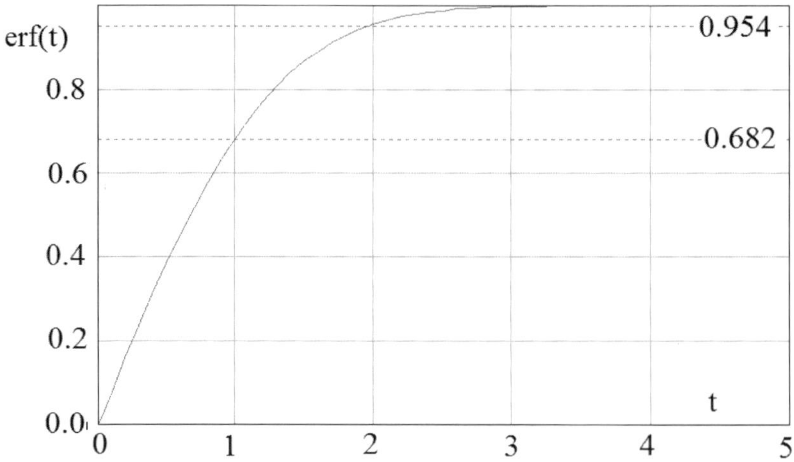

Fig. 5.2 Error function determined by (5.8)

Moreover few values of the error functions are given below:

$$\begin{aligned} \text{erf}\,(1) &= 0.68268949 \\ \text{erf}\,(2) &= 0.95449976 \\ \text{erf}\,(3) &= 0.9973002 \\ \text{erf}\,(4) &= 0.99993668 \end{aligned} \tag{5.9}$$

What has to be noted here concerns the possibilities of solving the integral (5.8) which cannot be done precisely and the above given values have been found numerically. Geometric interpretation of the definite integral is the area under the integrated function in certain limits. Therefore, to interpret for instance what it means that erf(1) = 0.68268949 requires resorting to Fig. 5.3 which shows the curve $\sigma = 1$ (see Fig. 5.1) colored grey between the limits (-1, +1).

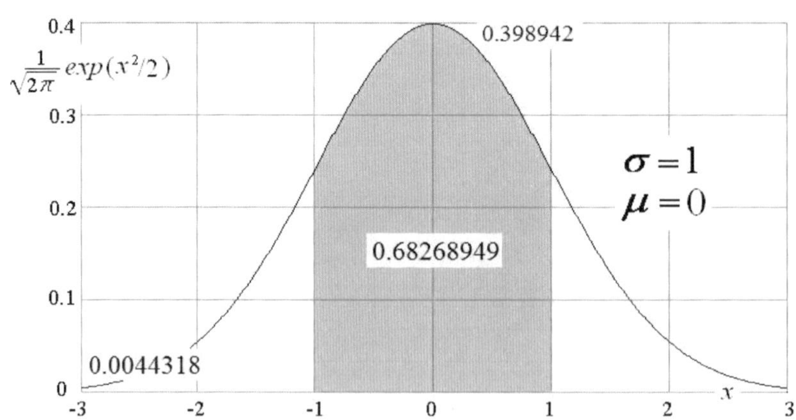

Fig. 5.3 The meaning of the error function seen in Fig. 5.2

Subchapter 5.3 follows up on the idea shown in Fig. 5.3 once more. In the next step of these preliminaries we provide the definition of the distribution function:

$$F(\xi) = \int_{-\infty}^{\xi} f(x)dx \tag{5.10}$$

Which for the normal variable has the following form :

$$F(\xi) = \frac{1}{\sqrt{2\pi}} \int_{-\infty}^{\xi} \exp\left(-x^2/2\right) dx \tag{5.11}$$

Expression (5.11) takes into account that $\sigma = 1$ and $\mu = \overline{x} = 0$ as shown in Fig.5.4 and the Student is warned about a possible confusion of the distribution function with the error function.

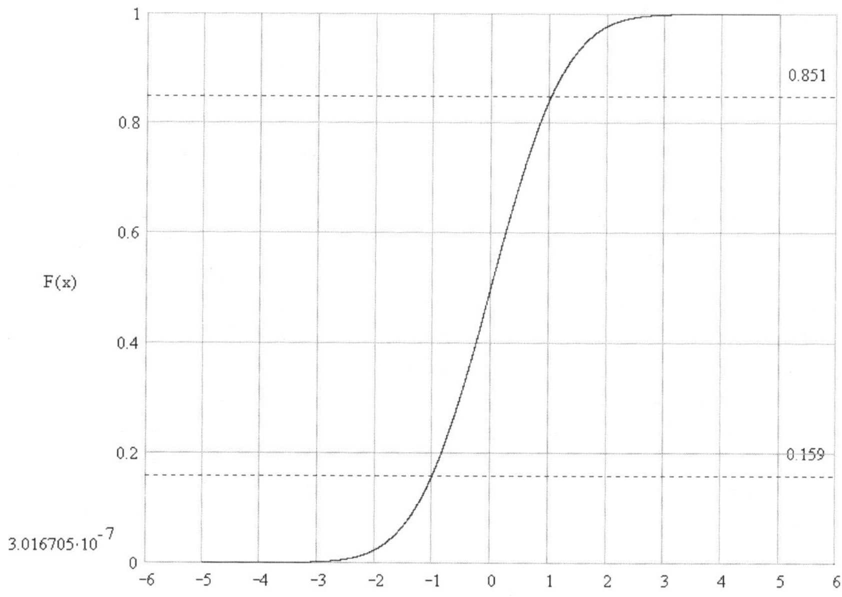

Fig. 5.4 The distribution of the normal variable

To emphasize our warning we provide below a new set of numerical values:

F(0)=0.5	F(3)=0.9986501	
F(1)=0.84134475	F(4)=0.99996833	(5.11a)
F(2)=0.97724987	F(5)=0.99999971	

And here is an example showing mutual correspondence between the error function and the distribution function:

$$erf(2) = 2 \cdot (F(2) - F(0.5)) \tag{5.12}$$

Closing this subchapter – we shall present another graphic idea related to the normal statistics. What we are going to present is frequently called *probabilistic paper* – and in particular a paper on the normal distribution. In order to do this we must define $T(u)$ the function which is the reverse function with respect to the distribution function given by:

$$\frac{1}{\sqrt{2\pi}} \int_{-\infty}^{T} \exp\left(-v^2/2\right) dv = u \tag{5.13}$$

The horizontal scale of Fig. 5.5 shows the values of this function – while the vertical scale represents the regular pattern. The vertical scale in Fig. 5.5 is linear while the horizontal scale corresponds to the non-linear function 5.13. Depicting the normal distribution function using this *paper* leads to a straight line. Probabilistic paper serves as an auxiliary tool to help recognize the type of the statistical distribution. If the empirical values show a pattern close to a straight line – by using a probabilistic paper of a particular distribution - it confirms the right strike. To illustrate the application of such a tool Fig. 5.5 shows grades of students from one of the classes taught by the Author of this book. These are the final examination results. Traditional grades from "C" to "A" – given as 3.0, 3.5, 4.0, 4.5 and 5.0 are supposed to belong to normal distribution.

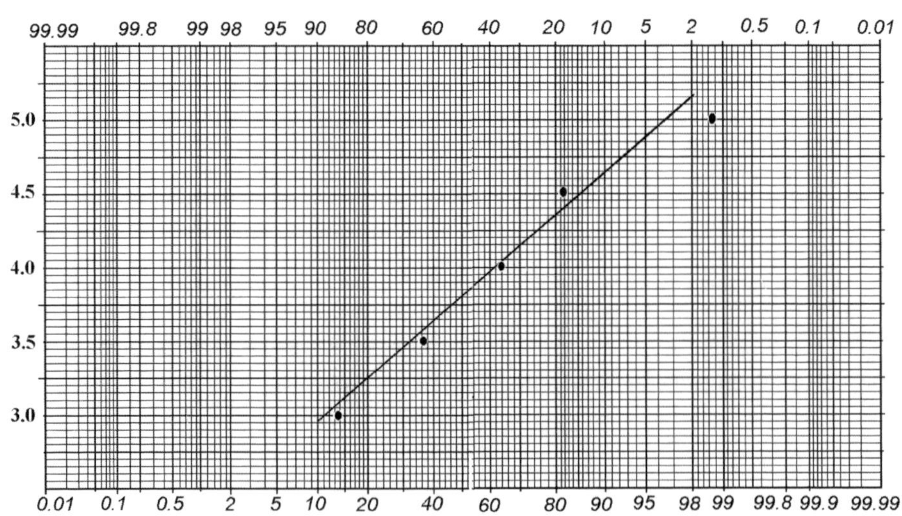

Fig. 5.5 Normal probability graph paper to test students' grades

5.2 Four Properties of the Normal Distribution

The properties refer to the normal curve depicted in Fig. 5.3 remarking that the independent variable of this curve has to be identified with the concept of the *z-score statistics*. Therefore let us recall an appropriate definition which is based upon both parameters of the normal distribution: its mean and its variance:

$$z_x = \frac{x_i - \overline{x}}{\sigma_x}$$

1. Normal distribution is concentrated at its origin corresponding to its mean value equal to zero. According to the exponential law expressed by (5.7) the rate of decrease is so high that practically the values of this function diminish outside the range (-3, +3).
2. The bell shaped curve is symmetric regarding the origin. Therefore determining its values for the positive arguments simultaneously determines them for the negative arguments, therefore this property has an exact mathematical formulation – it is called the *even function*:

$$f(-x) = f(x) \tag{5.14}$$

3. Normally distributed statistics have an infinite number of entities belonging to the continuous distributions. This property sometimes misleads beginners who are discouraged by the infinite span-wise character of the normal distribution.
4. The normal curve belongs to the class of single-mode curves that posses a single extreme (maximum) value. The middle value – called median - of the normal statistics is identical with its mode value and both are identical with its mean value.

5.3 Making Use of the Statistical Tables of the Normal Distribution

A specific feature of the normal statistics belonging to the class of continuous distributions is the fact that we do not have this information which is also related to the discrete distributions examined in Chapter 4 - namely we do not have the number of its terms and have to make conclusions about them by determining the area under the curve such as given by (5.7). Having the probability (density) distribution we have to resort to mathematics in order to find an appropriate value of the definite integral within certain limits. In general it can be done precisely, but not in this particular case (we have already mentioned it). Therefore we have to resort to the numerical methods of Analyses - which in numerical practice means that we are willing to use intermediate tools available on the market such as *Excell*, *MathCad*, *Statistica* , etc. In fact this practice is easier thanks to the popular Maths Tables and using them the Student can obtain the desired solution almost at once. Below is presented the first Example of this kind.

Example 5.1. Determine the fraction of the normal statistics which corresponds to the values of ***z-scores*** greater then 0.0 and smaller than 0.7.

Solution Fig. 5.6 suggests that we have to determine the highlighted area under the curve. Analytically it is expressed in the following form:

$$\frac{1}{\sqrt{2\pi}}\int_{0}^{0.7}\exp\left(-x^2/2\right)dx \approx 0.258036$$

Therefore the point is: how to get the above result? Looking at the upper part of Table 5.1 we propose a section – slightly cut down after removing some unnecessary parts to show how to easily obtain the desired result:

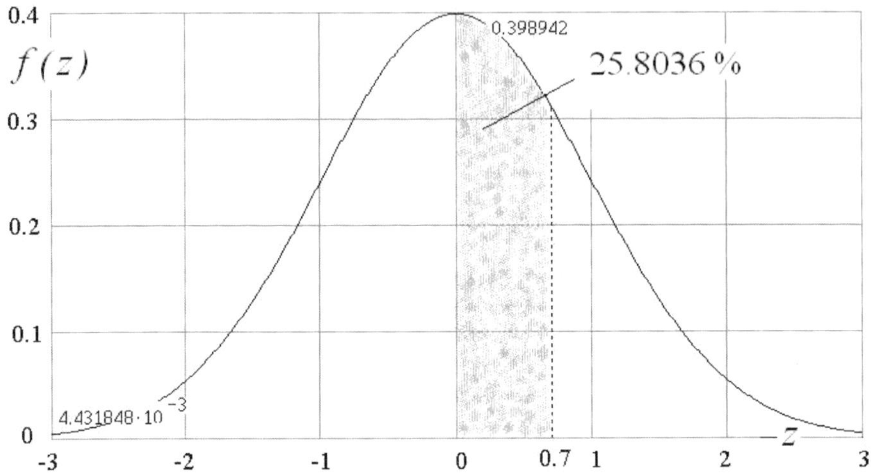

Fig. 5.6 For the solution of Example 5.1

Table 5.1 Initial part of the standard normal distribution table

z	.00	.01	.02	.03	.04	.05	.06	.07	.08	.09
0.0	.0000	.0040	.0080	.0120	.0160	.0199	.0239	.0279	.0319	.0359
0.1	.0398	.0438	.0478	.0517	.0557	.0596	.0636	.0675	.0714	.0753
0.2	.0793	.0832	.0871	.0910	.0948	.0987	.1026	.1064	.1103	.1141
0.3	.1179	.1217	.1255	.1293	.1331	.1368	.1406	.1443	.1480	.1517
0.4	.1554	.1591	.1628	.1664	.1700	.1736	.1772	.1808	.1844	.1879
0.5	.1915	.1950	.1985	.2019	.2054	.2088	.2123	.2157	.2190	.2224
0.6	.2257	.2291	.2324	.2357	.2389	.2422	.2454	.2486	.2517	.2549
0.7	**.2580**	.2611	.2642	.2673	.2704	.2734	.2764	.2794	.2823	.2852
0.8	.2881	.2910	.2939	.2967	.2995	.3023	.3051	.3078	.3106	.3133

The suitable upper limit of the integral under consideration is composed of the components of the first row and the first column of Tab. 5.1. This is a frequent trick used in creating such tables. In our example we combine 0.7 from the first column and 0.00 from the first row giving us 0.70 which leads to the value shown in bold in the above section, that is number **0.2580**. This value is given as a *fraction* – of the value 1.0. The table included in this book shows similar values but with higher accuracy and displays them as percentage.

Similar procedures allow to solve much more advanced problems – as seen in Part Two of this book. We do not provide the Student with an exhausting survey of formal matters related to the mathematical tables which usually are to be found in selected tools accompanying statistical books, but also in special publications.

5.4 Two Proofs

We mentioned earlier the presentation of the proofs justifying two important results regarding the mean and the variance. We present them in the same order.

In the first step the following substitution will take place:

$$z = \frac{x - \overline{x}}{\sigma} \qquad \text{or} \qquad x = \sigma \cdot z + \overline{x} \tag{5.15}$$

Then the defining formula (5.1) is substituted into (5.2) initiating the following chain of equivalences:

$$\mu = \int_{-\infty}^{+\infty} x \cdot f(x)dx \rightarrow \mu = \int_{-\infty}^{\infty} x \cdot \frac{1}{\sigma\sqrt{2\pi}} \cdot e^{-\frac{(x-\overline{x})^2}{2\sigma^2}} dx \rightarrow \mu = \frac{1}{\sigma\sqrt{2\pi}} \int_{-\infty}^{\infty} (\sigma \cdot z + \overline{x}) \cdot e^{-\frac{z^2}{2}} \cdot \sigma \cdot dz$$

$$\mu = \frac{\sigma}{\sqrt{2\pi}} \int_{-\infty}^{\infty} z \cdot e^{-\frac{z^2}{2}} dz + \frac{\overline{x}}{\sqrt{2\pi}} \int_{-\infty}^{\infty} e^{-\frac{z^2}{2}} dz$$

The Student should check that the first integral leads to zero, while the second to the one, therefore in the last instant it closes the proof:

$$\mu = \frac{1}{\sqrt{2\pi}} \cdot 0 + \overline{x} \cdot 1 = \overline{x} \tag{5.16}$$

Coming to the second procedure we commence by a formal proposal, instead of (5.5), we propose the following purely symbolic alteration of the problem:

$$\text{Var}\,(x) = \int_{-\infty}^{+\infty} (x - \overline{x})^2 \cdot f(x)dx = \sigma^2$$

Once $f(x)$ is replaced by (5.1) the initial step takes the shape:

$$\mathrm{Var}\,(x) = \int_{-\infty}^{\infty} (x-\overline{x})^2 \cdot \frac{1}{\sigma \cdot \sqrt{2\pi}} \cdot e^{-\frac{(x-\overline{x})2}{2\sigma^2}} dx$$

To proceed further with the proof requires application of (5.15) resulting in the form:

$$Var\,(x) = \frac{1}{\sigma \cdot \sqrt{2\pi}} \int_{-\infty}^{\infty} \sigma^2 \cdot z^2 \cdot e^{-\frac{z^2}{2}} \cdot \sigma\, dz$$ and then we get the desired result:

$$Var\,(x) = \frac{\sigma^3}{\sigma \cdot \sqrt{2\pi}} \int_{-\infty}^{\infty} z^2 \cdot e^{-\frac{z^2}{2}} dz = \frac{\sigma^2}{\sqrt{2\pi}} \cdot \sqrt{2\pi} \tag{5.17}$$

The above equivalence applies one more well known result on the value of the integral describing the exponential function under the integral.

5.5 The Central Limit Theorem – An Intuitive Approach

Below we propose a somewhat extensive procedure (as for the proportions of this book) which will result in a special kind of the normal distribution called the distribution of the sample mean. This distribution will be the only distribution of this kind belonging to the distributions of *mathematical statistics.*

To ensure appropriate general population for the foregoing theorem, open access to infinitely large statistics, that is to all its terms, should be assumed. Sometimes this statistics is called the *general population*. We shall call it the *population.* We do not know the probability distribution of this statistics, but we shall assume that the distribution allows to determine the mean and the variance which remain unknown. Imagine a statistical experiment allowing us to choose as many terms as we like from this population. Performing this procedure we will assume that we choose very time the same number of terms which we call *samples.*

Theorem 5.1. The Central Limit Theorem – Intuitionally. *Let us assume that as the first step M samples have been drawn from an infinite population . Each sample has N terms. As the second step the terms of each sample are summarized. Therefore, at the beginning of the third step we have only M numbers which we consider the terms of a new distribution which we call the "distribution of sample sums". It follows that this distribution is approximately normal.*

Below we shall try to present an approach offering necessary formal components to this Theorem. Let us denote the terms of the first sample as follows:

$$x_1^{(1)}, x_2^{(1)}, x_3^{(1)}, \ldots, x_N^{(1)} \tag{5.18}$$

accordingly, the *j-sample* is as follows:

$$x_1^{(j)}, x_2^{(j)}, \ldots, x_N^{(j)} \tag{5.19}$$

Concluding this stage it is apparent that both sequences (5.18) and (5.19) – present a single sample. As far as the number of samples M, its value remains open so far.
The next step presents the system of equations given below:

$$\begin{aligned} x_1^{(1)} + x_2^{(1)} + \ldots + x_N^{(1)} &= y_1 \\ x_1^{(2)} + x_2^{(2)} + \ldots + x_N^{(2)} &= y_2 \\ \vdots \qquad \vdots \qquad & \vdots \\ x_1^{(M)} + x_2^{(M)} + \ldots + x_N^{(M)} &= y_M \end{aligned} \tag{5.20}$$

Following the above given theorem CLT (*Central Limit Theorem*) – statistics y_i is normally distributed. The result is called the *distribution of sample sums.* There is close relation between the number N and the *quality* of the sample sums distribution.

To come to the kernel of the CLT we suggest the following virtual experiment. Assume first that we have a crate filled with a huge number of coins of the same dimension. Assume that each coin on one face has the digit "1" and on the other "0". Assume that a single coin is released at a time. Assume that the coins are well mixed – so we can be certain that we can get each coin in a purely random fashion. We read its designate and return it back to the crate. This experiment is repeated N=100 times. Remark that the number of samples M may be as great as we wish. Then there comes the question what the sample sum distribution will be.

In these circumstances it should be seen that the probability distribution is a binominal distribution with n =100 and p = 0.5. According to the CLT its limiting case would be normal distribution. This result has been discussed in Chapter 4 while here it is seen as a particular case of the CLT.

Nevertheless we may consider the above described experiment closer. From the binomial distribution follows for instance the probability of k = 50 successes corresponding to the mean value of the binomial distribution, thus we get:

$$P(k=50; n=100, p=0.5) = \frac{100!}{50! \cdot 50!} p^{50} \cdot q^{50}$$

The Word accessories include a calculator which can compute such values. The results close to the mean value are presented in Tab. 5.2.

Table 5.2 Binomial distribution – the right tail probabilities

50	0,07958923738717876149812705024217	49	0,078028664105077217155026519845265
48	0,073527010406707377703774989854192	47	0,066590499990980266599645273830212
46	0,05795839814029763944783940500037	45	0,048474296626430752992738411454854
44	0,038952559789096140797736223490508	43	0,030068642644214564826322698834778
42	0,02229226954657286702641165603268	41	0,015869073236543396866259144972417
40	0,010843866711637987858610415731151	39	0,0071107322699265494154822398237058
38	0,0044728799762441197936097960181375	37	0,0026979276047186754310662261696702
36	0,0015597393964779842335851620043406	35	0.00086385566574165280629332049471172

The numerical results presented in Tab. 5.2 may be used for many purposes. Here we suggest assessing to what extent they can be approximated by the normal distribution with the mean value 50 and the variance 25. It can be done directly and indirectly. For instance we can determine the ratio of the reduction of the distribution once it reaches $z = 1$, 2 and 3.

5.6 Distribution of Sample Means

The below Theorem 5.2 presents a particular case of the Theorem 5.1. Therefore we copied the formulation of the first and made necessary replacements/amendments.

Theorem 5.2. The Central Limit Theorem for Sample Means. *Let us assume that as the first step M samples have been drawn from an infinite population . Each sample has N terms. As the second step the terms of each sample are used to compute the mean of each sample. Therefore, at the beginning of the third step we have again M numbers which we consider the terms of a new distribution which we shall call the "distribution of sample means". Therefore it follows that also this distribution is approximately normal.*

If the Theorem 5.1 is true – then it guarantees the truth of the Theorem 5.2. Technical facilities at our disposal are too modest to present a rigorous proof of both theorems under our consideration. Therefore with respect to the Theorem 5.2 we limit ourselves to pointing out one possible proof. We have in mind a procedure resorting to the fact (which also should be taken for granted) that the normal distribution is invariant regarding linear transformations (of its variables). Therefore it seems enough to acknowledge that variables shown below as (5.21) are linearly dependent regarding variables given by (5.20).

Therefore our first step is given by the equations (5.21) shown below:

$$\begin{aligned} \frac{x_1^{(1)} + x_2^{(1)} + \ldots + x_N^{(1)}}{N} &= \mu_1 \\ \frac{x_1^{(2)} + x_2^{(2)} + \ldots + x_N^{(2)}}{N} &= \mu_2 \\ \vdots \qquad \vdots \qquad \vdots \quad & \quad \vdots \\ \frac{x_1^{(M)} + x_2^{(M)} + \ldots + x_N^{(M)}}{N} &= \mu_M \end{aligned} \tag{5.21}$$

According to the Theorem 5.1 probability distribution of the variables y_i is normal – therefore distribution of variables μ_i is also normal - because statistics μ_i and y_i are interrelated with the linear transformation:

$$\mu_i = \frac{y_i}{N} \tag{5.22}$$

Which concludes if not the proof then at least the procedure indicating a formal reason justifying the Theorem 5.2.

5.7 Properties of the Distribution of Sample Means

Distribution of the Sample Means possess some important properties from the point of view of its numerous applications. To satisfy the requirements imposed by the Probability Theory it is expected that the value of M can be infinite.

The *distribution of sample means* has the same mean value as the *general population*, due to the procedure described below:

Theorem 5.3. *As the first step an infinite number of samples is drawn from the original population. Each sample has N terms. As the second step the terms of each sample are used to compute the mean. They form the distribution of sample means. The mean value of this distribution is the same as the mean value of the original population.*

To formalize the sentence of the Theorem 5.3 let us denote the mean of the original population with the symbol μ, while the mean of the distribution sample means with the symbol $\mu_{\overline{X}}$. In this circumstances the thesis of the Theorem 5.3 will state what follows:

$$\mu_{\overline{X}} = \mu \tag{5.23}$$

Let us now come to the details expressed by the Theorem 5.3 by using the already developed formalism, first in the form :

$$\lim_{M \to \infty} \frac{\mu_1 + \mu_2 + \ldots + \mu_M}{M} = \mu \tag{5.24}$$

But in the second step we substitute the explicit terms μ_i given by (5.22) into (5.24) to get:

$$\lim_{M\to\infty} \frac{x_1^{(1)} + x_2^{(1)} + ... + x_N^{(1)} + x_1^{(2)} + x_2^{(2)} + ... + x_N^{(2)} + ... + x_1^{(M)} + x_2^{(M)} + ... + x_N^{(M)}}{N \cdot M} = \mu \qquad (5.25)$$

The result given by (5.25) contains evidence that the mean value of the sample means with the number of samples tending to infinity is calculated with respect to all terms of the original population, disregarding the sample volume, therefore it becomes identical with the mean value of the original population.

The last theorem in this sequence will explain the behavior of the *standard deviation of means.* It is described by the following procedure:

Theorem 5.4. *As the first step an infinite number of samples is drawn from the original population. Each sample has N terms. In the second step the terms of each sample are used to compute the mean. They form the distribution of sample means. The standard deviation of means equals the standard deviation of the original population divided by the square root of N.*

Denoting the standard deviation of the original population with σ and denoting the *standard deviation of means* with $\sigma_{\overline{X}}$ - we find that the Theorem 5.4 states

$$\sigma_{\overline{X}} = \sigma / \sqrt{N} \qquad (5.26)$$

To prove (5.26) we propose two steps. In the first step we shall prove that:

$$\sigma_y^2 = \sigma^2 \qquad (5.27)$$

To justify (5.27) claiming that statistics y_i retains the variability of the original population it will be sufficient to recall the Theorem 5.1. Then, let us consider the nature of transformations given by (5.20). Statistics y_i belongs to the class of statistics the values of which are *shifted* with respect to the values of the statistics x_i . We investigated this kind of a linear transformation in Chapter 1 concluding that such a transformation does not change the variability of the initial statistics. So, (5.27) is true.

In the second step let us notice that statistics y_i and statistics μ_i are in a linear relation:

$$\mu_i = \frac{y_i}{N} \qquad (5.28)$$

Let us recall once again the content of Chapter 1 regarding the transformation such as (5.28) denoted there with (1.13), while its consequences are presented by (1.15) and (1.17) - which are exactly the same as (5.26). Theorem 5.4 opens a wide field of applications. To support this suggestion a suitable example is provided below.

Example 5.2 *The parent population collects IQ test results describing the normal distribution with mean value equal to 100, and the standard deviation 15. Suppose we gather an infinitude of random samples of 10 scores each. Determine the distribution of the sample means obtained in this way and examine it numerically.*

This example offers a temptation which we cannot resist. It gave rise to the famous result whose author was for long time known as "Student". We know from the closing part of Chapter 1 that it was ***William Sealy Gosset*** who published paper [11] introducing his original idea – of determining the distribution of the sample means for *small samples* which is now known as *t-Student* distribution. This distribution occurs when the mother population has normal distribution with unknown variance. The point is that the *t-Student* distributions for samples of $N > 30$ become close to the normal distribution. Returning to the considered example according to Theorem 5.3 the sample means mean value is equal to the mean of the parent population:

$$\mu_{\overline{X}} = \mu = 100$$

On the other hand Theorem 5.4 will give us the standard deviation of means as:

$$\sigma_{\overline{X}} = \frac{\sigma}{\sqrt{N}} \quad \Rightarrow \quad \frac{15}{\sqrt{10}} \cong 4.7434 \tag{5.29}$$

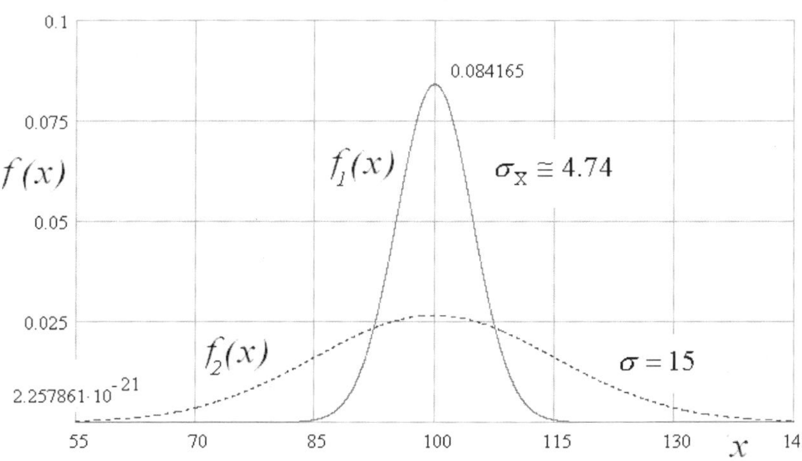

Fig. 5.7 Parent distribution and sample means distribution of Example 5.2

The best comment for the mutual interrelations between the two investigated distributions is offered by Fig. 5.7. According to Theorem 5.3 they have common means, and according to Theorem 5.4 their standard deviations are defined in (5.29). And using the *MathCad* package we get Fig. 5.7 which presents diagrams of both distributions (see how the rule of three-sigma is obeyed).

To complement the obtained solution we also include both analytical forms of both distributions as shown in (5.30). Some simple calculations are left for the Student.

$$f_1(x) = \frac{1}{15\sqrt{2\pi}} e^{-\frac{(x-100)^2}{2\cdot(15)^2}} \quad \text{and} \quad f_2(x) = \frac{1}{(4.7434)\sqrt{2\pi}} e^{-\frac{(x-100)^2}{2\cdot(4.7434)^2}} \tag{5.30}$$

A brief comment to the above presented results indicates a conclusion on sampling with an increasing number of terms. The foregoing distributions gradually concentrate closer and closer to the mean value. To support this obvious conclusion we add one more result presenting three sample means distributions – depicted in Fig. 5.8. The third one describes samples with the number of entries 100:

$$N = 100 \quad \Rightarrow \quad \sigma_{\overline{X}} = \frac{\sigma = 15}{\sqrt{N}} \quad \Rightarrow \quad \sigma_{\overline{X}} = 1.5 \tag{5.31}$$

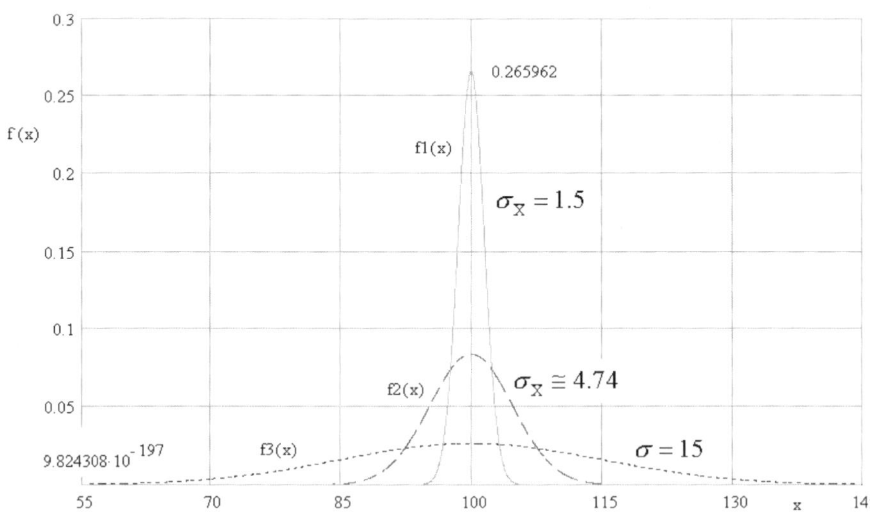

Fig. 5.8 Two sample means distributions with their parent distribution

5.8 To Initiate the Monte Carlo Simulation

Commencing with this exciting and important tool of Statistics and/or Probability let us recall Chapter 2 and the procedure of grouping variables which will be used afterwards. The point is that here we can use a very special kind of raw statistical data , i.e. pseudo random numbers uniformly distributed along the interval (0, 1) which will be grouped afterwards. We have in mind a simple tool in the form of a *scientific calculator* which is usually supplied with the *RND generator*. In fact it presents a secret of the manufacturers who supply the user with an instruction recommending extremely simple handling: just press a button to get the successive digit. Students today usually cannot imagine what sophisticated tools are accessible this way. Not so long ago – to make use of the random numbers the only way was to read them from special mathematical tables. Returning to the point it must be stated that the smallest random number seen on the display is 0.000, and the biggest 0.999. To come to the uniform probability distribution on the interval (0, 1) we provide formal steps leading to its mean and the variance:

$$\overline{x} = \int_0^1 x \cdot f(x) \cdot dx \qquad \sigma_x^2 = \int_{-\infty}^{\infty} (x - \overline{x})^2 f(x) dx \quad \text{here } f(x) \equiv 1 \text{ so: } \overline{x} = \tfrac{1}{2} \tag{5.32a}$$

$$\sigma^2 = \int_0^1 (x - \tfrac{1}{2})^2 dx = \tfrac{1}{3} - \tfrac{1}{2} + \tfrac{1}{4} = \tfrac{1}{12} \quad \text{then} \quad \sigma = \sqrt{\tfrac{1}{12}} = \frac{\sqrt{3}}{6} \approx 0.288675134... \tag{5.32b}$$

Commenting both results, the first one stating $\overline{x} = \frac{1}{2}$ seems to be quite obvious, but the second result stating that $\sigma^2 = \frac{1}{12}$ does not look so obvious. An *RND generator* has been used to produce a sample with 50 terms collected in Tab. 5.3. Maybe it will be reasonable to say that the succession shown in Tab. 5.3 omits the fact whether the terms have been generated preserving rows or columns.

Table 5.3 Random numbers generated by a RND generator

.993	.953	.982	.835	.327	.746	.564	.039	.029	.222
.521	.704	.126	.180	.459	.055	.186	.779	.714	.768
.152	.270	.724	.165	.333	.000	.276	.987	.709	.889
.229	.443	.898	.027	.360	.397	.778	.465	.489	.298
.586	.412	.063	.628	.556	.506	.998	.825	.450	.131

The statistics collected in Tab. 5.3 has been evaluated to determine the frequency histogram shown in Fig. 5.9. To establish the number of classes the rule of the thumb described in Chapter 2 has been used. Therefore, we have chosen 5 classes – so the class interval is equal to 0.2. All particular limits (preserving the left continuous) are also shown in Fig. 5.9.

The evidence for the left continuous can be found in the first interval due to the appearance of the term 0.000. Now, let us ask whether the frequency histogram shown in Fig. 5.9 documents sufficiently or rejects the hypothesis of the uniform distribution. Mathematical statistics offer in this respect some special tools to

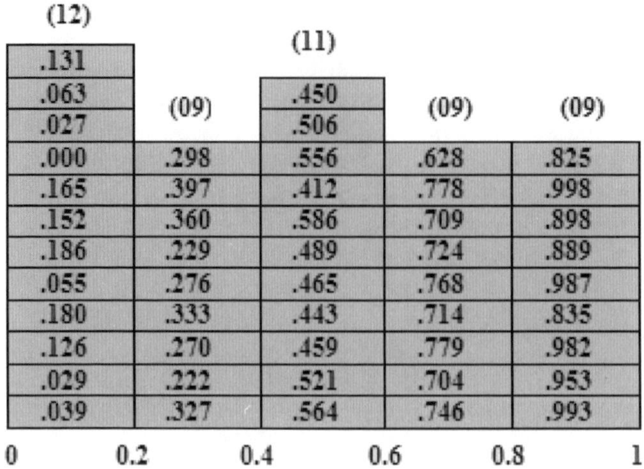

Fig. 5.9 Frequency histogram for data given in Tab. 5.3

resolve such problems (at least a part of them) but here we have no access to them, therefore we can – again – resort to the concept of the *Probability Graph Paper* (see [12] pp.212-215) already mentioned. In this case we present Fig. 5.10.

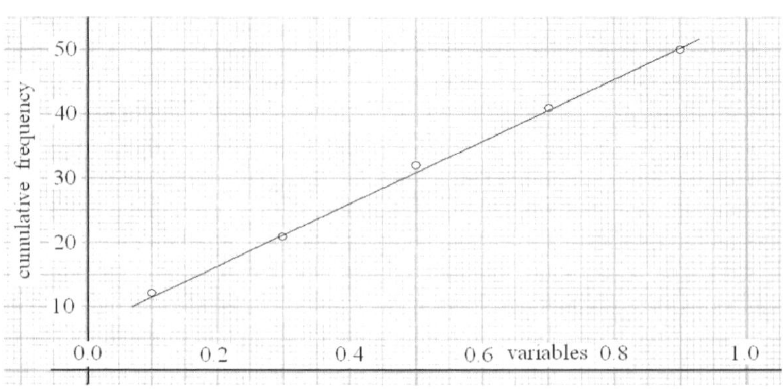

Fig. 5.10 Test on uniform probability graph paper

It seems that a visual examination of the hypothesis about the uniform distribution with respect to the sample given in Tab. 5.3 carried out with the help of Fig. 5.10, does not reject it. In other words – apparently the quality of the RND generator which has been used to generate this sample is rather good. With this example we came closer to concept in the title. However, such a tool as a *scientific calculator* cannot ensure full access to the matter. As the next step in the desired direction let us now present the idea of the generator which makes use of the rule of the thumb mentioned at the beginning of this Chapter in the footnote. This idea makes direct use of the sample means distribution although this time we resort to the uniform parents distribution. Therefore we recall first of all the system of equations given by (5.20) which now has the designate:

$$\begin{aligned} x_1^{(1)} + x_2^{(1)} + \ldots + x_N^{(1)} &= y_1 \\ x_1^{(2)} + x_2^{(2)} + \ldots + x_N^{(2)} &= y_2 \\ \vdots \qquad \vdots \qquad \vdots \\ x_1^{(M)} + x_2^{(M)} + \ldots + x_N^{(M)} &= y_M \end{aligned} \tag{5.33}$$

The difference lies in the fact that samples $x_1^{(j)}, x_2^{(j)}, \ldots, x_N^{(j)}$ are gathered from pseudo-random numbers uniformly distributed over the range (0, 1). Nevertheless for the foregoing purposes every term x_i has to undergo a linear transformation which will transpose it into a *z-scored* number – of which the mean is zero and the variance is equal to one. Having in mind fresh results given by (5.32a) and (5.32b) the desired transformation is determined by:

$$z_i = \frac{x_i - \overline{x}}{\sigma_x} = \frac{x_i - 0.5}{\sqrt{1/12}} \tag{5.34}$$

Now let us state clearly what we are going to obtain from the procedure defined by (5.33) regarding the terms determined by (5.34). Our target result has to be the standard normal distribution. Therefore we have to resort to the properties guaranteed by the sample sums theorem. Therefore there must appear the next new variable determined by:

$$\zeta_j = \frac{\sum_{i=1}^{N} z_i^{(j)}}{N} \tag{5.35}$$

Following the Theorem 5.3 this new variable ζ_j has the same mean value as the parent population made of variables z_i – it means that it is zero. The last step results from the Theorem 5.4 – which via (5.26) gives the relation between the appropriate variances. Therefore, to assess that the target variables λ_j will conform to the *z-score* requirements, values ζ_j and $\sqrt{N}$ need to be divided. Combining all the above statements will result in the final formula determining statistics λ_i from statistics x_i :

$$\lambda_j = \frac{\sum_{i=1}^{N} x_i^{(j)} - 0.5}{N \cdot \frac{\sqrt{1/12}}{\sqrt{N}}} \tag{5.36}$$

For $N = 12$ - formula (5.36) gives an extremely simple result:

$$\lambda_j = \sum_{i=1}^{12} x_i^{(j)} - 0.5 \tag{5.37}$$

Moreover (5.37) explains why we speak about the *rule of dozen*. If the above procedure despite all our efforts still leaves some uncertainty, we have reserved the most convincing procedure: by showing this result evidently providing the appropriate simulation. The point is that to present this procedure we no longer can use the *scientific calculator*. To calculate the sample of the desired standard normal statistics counting 50 terms as shown in Tab. 5.4 we have to use a computer code. The results collected in Tab. 5.3 could ensure getting only the first four terms of the desired statistics which (as we know from Chapter 2) cannot be considered even as a small sample. The Author of this book published a paper [5] where the inquiring Student may find many more details referring to such simulations.

Table 5.4 Simulated statistics λ $N = 12,\ M = 50$

0.10	0.77	1.17	-0.59	-0.99	-0.45	0.40	-0.63	-0.56	-0.61	-1.46	-0.59	-0.58	-2.42	1.13
0.41	-0.68	1.02	-0.27	-1.09	0.26	0.20	-1.11	-0.03	0.21	1.22	-0.86	0.29	-0.10	1.04
0.39	-0.33	-0.21	-0.80	-0.87	-0.08	-0.28	1.73	-2.04	0.83	-0.06	0.70	0.07	-1.81	0.08
1.40	-0.98	-0.20	1.22	0.12										

Usually quite a satisfactory frequency histogram for the raw statistics λ requires to have a number of terms M in the order of hundreds. Such data can be seen in [5]. A sample gathering only 50 terms here given in Tab. 5.4 allows to be grouped into not more than 5 classes. Therefore this stage has been here skipped. Instead Tab. 5.5 presents a cumulated histogram in order to test it using the *Probability Graph Paper* of the normal distribution. The result is seen in Fig. 5.11. A cautious conclusion is that the test does not reject the normal distribution hypothesis.

Table 5.5 Grouped data for data given in Tab. 5.4

Middle values	-2.5	-1.5	-0.5	0.5	1.5
Cumulated class	0.04	0.12	0.54	0.82	0.98

Having access to the *Statistica* package we cannot resist the temptation to apply the *test chi-square* called "goodness of fit" which does not reject this hypothesis at the confidence level $\alpha = 0.13$.

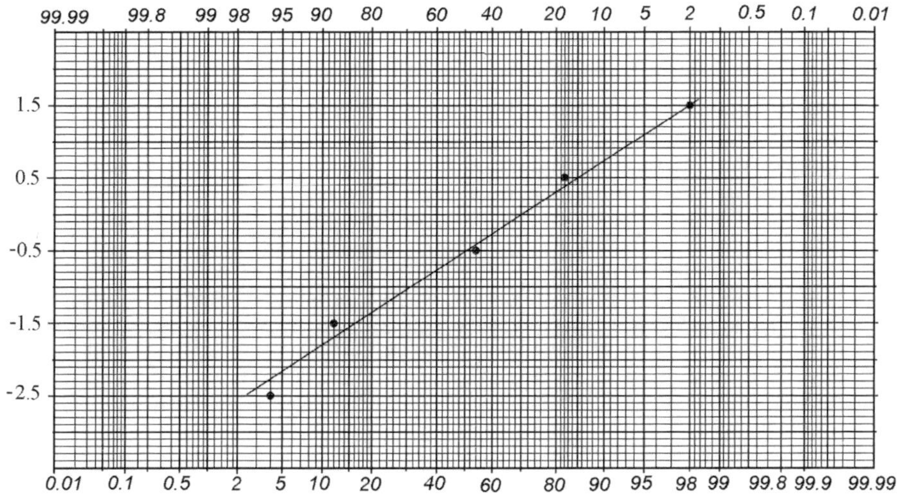

Fig. 5.11 Testing on normal probability graph paper of Tab.5.5 data

5.9 De Moivre–Laplace Limit Theorems

Regarding the fact who of the first few Christiaan Huygens, Blaise Pascal, Jacob Bernoulli or Abraham de Moivre may be credited as being the first to write the first tract on *Probability* of which even the name varies from one proposal to another, from *"Ratiociniis in Ludo Aleae"*, through *"Ars Conjectandi"* to *"The Doctrine of Chances"*, historians are divided, nevertheless the position of the book listed as [8] – is unquestionably important and significant. In this subchapter we present strong evidence in favor of the above claim. We have in mind two limit theorems bearing the names of Abraham de Moivre and Pierre Simon, Marquis de Laplace. Having in mind that now the Student has open access to *"The Doctrine of Chances*" (see [8]) we commence this subchapter with a few lines taken directly from p.243 of [8] in finding which we have to acknowledge the help of S. Stigler [7]):

>> *Although the Solution of Problems of Chances often require that several Terms of the Binomial* $(a+b)^n$ *be added together, nevertheless in very high Powers the thing appears so laborious, and so great of difficulty, that few people have undertaken that Task; for besides James and Nicolas Bernoulli, two great Mathematicians, I know of no body that has attempted it;* <<

The above quotation is considered the best introduction to the following material. Stigler ([7], p.82) – on his side – extracted from the quoted excerpt of *"The Doctrine of Chances*" three numbers which he inserted in a table and this table is partially reproduced here in the second column of our Tab. 5.6. For an inquiring Student we have to add that the first number 0.682688 is to be found on p.246 of [8], while the two others i.e. 0.95428 and 0.99874 on p.248 of [8]. The details of the numerical procedure – as it is seen – take up several pages and we can warn the Student – that their study is not easy! Also, it seems, that the contemporary account of Stigler – does not help significantly. Then let us add that the position of the pages 243-254 in the book [8] is quite special: they follow after PROBLEM LXXIII, and before PROBLEM LXXIV. De Moivre explains that he repeats there his own paper which was ready on November 12, 1733. With everything so far in this Chapter the Student should easily recognize the value of the results derived by Abraham de Moivre and give the right comment to the content of Tab. 5.6

Table 5.6 Error integral accounts

$n\,\sigma$	*De Moivre*	*Contemporary* [9]
$n = 1$	0.682 688	0.682 689 49
$n = 2$	0.954 28	0.954 499 76
$n = 3$	0.998 74	0.997 300 20

De Moivre-Laplace Local Theorem *probability of the appearance of exactly "k" successes when tossing a coin "n" times exactly expressed by the binomial* (4.27) *can be approximated by the normal distribution given below for every value of "p" only if "n" is sufficiently high:*

$$\frac{1}{\sqrt{2\cdot\pi\cdot n\cdot p\cdot(1-p)}}\exp\left\{-\frac{(k-n\cdot p)^2}{2\cdot n\cdot p\cdot(1-p)}\right\} \tag{5.38}$$

It is seen from (5.38) that the normal distribution has the same mean and the same variance as the binomial distribution (4.27). Having at hand a copy of De Moivre's book [8] we cannot resist the temptation to add a remark acknowledging the appearance of the symbol π in the square root of denominator (5.38). De Moivre there had to mention "*the Circumference of a Circle whose Radius is One*" (see p.244 of [8]). Returning to the point we have to note that (5.38) allows to determine the required probability without resorting to the troubling factorials demanded by (4.27). Before we draw the Student's attention to the numerical examples supporting these theorems we will present the second theorem bearing the same two names:

De Moivre-Laplace Integral Limit Theorem. *To make use of the normal approximation given below which approximates the number of successes* $\left(k_1, k_2\right)$ *as accurately determined by the binominal distribution, the condition* n $\rightarrow \infty$ *must be:*

$$\frac{1}{\sqrt{2 \cdot \pi}} \int_{z_1}^{z_2} \exp\left[-\frac{z^2}{2}\right] dz \tag{5.39}$$

of which z_1 lower and z_2 upper limits, *z-scored* auxiliary variables, are determined by:

$$z_1 = \frac{k_1 - n \cdot p}{\sqrt{n \cdot p \cdot (1-p)}} \text{ and } z_2 = \frac{k_2 - n \cdot p}{\sqrt{n \cdot p \cdot (1-p)}} \tag{5.40}$$

Note, that regarding the first theorem – the last subchapter 5.10 is thought to bring closer the quantitative convergence of the binomial to normal distribution. Nevertheless it can be instructing to see what the examination of the binomial diagrams presented in Fig. 5.12 offers in this respect.

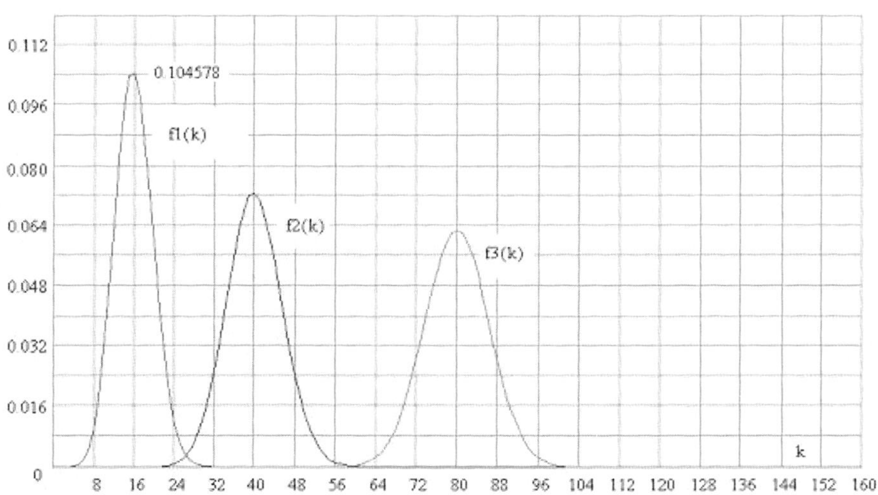

Fig. 5.12 Three binomials n = 160 for p = 0.1, 0.25 and 0.5

As we know the normal distribution is a symmetrical one. Therefore when examining Fig. 5.12 it is recommended especially to see how such comparatively high value of n converts unsymmetrical binomial distribution into a shape close to the symmetrical one. This fact qualitatively supports the essential meaning of the convergence . Numerical aspects of this convergence depicted in Fig. 5.12

are documented by the provided below coupled values of both binomial distributions. Let us first take *f1(k)* i.e. the binomial with $p = 0.1$ and the mean value 16 to examine the three couples with arguments indicating the distance of each argument from the mean value:

$$\begin{array}{ll} f1\,(-7) = 0.018554 & f1\,(+7) = 0.019593 \\ f1\,(-5) = 0.047166 & f1\,(+5) = 0.041864 \\ f1\,(-3) = 0.082313 & f1\,(+3) = 0.072148 \end{array}$$

The comparison shows rather unexpected behavior justifying the need of further investigations. Then let us take *f2(k)* corresponding to $p = 0.25$ with the mean value 40 and examine the next three couples symmetrically positioned regarding the mean:

$$\begin{array}{ll} f2\,(-12) = 0.006\,183 & f2\,(+12) = 0.006\,986 \\ f2\,(-8) = 0.025\,648 & f2\,(+8) = 0.024\,664 \\ f2\,(-4) = 0.057\,345 & f2\,(+4) = 0.054\,291 \end{array}$$

It is evident that not only do they differ but that those differences appear in a different pattern once their position with respect to the mean value of the binomial is changed. To this point we return on p.5.10 but here it is worth mentioning that Feller (see [13], p.182, Table 2) mentioned some details of the convergence which can be seen if we compare coupled values for the same argument but for both binomial and normal distributions under consideration. So far we will present two examples to illustrate both DeMoivre-Laplace theorems (the spelling "Demoivre" is sometimes used even by experts in this field).

Example 5.3. *Determine probability of gathering exactly $k = 4950$ successes while tossing a fair coin $p = 0.5$ for $n = 10000$ times.*

Coming to the numerical results and a possibility of deriving them by using a *scientific calculator* currently (i.e. in 2011) accessible in Word 7 it comes as a pleasant surprise that this calculator can calculate the below given expression:

$$f(k) = \frac{10000!}{4950! \cdot 5050!} \cdot \frac{1}{2^{10000}} \tag{5.44}$$

The result which we got in this way is as follows:

$$f(k) = 0.0048394951423164156764058974813258 \tag{5.45a}$$

The result obtained by using the local theorem has been obtained by the *MathCad* tools and is presented below:

$$n = 10\,000 \quad p = 0.5 \quad q = 0.5 \quad k = 4950$$

$$f(k) = \frac{\exp\left[-(k - np)^2 / 2\,n\,p\,q\right]}{\sqrt{2\pi\,n\,p\,q}}$$

$$f(k) = 0.004839414490382867$$

If we use an ordinary 10-digit calculator we get: 0.004 839 414 49 – and we see that the accuracy of this result is sufficient for the account of accuracy of the normal approximation which does not come above the first five digits. For the Polish Student it is reasonable to remark that in the popular book on Probability by T. Czechowski [6] the above example was also solved but the numerical result of 0.007042 is completely wrong, although it was repeated in both editions mentioned in our Literature (Edition 1, p.71, Edition 2, p.72).

In statistical practice especially the second theorem plays a significant role. Therefore comments, advice, examples usually concern this one. Before resorting to its numerical illustration we provide two valuables rules, following Jerzy Neyman's book [2] – both deal with the second theorem.

Rule 1.

Apply the integral theorem if the variance is greater than 3.

$$\sigma^2 = n \cdot p \cdot (1 - p) > 3 \tag{5.46}$$

Rule 2.

For the cases when n is comparatively small, broadening the limiting values (k_1, k_2) by the value 0.5 ensures higher accuracy of the normal approximation:

$$z_1^* = \frac{(k_1 - 0.5) - n \cdot p}{\sqrt{n \cdot p \cdot (1 - p)}} \qquad z_2^* = \frac{(k_2 + 0.5) - n \cdot p}{\sqrt{n \cdot p \cdot (1 - p)}} \tag{5.47}$$

Example 5.4 *Determine probability of the compound event regarding the number of successes from* $k_1 = 4950$ *up to* $k_2 = 5100$ *while tossing* $n = 10000$ *times the fair coin of* $p = 0.5$.

Solution of this example can be very simply obtained only by the integral theorem and in defined conditions partly without any calculations. We have in mind the procedure to determine suitable *z-scores* corresponding to $k1$=4950 and $k2$=5100 - in a view of $z1$ = -1, and $z2$ = +2. In the next step we have to make use of the normal distribution tables and read from there the appropriate values of probabilities: 0.3413 for $z1$ and 0.4772 for $z2$. Then we have to add both of them getting the desired final answer of 0.8185. Having in mind that both numbers of successes were so high there was no necessity to use the Rule 2 – and the related formula (5.42). It is easy to check that (5.40) gives satisfactory accuracy in determining $z1$ and $z2$.

5.10 Remarks on the Binomials Convergence [10]

Apparently the simplest idea of convergence will require determining the difference between the target, normal, and the initial binomial distributions. In Fig. 5.13 this idea is illustrated by using a numerical example of the binomial defined by $n = 160$, and $p = 0.5$. The maximum value of the normal distribution appearing at $k = 80$ can be obtained by the ordinary scientific calculator giving us the result $1/\sqrt{80 \cdot \pi} = 0.063078313$. With a help of the Word7 calculator we can obtain a corresponding value of the binomial shown here with the same accuracy as 0.062979831. Calculating the difference between them it becomes obvious that the first five digits depicted in Fig. 5.13 are confirmed.

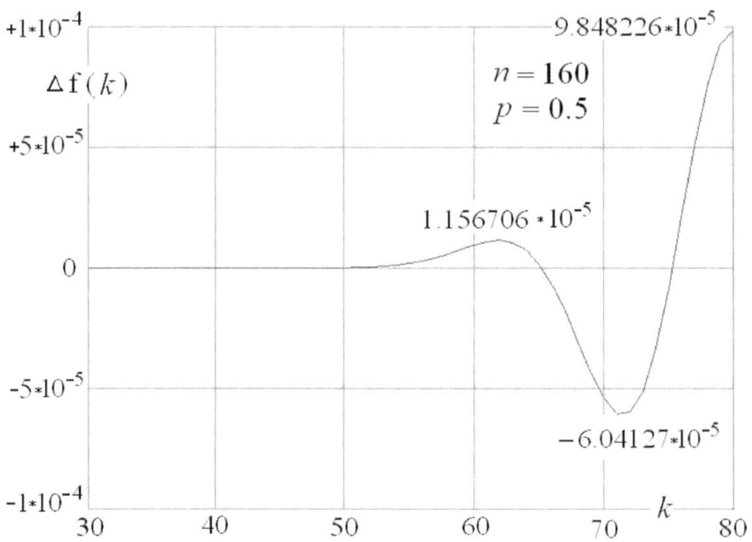

Fig. 5.13 Differences between normal and binomial distributions

Moreover, this first result tells us about the true sign of such differences – which allows to see further details, that is, in fact *the strange behavior* of these differences which afterwards diminish if we proceed far enough along the left tail of both distributions. Due to the symmetry of the considered binomial – exactly the same pattern is justified regarding the right tail of both distributions. But according to numerical results shown above it is not the case for all binomials covering the continuous range values regarding p.

Even more surprising is the behavior of relative differences which we define below with numerical and graphic results of appropriate calculations for two binomials – still corresponding to *fair dice* and the number of tossing $n = 100$ and $n = 160$. Appropriate pictorial illustrations follow in both cases after the numerical results.

$$s(k,k) = (f(k) - f_3(k))\frac{100}{f_3(k)}$$

$s(50,50) = 0.250308585417012$ $s(49,49) = 0.23052386349894$

$s(48,48) = 0.172749652552562$ $s(47,47) = 0.081724252467638$

$s(46,46) = -0.034658630703465$ $s(45,45) = -0.165348911667922$

$s(44,44) = -0.296126345798759$ $s(43,43) = -0.409561276492595$

$s(42,42) = -0.484933174430141$ $s(41,41) = -0.498085647225396$

$s(40,40) = -0.421191167879651$ $s(39,39) = -0.222392012222813$

$s(38,38) = 0.134724893591486$ $s(37,37) = 0.691873348440656$

$s(34,34) = 4.084729851476853$ $s(35,35) = 2.606224344119691$

$s(30,30) = 15.51682096306099$ $s(25,25) = 55.42194955676526$

$s(20,20) = 187.4005787083362$ $s(15,15) = 814.1630757492492$

$s(10,10) = 7.2996533542849151\ 10^3$ $s(5,5) = 3.4607055093653\ 10^5$

$s(0,0) = 1.95081241355982\ 10^9$

Graphic results shown in Fig. 5.14 –cover the range $50 > k > 35$.

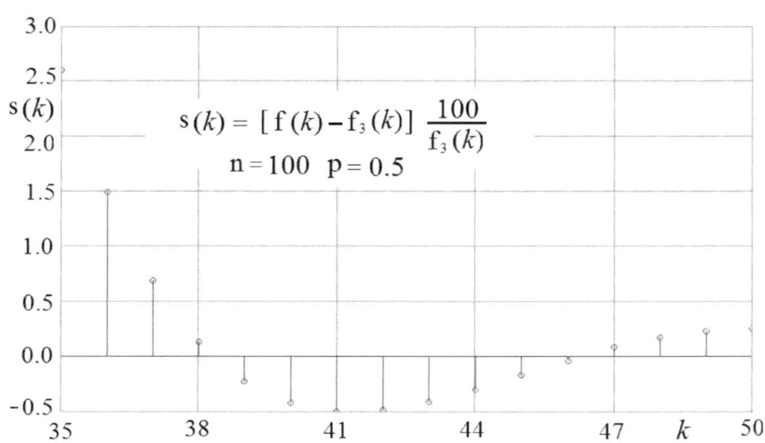

Fig. 5.14 Relative "errors" for n = 100 close to the mean value $\mu = 50$

Fig. 5.14 allows to see that the relative differences regarding the mean value $k = n\,p$ are comparatively small – and change their sign twice. However, once we cross $k = 38$ tending towards $k = 0$ the differences constantly increase reaching the maximum at $k = 0$. The same is seen for $n = 160$ - although the maximum value of the difference significantly exceeds the value shown in Fig. 5.14 for $n = 100$. We ask ourselves whether it is intuitive behavior.

$$s(k,k) = (f(k) - f_3(k))\frac{100}{f_3(k)}$$

$s(80,80) = 0.156371116551419$ $s(79,79) = 0.148611301391905$

$s(78,78) = 0.125719424499159$ $s(77,77) = 0.088858029879923$

$s(76,76) = 0.039964257048871$ $s(75,75) = -0.018250440539243$

$s(74,74) = -0.082299949413906$ $s(73,73) = -0.147921777104658$

$s(72,72) = -0.210072993569295$ $s(71,71) = -0.262922636176812$

$s(70,70) = -0.299839137767474$ $s(69,69) = -0.313371073790324$

$s(68,68) = -0.295219232426921$ $s(67,67) = -0.236197679504216$

$s(66,66) = -0.126181102958524$ $s(65,65) = 0.045964739619241$

$s(64,64) = 0.292473207053715$ $s(63,63) = 0.626792022025441$

$s(62,62) = 1.063728870512691$ $s(61,61) = 1.619633731454355$

$s(60,60) = 2.312625695756684$ $s(59,59) = 3.162873797447618$

$s(58,58) = 4.192943643678936$ $s(57,57) = 5.428224549508638$

$s(56,56) = 6.89745566433508$ $s(55,55) = 8.633374495893561$

$s(54,54) = 10.6735176676603$ $s(53,53) = 13.06121219265365$

$s(52,52) = 15.84680670352936$ $s(51,51) = 19.08920689959556$

$s(50,50) = 22.85779927982236$ $s(49,49) = 27.23487387805521$

$$s(0,0) = 1.66387548922877 \cdot 10^{14}$$

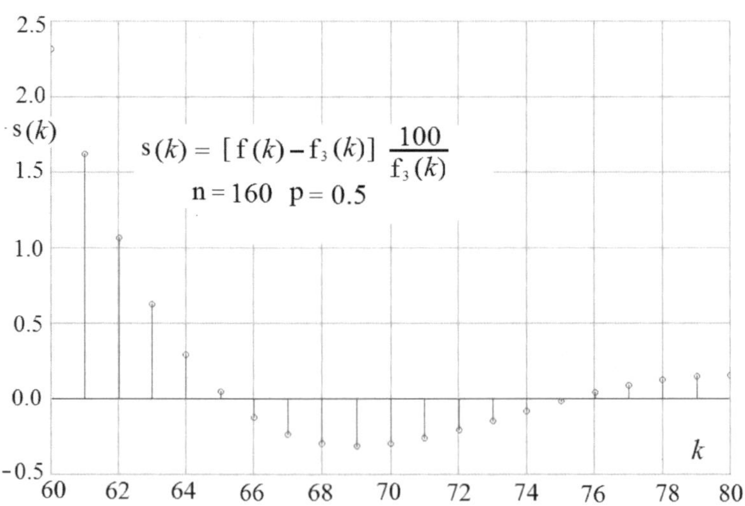

Fig. 5.15 Relative "errors" for $n = 160$ close to the mean value $\mu = 80$

References

[1] Weinberg, G.H., Schumaker, J.A., Oltman, D.: STATISTICS – An Intuitive Approach, 4th Edition. Brooks/Cole Publishing Company, Monterey (1981)

[2] Neyman, J.: First Course in Probability and Statistics. HR&W, New York (1950); Polish translation Zasady Rachunku Prawdopodobieństwa i Statystyki Matematycznej, PWN, Warszawa (1969); Russian translation – Nauka, Moskwa (1968)

[3] Fisz, M.: Rachunek Prawdopodobieństwa i Statystyka Matematyczna (in Polish: Probability and Mathematical Statistics), 3rd edn. PWN, Warszawa (1967)

[4] Zubrzycki, S.: Wykłady z Rachunku Prawdopodobieństwa i Statystyki Matematycznej (in Polish: Letures on Probability and Mathematical Statistics). PWN, Warszawa (1966)

[5] Laudański, L.M.: Pre-stochastic Models of Computational Probability. Transactions of the Aviation Institute 159, 4/99, 30–36 (1999)

[6] Czechowski, T.: Elementarny Wykład Rachunku Prawdopodobieństwa (in Polish: Introductory Course to Probability). PWN, Warszawa - Edition 1 – 1958, Edition 2 – 1968

[7] Stigler, S.M.: The History of Statistics – the Measurement of Uncertainty before 1900. The Balknap Press of Harvard UP, Cambridge (1986)

[8] de Moivre, A.: The Doctrine of Chances or A Method of Calculating the Probabilities of Events in Play, 3 edn., Fuller, Clearer, and more Correct than the Former. London, A. Millar, pp. 1–378 (1756), Digitized by Google, Internet

[9] Laudański, L.M.: Statystyka nie tylko dla licencjatów (in Polish: Statistics not only for the undergraduates), Part 2, 2nd edn. Publishing House of the Rzeszow TU (2009)

[10] Laudański, L.M., Dłubis, E.: Zbieżność i rozbieżność rozkładu dwumianowego względem rozkładu Gaussa w praktyce statystycznej (in Polish: The convergence and divergence between the binomial and Gaussian distribution as used in statistical practice). In: Proceedings of the IV-th International Scientific Conference in Jarosław: Prospect of the development of the Rzeszow Province due to European integration processes. PWSZ Jarosław, pp. 263–272 (2000)

[11] Student: The probable error of a mean. Biometrika 6, s.1– s.25 (March 1908), accessible on the Internet

[12] Lloyd, E.: Handbook of Applicable Mathematics. Probability, vol. II, pp. 1–450. J. Wiley & Sons, New York (1980)

[13] Feller, W.: An Introduction to Probability Theory and Its Applications, 3rd edn., vol. I. Posthumous Edition, p. 509. John Wiley and Sons, New York (1971)

Les Gross Poissons

Peter Bruegel (c. 1525-1569), “Big fish eat little fish”, engraving. © Trustees of the British Museum.

First Approach

One of the most haunting of Bruegel's images, *Big Fish Eat Little Fish* is among the first of the artist's many treatments of proverbs in paintings or prints. The image reveals many small and large fish tumbling out of the mouth of an enormous beached fish. A small, helmeted figure with an oversized knife slices open the big fish's belly, revealing even more marine creatures. Land, air, and water seem to be overrun by an odd assortment of real and fantastic fish, while in the foreground a man, accompanied by his son, gestures toward the scene. The meaning of his gesture is conveyed in the Flemish inscription below, which translates: "Look son, I have long known that the big fish eat the small." This vernacular form of the ancient Latin proverb, which appears in majuscule lettering just above, relates to the theme of a senseless world in which the powerful instinctively and consistently prey on the weak. That the son understands the lesson is apparent from his gesture toward the other man in the boat, who has extracted a small fish from a larger one. Bruegel's brilliant visualization of the proverb was first conceived as a drawing (Vienna, Graphische Sammlung Albertina, pen and brush drawing 22 x 31 cm) that is signed by the artist and dated 1556. This engraving by Pieter van der Heyden (22.9 x 29.6 cm), however, is signed in the lower left corner with the name Hieronymus Bosch, who had died in 1516. The print's publisher, Hieronymus Cock, was probably responsible for replacing Bruegel's name with that of the more famous and salable Bosch, who had, not coincidentally, a major influence on Bruegel. Depicted engraving is owned by the British Museum, very similar one is in a possession of the Metropolitan Museum, and the third one stores the Royal Library in Bruxelles.

[Wikipedia, unknown author]

Second Approach

Here, like Jheronimus Bosch before him, Bruegel illustrates a proverb. The saying 'Big fish eat little fish', meaning that the rich get richer while the poor get poorer, had been widely known throughout Europe since ancient times. Bruegels's representation has generally been interpreted as an allusion to greed, yet surprisingly enough, the most significant detail has often been overlooked: the biggest and greediest fish lies stranded and gutted, while small fish spill from its entrails. In other words, all it accumulated is now lost, the moral of the story being that greed does not pay. We see, in the detail below, the father showing his child the fish and teaching him a valuable lesson. The text that accompanied the original artwork stated "Here son, I have long known that big fish eat little fish." Also of interest, is the hallmark born by the cutter's knife, what appears to be an orb of State. Hence this big fish, the biggest in the sea, is still subject to the rule of Princes. However, this proverb seems not to hold in our current times: some Big Fish are now 'too big to fail', think on the Banks & I'm sure you will agree that something has changed in the grand order of things. The knife of State is now a blunt instrument & the Big Fish roam free. [Wikipedia, unknown author]

Third Strike

Bruegel' drawing can be also associated with the famous quote of Isaac Newton (1642-1727):

If I have seen further it is only by standing on the shoulders of giants.

In other words - it is allegory regarding the processes of accumulating the human knowledge: "the big fish" stands for the next generation of scientists, "the small fish" – symbolize the past generations – a symbolic cascade of fruitful efforts of the creativity.

Book Two
Exercises

Unit 1
Descriptive Statistics

Problem 1.1 (see: [1], Prob. 4.10, p.59)

Salesperson Sarah sells 50 units on day one when the mean number of units sold is 65 and the standard deviation is 12, and 18 units on day two when the mean is 15 with a standard deviation of 3. Salesperson Saul sells 45 units the first day and 21 units on the second day. Which salesperson is the most successful on the basis of these two days' sales?

Solution

We have to assume that the problem requires understanding the sales conditions on the two days which apply to both salespersons. It is suggested that the *z-scored* results are used to determine the selling efficiency. Therefore, we have to calculate four values of the appropriate *z-scores*.

Sarah had $z_1 = \dfrac{50-65}{12} = -1.25$ and $z_2 = \dfrac{18-15}{3} = +1.00$ her total score - 0.25

Saul had $z_1 = \dfrac{45-65}{12} = -1.67$ and $z_2 = \dfrac{21-15}{3} = +2.00$ his total score + 0.33

Conclusion: Saul is the most successful salesperson.

Problem 1.2 (see: [1], Prob. 2.8, p.23)

The ages at which 20 emphysema victims who were pack-a-day smokers were diagnosed are: 68, 57, 35, 49, 57, 53, 70, 48, 59, 67, 63, 65, 48, 55, 63, 65, 41, 85, 60, 57. Compute the mean and median ages at which the disease was diagnosed for this population. Determine also the variance of the population.

L.M. Laudański: Between Certainty and Uncertainty, ISRL 31, pp. 165–187.
springerlink.com

Solution

Ordered statistics allows to determine the median (it is the medium value) as (57 + 59)/2 = 58. The calculations will partly be done with the help of the SD procedure. It will produce the following results: $\sum x_i = 1165$, $\sum x_i^2 = 70203$ - and in particular $\overline{x} = 58.25$ with $\sigma_x = 10.82069776$ and $\sigma_x^2 = 117.0875$. Therefore, the median and the mean value are almost identical.

However, it has to be pointed out that the variance is greater than the mean. As the Student will see (it will be discussed in Chapter 5) such cases correspond to the negative binomial distribution. The Student is advised to complete calculations of all the squared values and then the mean square value as 70203/20 = 3510.15 and the square of mean as 3393.0625. Their difference gives the variance 3510.15 – 3393.0625 = 117.0875, i.e. exactly the same as shown above. The Student who wants to broaden his/her skills by performing numerical calculations is advised to derive the variance applying the direct definition (1.4).

Table 1P.1 Age of smokers

x_i	x_i^2	x_i	x_i^2
35	1225	59	3481
41	1681	60	3600
48	2304	63	3969
48	2304	63	3969
49	2401	65	4225
53	2809	65	4225
55	3025	67	4489
57	3249	68	4624
57	3249	70	4900
57	3249	85	7225
---	---	$\sum$ 1165	$\sum$ 70203

Problem 1.3 (see [1], Prob. 2.20, p.25)

Ten mopeds tested for gas mileage yielded the following results in miles per gallon: 123, 85, 97, 92, 103, 114, 109, 91, 98, 83 mil/gal. Find the mean and the variance. If the antipollution device is installed on any moped, its gas mileage is decreased by 7.5 mil/gal – find the mean and the variance also for this case.

Solution 1

It is seen that $\sum x_i = 995$, $\overline{x} = 99.5$, then $\sum x_i^2 = 100487$, therefore the coefficients of variability $\sigma_x = 12.18400591$ and $\sigma_x^2 = 148.45$.

Table 1 P.2 US gas mileage

x_i	x_i^2	x_i	x_i^2
123	15129	114	12996
85	7225	109	11881
97	9409	91	8281
92	8464	98	9604
103	10609	83	6889
---	---	$\sum$ 1165	$\sum$ 100487

Now to the second question: there are two solutions but only one of them is straightforward. The mean value of the new statistics is less than above determined for 7.5 mil/gal and is equal to 92 mil/gal. But the variance in both cases remains the same.

The Student who does not understand these results has two choices. First, to justify the above results regarding the mean and the variance it is enough to recall the results (1.10) and (1.12).

There is also another approach which is less formal but perhaps more evident. To resort to the calculations given above with respect to statistics $y_i = x_i - 7.5$. This kind of proof is less general, we may even say that it is particular as gives the answer only for this numerical case.

Solution 2

The second approach follows the European standard to find out the gas mileage. First it uses SI Units, therefore 1 gallon = 3.785 liter, 1 state mile = 1609 meters. Secondly in Europe we determine the gas yield per 100 km. Let us apply the conversion procedure for the first moped which travels 123 miles on 1 gallon of gas, which is 197.907 km per 1 gallon, which gives 52.287186 km per 1 liter, therefore to travel 100 km it will use 1.9125144638643403214641220371184 liters of gas. With only four digits after the decimal point, the results in the SI units are presented below.

Table 1P.3 European gas mileage

x_i	x_i^2	x_i	x_i^2
1.9125	3.65771157	2.0635	4.25804235
2.7675	7.65917210	2.1582	4.65764821
2.4251	5.88133897	2.5850	6.68246811
2.5569	6.53798658	2.4004	5.76194033
2.2839	5.21609185	2.8342	8.03273602
---	---	$\sum$ 23.9872	$\sum$ 58.34513609

$$\overline{x} = 2.39872 \quad \sigma_x = 0.283999948 \quad \sigma_x^2 = 0.0806559706$$

Now, let us convert the above result for the mean mileage 99.5 mil/gal into SI units, it will give us 2.3642 - which compared to 2.39872 shows troubling discrepancy demanding an explanation. The answer is simple but unexpected at first glance: the example documents that the calculated average 2.39872 is wrong – and has to be replaced by the so called harmonic average defined below:

$$x_{harm} = N \Big/ \sum_{i=1}^{N} 1/x_i \tag{1P.1}$$

Let us apply the above definition to the example under consideration by substituting values given in the Table1P.3 for all x_i into the denominator of the formula (1P.1). Then we obtain the following result performing the necessary calculations with Word7:

$$\sum_{i=1}^{N} 1/x_i = 4.22975331894227094783900192049566$$

Therefore, we get the *harmonic average* as

$$x_{harm} = 2.36420406722458399232148030128444$$

Ignoring the unnecessary digits at the end of the procedure which cannot be considered as accurate as it does not provide more than the first four decimal places – we have exactly the same value as was obtained in Solution 1 as the mean value – sometimes called the *arithmetic average.*

Then the following question appears: why could the same procedure applied in Solution 2 not give the right answer? Again, the answer is trivial: the calculations cannot be done *mechanically*. In fact the procedure to determine the values given in Table1P.3 uses *reciprocal* values and applying a mechanical averaging procedure to them the gives the wrong result. We hope that the following example will give the Student a satisfying opportunity to consider this matter once again but in a simpler arrangement.

Problem 1.4

Determine the mean yield (i.e. the number of quintals of the crop from one hectare) taking into account the three neighbor farmers whose data is given in Table 1P.4.

Solution

It should be obvious that the answer (20 + 22 + 24)/3 = 22 is wrong. Such answer would suggest that the arable acreage for each farm is the same, but it is not. The desired answer is given below:

Table 1P.4 Farm yield

Owners of the farms	Yield from hectare in quintals	Total crop in quintals	Acreage arable in hectares
Malinowsky	20	1000	50
Nowacks	22	1100	50
Zelias	24	1440	60
------------------	----------	3540	160

$$x_{ave} = \frac{\sum 3540}{\sum 160} = 22.125 \;\; q/ha \tag{1P.2}$$

A closer analysis of this procedure allows to establish some details which deserve to be understood as they are more general. Let us examine the following calculations

$$x_{ave} = \frac{1000 + 1100 + 1440}{\dfrac{1000}{20} + \dfrac{1100}{22} + \dfrac{1440}{24}} \tag{1P.3}$$

Moving forward, algebraic formalism identifies (1P.3) by the general expression:

$$x_{harm} = \sum_{i=1}^{N} f_i \Big/ \sum_{i=1}^{N} f_i / x_i \tag{1P.4}$$

Going backwards, formula (1P.1) used for *grouped data* may be simplified to one which is known as *harmonic average* in the verbal arrangement and is used in descriptive statistics.:

$$x_{harm} = N \Big/ \sum_{i=1}^{N} 1/x_i \tag{1P.1}$$

Note, that (1P.4) has been already used above to solve Problem 1.4.

Ending this passage we supply our Student with one more example of this kind which can be found in Reichmann's book [2], see pp.56-57. As a point of interest we quote Reichmann' book stating this problem which is so puzzling that it deserves to be called a *brain-twister,* in Russian expressed by the term *головоломка*.

Business take-over. *There are two greengrocers in a vegetable market. A regularly sells new potatoes at a shilling for two pounds, while B sells slightly lower quality at a shilling for three pounds. Each of them sells 60 lb per day. Together they earn 30 + 20 shilling daily. One day they are both replaced by a third one who sells two and half pounds potatoes for one shilling. He sells 120 lb, but he earns only 48 shillings. Why?!*

We can add in the end that the Polish Student can find all these examples in this Author's book [3] presented in an alternative way.

Problem 1.5 (Follows Prob.4.25 [1])

Seventeen readings of the earthquakes registered at the same station at two periods of time "A" and "B" are given. Consider the consequences of merging these readings. "**A**": 5.3, 5.6, 3.7, 2.9, 5.1, 5.0, 4.6, 5.1 "**B**": 3.7, 3.9, 4.2, 7.1, 6.3, 3.7, 3.8, 4.5, 7.5

Solution

Regarding statistics "A" see Table 1P.5 given below.

Table 1P.5 Earthquakes records "A"

x_i	x_i^2	*z-score*
5.3	28.09	0.747498513
5.6	31.36	1.09926252
3.7	13.69	-1.128576187
2.9	8.41	-2.066613537
5.1	26.01	0.512989176
5.0	25.00	0.395734507
4.6	21.16	-0.073284168
5.1	26.01	0.512989176
$\sum$ 37.3	$\sum$ 179.73	$\sum$ 0

Earthquake statistics "A" shows $\mu_A = 4.6625$, and $\sigma_A = 0.852844505$. These values serve as a required constant to derive *z-scores* statistics given in the third column of Table1P.5.

Regarding statistics "B", the data are provided in Table1P.6 - to derive similar results as those in Table 1P.5.

Earthquake statistics "B" shows $\mu_B = 4.96(6)$, and $\sigma_B = 1.462873887$ – then used to calculate *z-score* statistics shown in the third column of Table1P.6.

Now let us consider the problem of merging data "A" and data "B". The first concept is shown in Table1P.7 – presenting concatenation of the set "A" with the set "B" - at the level of *z-scores* for both statistics. Table1P.7 presents ordered entries with the total number of entries determined by $8 + 9 = 17$.

The new statistics is *z-scored* statistics. The problem is how to prove it? It is very simple to claim that the mean value is *zero*. But how to prove that the variance remains equal to 1?

Table 1.P.6 Earthquake records "B"

x_i	x_i^2	*z-score*
3.7	13.69	-0.865875505
3.9	15.21	-0.729158320
4.2	17.64	-0.524082542
7.1	50.41	1.458316641
6.3	39.69	0.911447901
3.7	13.69	-0.865875505
3.8	14.44	-0.797516912
4.5	20.25	-0.319006764
7.5	56.25	1.731751011
$\sum$ 44.7	$\sum$ 241.27	$\sum$ 0

Table 1P.7 First merging, z-scores

-2.066613537	-0.729158320	0.512989176	1.458316641
-1.128576187	-0.524082542	0.512989176	1.731751011
-0.865875505	-0.319006764	0.747498513	-----
-0.865875505	-0.073284168	0.911447901	-----
-0.797516912	0.395734507	1.09926252	-----

Let us denote statistics "A" with the number of entries equal to 8 by x_i and the other with 9 entries by y_i . If each statistics has mean value equal to zero and the variance equal to 1, then it is true that:

$$\sum x_i^2/8 = 1 \quad \text{and} \quad \sum y_i^2/9 = 1 \quad \text{so} \quad \sum x_i^2 = 8 \quad \text{and} \quad \sum y_i^2 = 9$$

The above obtained results prove that the mean square value *z-scored* statistics is equal to its variance and then, equal to 1.

Therefore, for the concatenated statistics the variance will be also equal to the mean square which in the end will be equal to (8 + 9)/17 = 1. What ends the proof.

In the second approach let us concatenate both original statistics "A" and "B". The mean value of the new statistics can be calculated as follows with the help of the results given in Table1P.5 and 6:

$$\mu_{A+B} = \frac{37.3 + 44.7}{9 + 8} = 4.823529412$$

In a similar way the variance (of the concatenated statistics) can also be calculated. We prefer using the auxiliary form (1.5) i.e. calculating the mean square value and subtracting the square of the mean to derive the variance. Again, making use of the partial sums given in the above mentioned two tables we get:

$$\sigma^2_{A+B} = \frac{241.27 + 179.73}{9+8} - \left(\frac{82}{17}\right)^2 = 1.498269897$$

Using the Windows7 calculator we get:

1.4982698961937716262975778546 7128

Corresponding standard deviation:

$$\sigma_{A+B} = 1.224038357$$

Then we face the following question: did both concatenations give the same statistics or did we derive two different statistics in this way? In order to find the answer to this question we calculated ***z-scored*** statistics for the second merging statistics and obtained results presented in Table 1P.8.

The problem arising here can be re-stated as follows: what makes the difference between two *z-scored* statistics? The obvious answer is: they differ only in their probability distributions. So, to get the final answer here rather exceeds our possibilities at this stage. Nevertheless, we may make a step towards the methods developed in Chapter-2 "Grouped data" where among other tools an important place is occupied by the frequency histogram which can be considered a substitute for the probability density functions. So we suggest the Student bravely tackles the problems of drawing histograms.

In order to complete such a task in the simplest manner, we have to have ordered statistical data at hand. So we recall here Table1P.7 and insert its data into graph of the histogram – as seen in Fig.1P.1. To help the Student obtain such a histogram we present a few figures from Table1P.7 showing their place in the histogram.

Table 1P.8 Second merging, original statistics

x_i	x_i^2	*z-scores*	x_i	x_i^2	*z-scores*
3.7	13.69	-0.917887421	5.3	28.09	0.389261158
3.9	15.21	-0.754493848	5.6	31.36	0.634351517
4.2	17.64	-0.50940349	3.7	13.69	-0.917887421
7.1	50.41	1.85980331	2.9	8.41	-1.571461711
6.3	39.69	1.20622902	5.1	26.01	0.225867586
3.7	13.69	-0.917887421	5.0	25.00	0.144170799
3.8	14.44	-0.836190634	4.6	21.16	-0.182616345
4.5	20.25	-0.264313131	5.1	26.01	0.225867586
7.5	56.25	2.186590455	------	-------	-------------
			$\sum$ 82	$\sum$ 421	$\sum$ 0

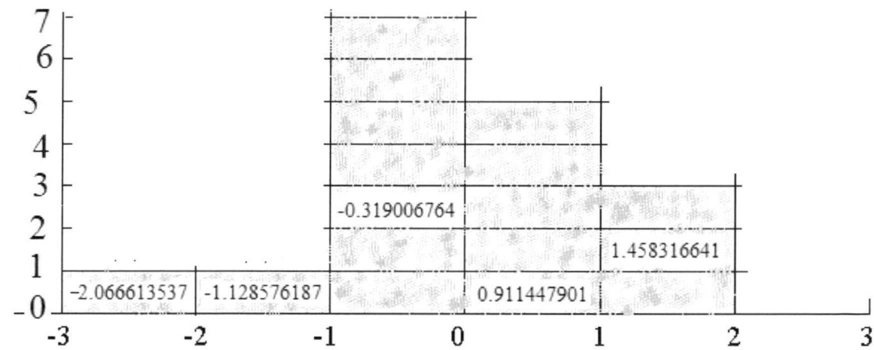

Fig. 1P.1 Histogram of the first merged statistics

To draw the second histogram we first collected all obtained *z-scores* given in Table1P.8 – into an ordered set given in Table1P.9.

Table 1P.9 Second merging – z-scores

-1.571461711	-0.754493848	0.225867586	1.85980331
-0.917887421	-0.50940349	0.225867586	2.186590455
-0.917887421	-0.264313131	0.389261158	-----
-0.917887421	-0.182616345	0.634351517	-----
-0.836190634	0.144170799	1.20622902	-----

Then the second frequency histogram – Fig.1P.2 was drawn in the same way as the first shown in Fig.1P.1.

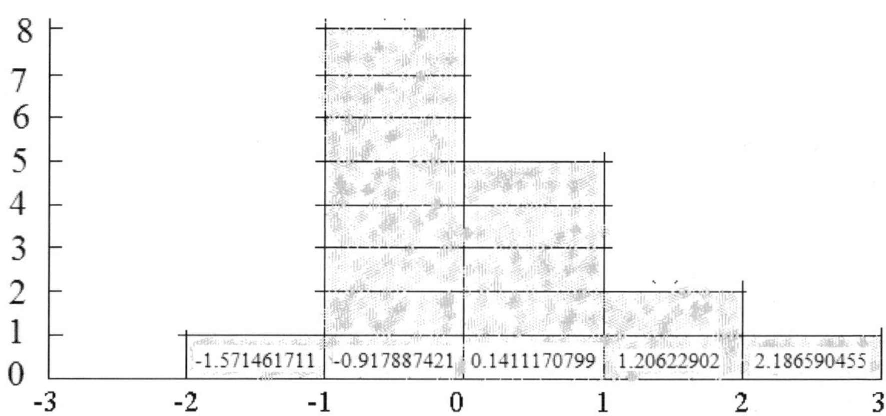

Fig. 1P.2 Histogram of the second merged statistics

And now we face directly a relatively difficult problem of the similarity of both histograms presented in Fig.1P.1 and Fig.1P.2. We cannot ignore the significance of the number of gathered records of the Earthquakes. So, with this reservation a cautious conclusion is that they are similar but not identical.

Problem 1.6 (see [1], Prob.2.29)

Fifteen haulers presently provide trash service for the city. The number of residences served by the haulers are 6050, 4750, 8093, 3857, 6247, 5190, 3025, 5520, 4055, 6385, 6700, 4350, 4470, 3960 and 8030. Find the mean, variance and standard deviation of residents served per hauler. If the city agrees to contract with a new company to take over 7% of each route then how many residents will the new company serve and what will the new mean, variance, and standard deviation per hauler be?

Answers: $\overline{x}_1 = 5378.8$, $\Delta x = 5648$, $\overline{x}_2 = 5395.625$; $\sigma_{x1} = 1467.765657$, $\sum x_1 = 80\,682$, $\sum x_1^2 = 466\,287\,382$; $\sum x_2 = 86\,330$, $\sum x_2^2 = 498\,187\,286$, $\sigma_{x2} = 1422.651127$.

Moreover $\sum x_1 = \sum_{i=1}^{15} x_i$ and $\sum x_1^2 = \sum_{i=1}^{15} x_i^2$ then $\sum x_2 = \sum_{i=1}^{16} x_i$, $\sum x_2^2 = \sum_{i=1}^{16} x_i^2$.

The above results have been obtained by using the procedure SD of scientific calculators on the market. But it can be verified that the rule denoted as (1.5) (see Part 1, Chapter 1) gives us the following results:

$$\sigma_{\mathrm{x}}^2 = \frac{466287382}{15} - \left[\frac{80682}{15}\right]^2 \text{ then } \sigma_{\mathrm{x}}^2 = 31085825.46(6) - 28931489.44$$

And in the two next steps we get the final answers:

$$\sigma_{\mathrm{x}}^2 = 2154336.026(6) \quad \text{and} \quad \sigma_x = 1467.765658$$

to satisfy the inquiring Student with higher accuracy of both results. Complementing the above we also present a direct solution of the problem in Table 1P.10.

It is left for the Student to make use of the results given in Table1P.10 – with a hint, that the results are accurate and as such can be used to continue with further *accurate* calculations as a challenge to an ambitious Student.

Table 1P.10 Direct solution

x_i	x_i^2	$x_i - \overline{x}$	$(x_i - \overline{x})^2$
6050	36602500	671.2	450509.44 #
4750	22652500	-628.8	395389.44 #
8093	65496649	2714.2	7366881.64 #
3857	14876449	-1521.8	2315875.24 #
6247	39025009	868.2	753771.24 #
5190	26936100	-188.8	35645.44 #
3025	9150625	-2353.8	5540374.44 #
5520	30470400	141.2	19937.44 #
4055	16443025	-1323.8	1752446.44 #
6385	40768225	1006.2	1012438.44 #
6700	44890000	1321.2	1745569.44 #
4350	18922500	-1028.8	1058429.44 #
4470	19980900	-908.8	825917.44 #
3960	15681600	-1418.8	2012993.44 #
8030	64480900	2651.2	7028861.44 #
$\sum$ 80682	$\sum$ 466287382	$\sum$ 0	$\sum$ 32315040.4

Problem 1.7 (see [1], Prob.2.26)

One of the priorities of the new state administration is to encourage foreign tourists to visit the state. Records for the past eight years show that the following numbers of foreigners have vacationed in the state (per year). 200 500, 185 000, 190 000, 210 000, 155 500, 145 000, 187 000, 165 000. What has been the mean, variance, and standard deviation of foreign visitors per year during the last eight years? By what percent would the mean yearly number have to be increased to reach the state goal of 300 000 visitors from other countries per year?

Answers: $\overline{x}_1 = 179\ 750$ tourist/year, $\sum x_1 = 1\ 438\ 000$, $\sum x_1^2 = 2.620245 * 10^{11}$,

$\sigma_{x1} = 21\ 047.565\ 18$; desired increase 67% to raise the new average to 300 thousand tourist/year.

There are also two direct solutions. The first solution uses numbers expressed by all digits:

Table 1P.11 Direct solution 1

x_i	x_i^2	$x_i - \overline{x}$	$(x_i - \overline{x})^2$
200 500	40200250000	20750	430562500
185 000	34225000000	5250	27562500
190 000	36100000000	10250	105062500
210 000	44100000000	30250	915062500
155 500	24180250000	-24250	588062500
145 000	21025000000	-34750	1207562500
187 000	34969000000	7250	52562500
165 000	27225000000	-14750	217562500
$\sum$ 1438000	$\sum$ 262024500000	$\sum$ 0	$\sum$ 3544000000

The second direct solution economizes on calculations by using thousands:

Table 1P.12 Direct solution 2 in thousands

x_i	x_i^2	$x_i - \overline{x}$	$(x_i - \overline{x})^2$
200.5	40200.25	20.75	430.5625
185	34225	5.25	27.5625
190	36100	10.25	105.0625
210	44100	30.25	915.0625
155.5	24180.25	-24.25	588.0625
145	21025	-34.75	1207.5625
187	34969	7.25	52.5625
165	27225	-14.75	217.5625
$\sum$ 1438	$\sum$ 262024.5	$\sum$ 0	$\sum$ 3544

Problem 1.8 (see: [1], Prob.2.23)

Hospital occupancy figures show the following totals for number of unoccupied beds in the metropolitan area.

Jan.	385	Apr.	250	July	629	Oct.	297
Feb.	400	May	227	Aug.	600	Nov.	318
Mar.	270	June	573	Sept.	340	Dec.	635

Find the mean, variance, and standard deviation of the monthly totals. If the construction of a new hospital increased the number of unoccupied beds by 25% as claimed by the hospital council, what would the new mean be?

Ans.: $\overline{x}_1 =$ 410.3(3); $\sum x_1 = 4924$, $\sum x_1^2 = 2\ 287\ 282$, $\sigma_{x1} = 149.1086479$);

$\overline{x}_2 =$ 512.91(6)

Problem 1.9 (see [1], Prob.3.9)

A dozen large banks report that sales of gold in ounces for the week were

685, 857, 973, 495, 453, 892, 1173, 733, 1244, 797, 852 and 971

a. Find the mean, variance, and standard deviation for gold sales at 12 banks for the week in question by subtracting 400 from each term.
b. If prices had gone up as expected, sales would have been down approximately 10%; what would the mean, variance, and standard deviation have been increased each term by 10%?

Solution

The problem is solved in the British customary units. The Student who wants to obtain the solution in SI units has to remember the following conversion factors. 1 lb = 16 ounces. Then 1 lb = 0.45359237 kg which is frequently rounded up to 0.4536 kg (regarding the mass).

Table 1P.13 Direct solution

x_i	$(x_i - 400)$	$(x_i - 400)^2$
685	285	81225
857	457	208849
973	573	328329
495	95	9025
453	53	2809
892	492	242064
1173	773	597529
733	333	110889
1244	844	712336
797	397	157609
852	452	204304
971	571	326041
$\sum$ 10125	$\sum$ 5325	$\sum$ 2981009

To determine the true mean of the statistics under consideration one has to perform the following calculations:

$$\bar{x} = \frac{\sum (x_i - 400)}{N} + 400 = 443.75 + 400 = 843.75$$

For the statistics of differences shown in Table1P.13 we can use a suitable notion for its mean:

$$\mu_{400} = \frac{\sum (x_i - 400)}{N} = 443.75 \quad \text{then write} \quad \bar{x} = \mu_{400} + 400$$

Using data given in Table1P.13 the mean value of the original statistics results from

$$\bar{x} = \frac{10125}{12} = 843.75$$

Regarding the variance and standard deviation we obtain:

$$\sigma_x^2 = \frac{\sum (x_i - 400)^2}{N} - \left[\frac{\sum (x_i - 400)}{N}\right]^2 = \frac{2981009}{12} - \left[\frac{5325}{12}\right]^2 = 248417.4166 - 196914.0625$$

Finally the variance, and standard deviation will be equal to

$$\sigma_x^2 = 51503.3541 \qquad \text{and} \qquad \sigma_x = 226.9435042$$

If an inquiring Student uses the Standard Deviation procedure in scientific calculators, they get $\sum x_i = 10125$, $\sum x_i^2 = 9161009$, $\sigma_x = 226.9435043$. It should serve as a practical reminder of the conclusion (1.11) known from Chapter 1: the shifted statistics retain the variability of the original statistics. The solution of point (b) regarding new statistics where all terms are reduced by factor 0.9 is left for the Student. Hint: it can be done very simply by recalling results in (1.15) and (1.17) from Chapter 1.

Short Note

Another aspect of the above given can be also seen in the following considerations presented below. First, it is seen that:

$$\sigma_{400}^2 = \sigma_x^2 + \mu_{400}^2$$

To be certain about the meaning of the notation σ_{400}^2 it is given explicitly below:

$$\sigma_{400}^2 = \frac{\sum (x_i - 400)^2}{N}$$

Let us make use of the initial formula to derive the variance of the original statistics:

$$\sigma_x^2 = \sigma_{400}^2 - \mu_{400}^2$$

Substitution of the numerical values from Table1P.13 will confirm that:

$$\sigma_x^2 = \frac{2981009}{12} - 443.75^2 = 248417.4166 - 196914.0625 = 51503.3541$$

The Student should see that the important formula (1.5) was used above although in somewhat different disguise.

Problem 1.10 (see: [1], Prob.3.7)

Fifteen sufferers from migraine headaches required the following number of milligrams of the active ingredient in a new remedy for relief.

325, 350, 270, 280, 295, 380, 400, 250, 420, 415, 315, 365, 385, 415, 505

(i) what is the variance, and standard deviation of these 15 terms? (ii) had each person required 50 milligrams less, what would be the variance and standard deviation?

Solution

Table 1P.14 Direct solution

x_i	$x_i - 358$	$(x_i - 358)^2$
325	-33	1089
350	-8	64
270	-88	7744
280	-78	6084
295	-63	3969
380	22	484
400	42	1764
250	-108	11664
420	62	3844
415	57	3249
315	-43	1849
365	7	49
385	27	729
415	57	3249
505	147	21609
$\sum$ 5370	$\sum$ 0	$\sum$ 67440

The numerical solution to the first question of Prob.1.10 proposed in Table1P.14 also presents a direct way to derive the answer. This approach makes use of the basic mean which has to be determined initially; this can be done by applying the data in the first column of the mentioned table. We get $\overline{x} = 358$. Then in the second column we derived all differences from the mean, to be squared in the third column. Therefore, finally the variance follows as the mean square deviation from the mean:

$$\sigma_x^2 = \frac{\sum (x_i - 358)^2}{N} = \frac{67440}{15} = 4496 \quad \text{and} \quad \sigma_x \cong 67.05221845696084051$$

The obvious answer to the second question is left for the Student. But we have one more similar problem to give an opportunity to consider how ambiguous wording in presenting the problem may lead us in the wrong direction.

Problem 1.11 (see: [1], Prob.3.21)

Ten standard homeowner insurance policies from different companies were compared with the companies' standard policies for the previous year. In each case the newer policy was easier to understand and contained fewer words. The word reduction (in numbers of words) for the 10 policies were 897, 513, 400, 1057, 615, 299, 753, 1184, 387, and 350. (a) Determine the variance and standard deviation of the 10 terms. (b) If each policy were shortened by an additional 150 words, what would the variance and standard deviation be?

Solution by using the SD procedure:

Answers: (a) $\sum x_i = 6455$, $\sum x_i^2 = 5053787$, $\sigma_x^2 = 88708.45, \sigma_x = 297.8396381$

(b) except the mean reduced to 495.5, the variability remains the same!

Problem 1.12 (see: [1], Prob.3.13, p.42)

The numbers of repairs necessary for over five-year period for 25 automobiles of the same make are given here: 37, 43, 21, 28, 29, 17, 42, 33, 25, 26, 29, 24, 37, 35, 32, 32, 18, 15, 24, 39, 24, 29, 18, 41, 12. How many of the above terms are within (a) one, (b) two, and (c) three standard deviations of the mean? Answers: $\sum x = 710$, $\sum x^2 = 21998$, $\sigma = 8.741662695$, $2\sigma = 17.48332539$, $3\sigma = 26.22498808$, $\overline{x} = 28.4$.

<u>Solution</u>

A simple way to obtain the answer is to order these statistics from the lowest to the highest value: 12, 15, 17, 18, 18, 21, 24, 24, 24, 25, 26, 28, 29, 29, 29, 32, 32, 33, 35, 37, 37, 39, 41, 42, 43; and then establish the limits for all the three classes with (a) 19.66-37.14; (b) 10.92 – 45.88; and (c) 2.18 - 54.62. Therefore, all 25 are within (c), also all 25 are within (b), and 16 of them are within (a) [the last answer in Weinberg's book shows 14, which is incorrect]. Note: if all the terms belong to class (b) it is obvious that they will all be also in class (c).

A very similar problem has been provided below, the only difference is that there is no answer.

Problem 1.13 (see [1], 3.28, p.45)

Consider the following 17 terms, which indicate miles walked or ridden to school by the first-graders in an experimental school designed to integrate the community culturally and economically: 3.2, .1, 6.8, 9.0, 2.1, 4.4, 7.8, 2.1, 5.5, 2.3, 14.0, 3.7, 12.3, 8.7, 6.2, 9.7, 11.2. How many children travel distances within (a) one, (b) one and half, and (c) three standard deviations of the mean?

Solution

As in the previous problem we use the Standard Deviation procedure included in scientific calculators on the market. We get the following results: $\sum x = 109.1$, $\sum x^2 = 956.69$, $\sigma = 3.884544836$, $1.5\sigma = 5.826817254$, $3\sigma = 11.65363451$, $\bar{x} = 6.417547059$.

Also, as in Problem 1.12, to solve this Problem we order the statistics in the following way: 0.1, 2.1, 2.1, 2.3, 3.2, 3.7, 4.4, 5.5, 6.2, 6.8, 7.8, 8.7, 9.0, 9.7, 11.2, 12.3, 14.0. Initial data from the SD procedure make it possible to determine class limits. We find limits (2.53, 10.3) for class (a) ; the limiting values (0.59, 12.24), for the class (b) and finally limits for class (c) of (0, 18.07) . Having in mind the fact that distances can not be negative we established the lowest limit as zero. Looking at the ordered statistics we notice that there are 10 terms in class (a), 14 terms in class (b), and all 17 in class (c). It may be said here that taking the class of two sigma will classify all terms, as its limits will be 0 - 14.19 . But how this information may support community integration culturally and economically, the Author of this book cannot say.

***

As the next we propose a problem which is formulated and solved in Hawkins/Weber's [6] in order to introduce such statistical measures which are not so frequently used: Pearson's coefficient of skewness, and two central moments to measure asymmetry and flatness (called also peakedness) – skewness, and kurtosis.

Problem 1.14 (see [6], 2.21)

The Federal Statistical System for 1975 of federal construction jobs in 13 western states ($ million) is given in the following, ordered statistics: 13.4, 22.4, 25.6, 27.0, 29.3, 56.8, **72.5**, 102.4, 107.9, 135.6, 143.6, 206.9, 370.6 . Examine it from the point of view of asymmetry and flatness.

Solution

Before we start answering/solving the Problem, we will define the above parameters. But in order to do that we have to note that in fact they are rooted in grouped data and probability density functions. If we apply them in this statistics which we analyze in Unit 1 it can be done purely by using the core idea of the appropriate definitions.

The *Pearsonian coefficient of skewness* SK is defined as

$$SK = \frac{3\times(\text{mean - median})}{\text{standard deviation}}$$

Coefficient of skewness α_3 :

$$\alpha_3 = \frac{\sum (x_i - \overline{x})^3 / N}{\sigma_x^3}$$

Coefficient of kurtosis α_4:

$$\alpha_4 = \frac{\sum (x_i - \overline{x})^4 / N}{\sigma_x^4}$$

In the second step it is justified to make use of the Standard Deviation procedure to derive the basic averages: mean and the standard deviation. In this way we get the following

$$\sum x = 1314, \ \sum x^2 = 252695.12, \ \overline{x} = 101.0769231, \ \sigma_x = 96.02885907$$

The above values have been verified using a Word 7 calculator. The two essential sums are exactly the same as the ones above. Then the mean, the mean square, and the standard deviation were calculated using the Word 7 calculator. The results are presented below.

$$\overline{x} = 101.0769230769230769230769230769 2$$

$$\overline{x^2} = 19438.086153846153846153846153846$$

$$\sigma^2 = 9221.5417751479289940828402366864$$

$$\sigma = 96.028859074488274345394113355714$$

Subsequently to determine the successive parameters we start the calculations with the simplest *Pearsonian skewness*. By inserting the median value in bold of 72.5 we get

$SK = 0.29758682200151225164352702384652$ rounded to $SK = 0.2976$

In order to determine the coefficient of skewness we propose the following procedure. It is easily possible to justify that the following formula is true:

$$\sum\left(x_i - \overline{x}\right)^3 / N = \overline{x^3} - 3\cdot\overline{x}\cdot\overline{x^2} + 2\cdot\overline{x}^3$$

It means that to derive the skewness coefficient we have to determine the mean cube value $\overline{x^3}$ besides the known averages. In the procedure leading to this result, the entire sum of the cubic values was calculated first

$\sum x^3 = 68180756.118$, and then $\overline{x^3} = 5244673.5475384615384615384615385$

In the next step we derived the value of the negative term $3\cdot\overline{x}\cdot\overline{x^2} = 5894225.8168047337278106508875708$. Then we collected the two positive terms $\overline{x^3} + 2\cdot\overline{x}^3 = 7309987.2880937642239417387346367$. In the end of this essential step, the third central moment was derived as

1415761.4712890304961310878470667.

The above value was normalized by the cube of standard deviation to get the desired result of the skewness coefficient:

$\alpha_3 = 1.5987655522386401022098157062012$ rounded to $\alpha_3 = 1.60$

The same rounded value was obtained by Hawkins/Weber, although it was derived by direct evaluation of the presented formula of the central moment. It is obvious that the procedure shown here saves time and may lead to the final result of higher accuracy.

Intending to determine kurtosis we derived the formula again using an indirect approach. In doing so, the essential formal result was obtained first:

$$\sum\left(x_i - \overline{x}\right)^4 / N = \overline{x^4} - 4\cdot\overline{x^3}\cdot\overline{x} + 6\cdot\overline{x^2}\cdot\overline{x}^2 - 3\cdot\overline{x}^4$$

To apply the above formula, the average of the mean fourth power $\overline{x^4}$ has to be determined. The four other components can be calculated by using already known results. So, we calculated first

$\sum x^4 = 21744799306.4996$ and then the desired

$\overline{x^4} = 1672676869.7307384615384615384615$.

To determine the final result we separated negative and positive terms and then added them. The remaining positive term value is

$$6\cdot\overline{x^2}\cdot\overline{x}^2 = 1191540418.9663723258989531178871$$

These two give 2864217288.6971107874374146563486. Two negative terms are calculated as

$$4\cdot\overline{x^3}\cdot\overline{x} = 2120461858.9124733727810650887567 \quad \text{and}$$

$$3 \cdot \overline{x}^4 = 313133337.1257308917754980567904 4 .$$

And the final result is 430622092.65890652288085151080087. Dividing it by the fourth power of standard deviation gives us kurtosis of 5.0639479262287234666531003039779 rounded to $\alpha_4 = 5.06$. The result shown by Hawkins/Weber is the same as the above rounded one. To finish this solution we decided to present own results in view of the Table 1P.15. The main tool used in order to fill this Table could not be a *scientific calculator* available on the market although it allows to obtain accurate results. Then for comparison we also present Table 1P.16 which shows the solution given by Hawkins/Weber.

Table 1P.15 Determining skewness and kurtosis

x	$x\ x$	$x\ x\ x$	$x\ x\ x\ \mathrm{X}$
13.4	179.56	2406.104	32241.7936
22.4	501.76	11239.424	251763.0976
25.6	655.36	16777.216	429496.7296
27.0	729.00	19683.000	531441.0000
29.3	858.49	25153.757	737005.0801
56.8	3226.24	183250.432	10408624.5376
72.5	5256.25	381078.125	27628164.0625
102.4	10485.76	1073741.824	109951162.7776
107.9	11642.41	1256216.039	135545710.6081
135.6	18387.36	2493326.016	338095007.7696
143.6	20620.96	2961169.856	425223991.3216
206.9	42807.61	8856894.509	1832491473.9121
370.6	137344.36	50899819.816	18863473223.8096
1314.0	252695.12	68180756.118	21744799306.4996

To finish it will be reasonable to note that according to our standards the solution shown in Table 1P.16 is not a not recommended one, mainly because there is no the concluding sums which have to end each column.

Table 1P.16 Determining skewness and kurtosis following Hawkins/Weber

STATE	AMOUNT	$(x-\bar{x})$	$(x-\bar{x})^2$	$(x-\bar{x})^3$	$(x-\bar{x})^4$
1	13.4	−87.7	7,691	−674,526	59,155,942
2	22.4	−78.7	6,194	−487,443	38,361,796
3	25.6	−75.5	5,700	−430,369	32,492,850
4	27.0	−74.1	5,491	−406,869	30,148,994
5	29.3	−71.8	5,155	−370,146	26,576,499
6	56.8	−44.3	1,962	−86,938	3,851,367
7	72.5	−28.6	818	−23,394	669,059
8	102.4	1.3	2	2	3
9	107.9	6.8	46	314	2,138
10	135.6	34.5	1,190	41,064	1,416,695
11	143.6	42.5	1,806	76,766	3,262,539
12	206.9	105.8	11,194	1,184,287	125,297,577
13	370.6	269.5	72,630	19,573,852	5,275,153,207

$\bar{x} = \frac{1314.0}{13} = 101.1$ $s^2 = 9221.5$ median $= 72.5$ $\alpha_3 = 1{,}60$ $\alpha_4 = 5.06$

Problem 1.15 (see: [1], Prob.3.27)

Each of 20 members of the music faculty contributes the same amount of the coffee fund each month. The total number of cups consumed, however, varies from teacher to teacher as indicated by the totals for each week given here.

Table 1P.17 Coffee cups consumed per week

14	25	46	15	24
18	23	29	10	05
15	08	17	17	15
30	11	12	10	25

(1) find the variance and standard deviation for the number of cups consumed for the week in question; (2) assuming everyone cuts his or her coffee drinking exactly in half next week, find next week's variance and standard deviation; (3) if everyone increases their coffee drinking next week by five cups, what will the variance and standard deviation be?

Answers: $\bar{x} = 18.45$, $\sigma_x = 9.303090884$, $\sigma_x^2 = 86.5475$,

$$\sum x_i = 369, \quad \sum x_i^2 = 8539 .$$

Solution

To present the complete calculations of the statistics given in Table1P.17, its content has been ordered from the smallest term to the biggest one and the table was filled from the top to the bottom. To save space, data in columns 3 and 4 of Table 1P.18 are the continuation of columns 1 and 2.

Table 1P.18 Direct solution

x_i	x_i^2	x_i - cont.	x_i^2 - cont.
05	25	17	289
08	64	17	289
10	100	18	324
10	100	23	529
11	121	24	576
12	144	25	625
14	196	25	625
15	225	29	841
15	225	30	900
15	225	46	2116
---	---	$\sum$ 369	$\sum$ 8539

Following a method which economizes on calculations, the procedure applying the (1.5) rule has been proposed which requires only to determine the squared values of the statistics. Then it is seen that the results obtained in this way shown at the bottom of the third and the fourth columns confirm the SD procedure. We partially present it here.

$$\sigma_x^2 = \overline{x^2} - (\overline{x})^2 \quad \overline{x^2} = \frac{8539}{20} \quad \rightarrow \quad \overline{x^2} = 426.95 \quad \overline{x} = \frac{369}{20} \quad \rightarrow \quad \overline{x} = 18.45$$

$$\sigma_x^2 = 86.5475, \ \sigma_x = 9.303090884 .$$

We add to the following to a short discussion on the consequences of two *revolutionary* changes in the teaching staff habits of drinking coffee. Regarding the consequences of reducing each term of the original statistics by half – the new mean value and new standard deviation will be also reduced by half. The reduction of each term by 5 units will result in the reduction of the mean by 5 units giving $\overline{y} = 13.45$ and the variability of the new statistics will not change.

In the end we have a proposal rather similar to that in the solution of Problem 1.5. Let us fix six class intervals each with eight terms and class limits $(5 \div 12)$, $(13 \div 20)$, $(21 \div 28)$, $(29 \div 36)$, $(37 \div 44)$, and $(45 \div 53)$; then determine the number of class frequencies for each interval as 6, 7, 4, 2, 0, 1 in the end we obtain the frequency histogram shown in Fig.1P.3.

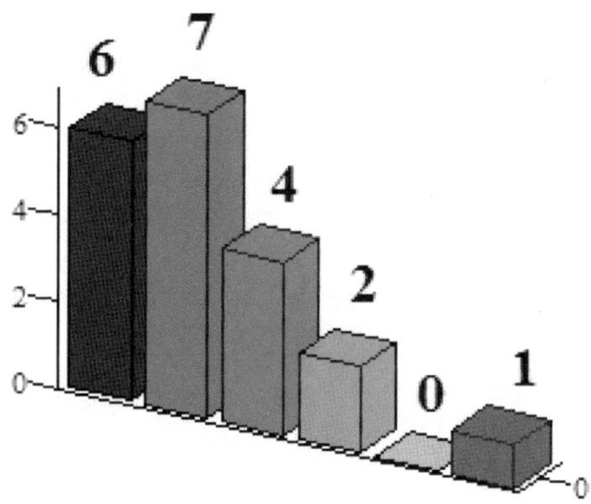

Fig. 1P.3 Frequency histogram of consumed coffee

Comments. (1) such distinct six classes of faculty teachers should equally contribute to the total coffee fund despite the fact that their coffee consumption is so different; this gives rise to the following question: how would a justified proposal of their contribution to the total coffee fund look like? (2) the Student is encouraged to discuss the consequences of the appearance in this statistics of an unusually high term "46"– the *z-score* of which is close to 2.96; so, how will the mean and the variance be changed if this term is rejected?

Answers: $\overline{x} = 17$, $\sigma_x^2 = 49.0526315$, $\sigma_x = 7.00378385$.

References

[1] Weinberg, G.H., Schumaker, J.A., Oltman, D.: Statistics – An Intuitive Approach, 4th edn., pp. 1–447. Brooks/Cole, Monterey (1981)

[2] Reichmann, W.J.: Use and Abuse of Statistics. Penguin Books, Middlesex (1961), p. 345 (1976); Polish translation by Robert Bartoszyński (1933-1998) Drogi i bezdroża statystyki, pp. 1–395. PWN, Warszawa (1968)

[3] Laudański, L.M.: Statystyka nie tylko dla Licencjatów (in Polish: Statistics not only for undergraduates), 2nd edn., vol. 1. Publishing House of the Rzeszow TU, Rzeszów (2009)

[4] Yule, G.U.: An Introduction to the Theory of Statistics. Charles Griffin and Co., London (1911); 2-nd Edition translated into Polish by Z. Limanowski: Wstęp do Teorji Statystyki, Gebethner i Wolff, Warszawa 1921; pp. 1–446. Vi-th Edition of 1922 accessible by Internet, pp. 1–415. 14-th edition, co-author M.G. Kendall, 1950 translated into Polish as Wstęp do Teorii Statystyki. PWN, Warszawa (1966)

[5] Spiegel, M.R.: Schaum's Outline of Theory and Problems of Statistics, pp. 1–359. McGraw-Hill, New York (1972), 870 solved problems

[6] Hawkins, C.A., Weber, J.E.: Statistical Analysis. Applications in Business and Economics, p. 626. Harper & Row, New York (1980)

Unit 2
Grouped Data

First, we offer our Students an opportunity to master the subject from the opening part of Chapter 2 related to the theory of attributes in view of three problems from Udny Yule's book [1]. The problems are stated and have complete solutions. In a preparatory stage we give a brief account of the specific definitions and major concepts on this subject.

Problem 2.1 (see [1], p.15)

The following are the numbers of boys observed with certain classes of defects amongst a number of school-children. A, denotes development defects; B, nerve signs; C, low nutrition.

(ABC)	149	(αBC)	204
$(AB\gamma)$	738	$(\alpha B\gamma)$	1,762
$(A\beta C)$	225	$(\alpha\beta C)$	171
$(A\beta\gamma)$	1,196	$(\alpha\beta\gamma)$	21,842

Find the frequencies of the *positive* classes.

Solution

To solve the problem we can follow a similar solution given in Chapter 2. Let us first evoke the concept of the positive classes by listing all of them for this case: (A), (B), (C), (AB), (AC), (BC), and (ABC). The term follows from the fact that each class collects the positive attributes. It also seems worth emphasizing that each collection discussed here includes only the attributes with a division by dichotomy: that is the object (individual) either possesses a chosen attribute or not. Then, it is important to determine the total number of non empty classes possessing the attributes under investigation. We also remind the Student that the selection of three attributes as is the case here results in six non empty classes of order one, twelve classes of order two, and eight (non empty) classes of order three. Therefore, the data given in Problem 2.1 describe the problem completely. So, adding all class frequencies for those eight classes, we get $N = 26\,287$, the totality of the objects under investigation, i.e. boys amongst the number of school-

L.M. Laudański: Between Certainty and Uncertainty, ISRL 31, pp. 189–215.
springerlink.com

children. According to the proposed terminology (consult Chapter 2) these eight classes form the ultimate set – that is the set of all attributes, moreover these classes possesses the property of completeness of statistical description. It also should be pointed out that it is never necessary to enumerate more than the ultimate frequencies. Chapter 2 show (see Example 2.2) how to determine all ultimate frequencies from the positive class frequencies. Requested *positive* classes also form the complete set. Now let us go back to the solution of our problem: determining all the positive class frequencies:

$(A) = (ABC) + (AB\gamma) + (A\beta C) + (A\beta\gamma) \rightarrow 149 + 738 + 225 + 1196 = 2308$
$(B) = (ABC) + (AB\gamma) + (\alpha BC) + (\alpha B\gamma) \rightarrow 149 + 738 + 204 + 1762 = 2853$
$(C) = (ABC) + (\alpha BC) + (A\beta C) + (\alpha\beta C) \rightarrow 149 + 204 + 225 + 171 = 749$
$(AB) = (ABC) + (AB\gamma) \rightarrow 149 + 738 = 887$
$(AC) = (ABC) + (A\beta C) \rightarrow 149 + 225 = 374$
$(BC) = (ABC) + (\alpha BC) \rightarrow 149 + 204 = 353$
$(ABC) = 149; \quad N = 26287$

Problem 2.2 (see [1], p.16)

The following are the frequencies of the positive classes for the girls in the same investigation :—

N	23,713	(AB)	587
(A)	1,618	(AC)	428
(B)	2,015	(BC)	335
(C)	770	(ABC)	156

Find the frequencies of the ultimate classes.

Solution

We proceed with calculations which in the beginning follow three ultimate class patterns given in the first paragraph of Chapter 2:

$$(AB\gamma) = (AB) - (ABC) \rightarrow 587 - 156 = 431$$

$$(A\beta\gamma) = (A\gamma) - (AB\gamma),\ (A\gamma) = (A) - (AC) \rightarrow 1618 - 428 - 431 = 759$$

$$(\alpha\beta\gamma) = (\beta\gamma) - (A\beta\gamma),\ (\beta\gamma) =$$

$$= N - (B) - (C) + (BC) \rightarrow 23713 - 2015 - 770 + 335 - 759 = 20504$$

$$(A\beta C) = (AC) - (ABC) \rightarrow 428 - 156 = 272$$

$$(\alpha BC) = (BC) - (ABC) \rightarrow 335 - 156 = 179$$

$$(\alpha B\gamma) = (B\gamma) - (AB\gamma),\ (B\gamma) = (B) - (BC) \rightarrow 2015 - 335 - 431 = 1249$$

$$(\alpha\beta C) = (\alpha\beta) - (\alpha\beta\gamma),\ (\alpha\beta) = N - (A) - (B) + (AB) \rightarrow 163$$

The above results, as our Student should remember, can satisfy the obvious condition: the totality of the ultimate class frequencies must be equal to the total number of individuals appearing in considered statistics. This time their sum should give the number of 23713 and the Student can easily check that it really gives this number.

Note: the meaning of the negative class $(\alpha\beta\gamma)$ is especially meaningful – those girls are healthy, free of the any of the listed defects.

Problem 2.3 (see: [1]), p.16)

Convert the Census statement as below into a statement in terms of (*a*) the positive, (*b*) the ultimate class-frequencies.

A = blindness, B = deaf-mutism, C = mental derangement.

N	29,002,525	$(AB\gamma)$	82
(A)	23,467	$(A\beta C)$	380
(B)	14,192	(αBC)	500
(C)	97,383	(ABC)	25

Solution

As we may guess the Census statement also presents a set of classes with the property of completeness. It can be called a mixed set. To complete the requirements of sub problem (a) we have to determine only the following three positive class frequencies:

$$(AB) = (ABC) + (AB\gamma) \rightarrow 25 + 82 = 137$$
$$(AC) = (ABC) + (A\beta C) \rightarrow 25 + 380 = 405$$
$$(BC) = (ABC) + (\alpha BC) \rightarrow 25 + 500 = 525$$

To complete the requirements of sub problem (b) the following ultimate classes must be determined:

$(A\beta\gamma)$, $(\alpha\beta\gamma)$, $(\alpha B\gamma)$, and $(\alpha\beta C)$. The details of the calculations may follow the above solution to Problem 2.2. Therefore, we obtain:

$$(A\beta\gamma)=(A\gamma)-(AB\gamma),\ (A\gamma)=(A)-(AC) \rightarrow 23467 - 405 - 82 = 22980$$
$$(\alpha\beta\gamma)=(\beta\gamma)-(A\beta\gamma),\ (\beta\gamma)=N-(B)-(C)+(BC)$$
$$\rightarrow 29002525 - 14192 - 97383 + 525 - 22980 = 28\,868\,495$$
$$(\alpha B\gamma)=(B\gamma)-(AB\gamma),\ (B\gamma)=(B)-(BC) \rightarrow 14192 - 525 - 82 = 13585$$
$$(\alpha\beta C)=(\alpha\beta)-(\alpha\beta\gamma),\ (\alpha\beta)=N-(A)-(B)+(AB)$$
$$\rightarrow 29002525 - 23467 + 137 - 28868495 = 110700.$$

***

In the next step we move towards the theory of variables regarding grouped data. The most frequent exposition, as in [1] for instance, does not pay special attention to the initial procedure – leading from the raw statistical data to the grouped data. Here the Student will find the discussion of this stage preceding further evaluation procedures of the grouped data. In this case we follow and make use of the book by Weinberg [2] and Spiegel [4], not to mention our own book [3]. There is a question of which case should be taken first, the continuous or the discrete one? Despite the common practice in mathematics courses which gives priority to the discrete data – here in grouping statistical data, an apparently easier case is presented by the continuous data, therefore we commence from this case.

Problem 2.4 (see [2], Prob.5.12) - The Continuous Case

According to the assessor's office the sizes (in hundreds of the square feet) of the 77 homes in the Fairfax addition are as given below. Represent the data using a grouped frequency table (continuous case), and a frequency histogram. Find the mean, variance, standard deviation for the grouped data and compare them with the raw statistics data.

16.83	17.30	23.90	15.21	18.75	19.31	15.95	25.72	18.40	15.00	19.30
22.00	21.70	16.45	23.43	19.30	17.41	17.42	18.75	18.30	28.00	16.02
15.10	16.32	13.05	15.47	16.81	16.11	18.00	28.70	18.20	25.20	15.90
29.30	30.04	18.10	19.40	17.93	15.50	19.00	19.60	19.76	16.50	16.70
16.03	19.51	16.30	28.40	15.75	27.20	14.30	16.00	23.93	30.00	15.00
23.00	14.70	14.25	17.11	19.50	15.21	17.15	17.80	15.50	18.00	16.50
15.00	16.00	14.03	18.03	21.08	18.50	18.70	24.10	15.50	17.08	21.40

Solution

SD procedure gives the following raw statistics results:

$$\sum x_i = 1455.7,\ \sum x_i^2 = 28\,834.1496,\ \overline{x} = 18.9051948,\ \sigma_x = 4.130748697$$

Grouped data results are gradually presented below with some comments.

Following (see Chapter 1 for a possible reference) an algorithm of designing the frequency histogram for the continuous case data we proceed as follows.

Selecting a number of classes (or intervals) it must be kept in mind that there is no common rule, therefore the end product *is not unique*. We recommend as a guideline the thumb rule which in this particular case $\left(2^6 = 64\right) \leq 77$ recommends to assume six classes.

To determine the total range the difference between the largest and the smallest terms must be found: (30.04 - 13.05) = 16.99.

Then the above results allow to determine the interval of each class as 16.99 : 6 = 2.8316 (6). But it should be noted that determining the highest class limits as (27.21, 30.04) the left band continuity will be broken. It means that the greatest term 30.04 would belong to the non existing seventh class. The problem can be solved by choosing a common interval width of 2.832. It will lead to the class intervals as shown in Table2P.1 and then to the frequency histogram shown in Fig.2P.1.

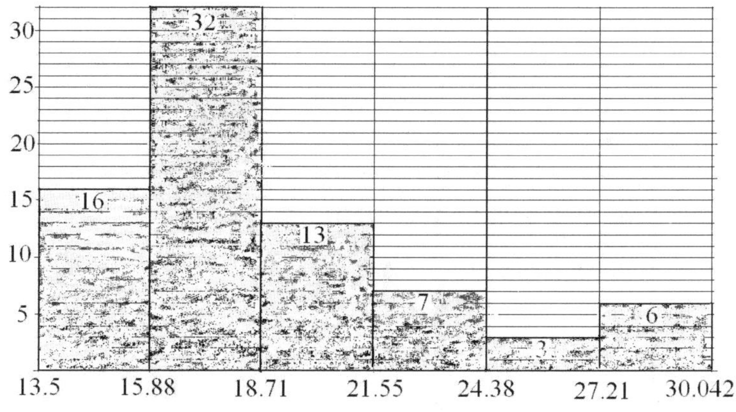

Fig. 2P.1 Frequency histogram - the continuous case

The histogram in Fig. 2P.1 presents the class frequencies (the third column of Table2P.1). With the midpoints shown in the second column of the table – the stage of grouping is completed. For the continuous data it is recommended to determine the midpoints as was shown for the highest class:

(i) 30.042 – 27.21 = 2.832 (ii) 2.832 : 2 = 1.416 (iii) 27.21 + 1.416 = 28.626 ≈ 28.63

The lower midpoint value can be obtained by subtracting from the class width, the exact midpoint value of the higher class. Rounding is always as the last step. The rounded value is inserted in the table. If we require higher accuracy, the subsequent numerical steps do not have to be rounded.

Table 2P.1 Direct method to determine mean values

Class limits	x	f	fx	x^2	$f\,x^2$
27.21 – 30.042	28.63	06	171.78	819.6769	4918.0614
24.38 – 27.21	25.79	03	077.37	665.1241	1995.3723
21.55 – 24.38	22.96	07	160.72	527.1616	3690.1312
18.71 – 21.55	20.13	13	261.69	405.2169	5267.8197
15.88 – 18.71	17.30	32	553.60	299.2900	9577.2800
13.05 – 15.88	14.47	16	231.52	209.3809	3350.0944
		N= 77	1456.68		28798.7590

Let us now proceed with the determining numerical values of the mean, variance, and standard deviation for this grouped data. We make use of the values determined in Table 2P.1.

$$\overline{x} = \frac{\sum x_i \cdot f_i}{\sum f_i} \rightarrow \overline{x} = \frac{1456.68}{77} \rightarrow \overline{x} = 18.91792208 \ [\overline{x} = 18.9051948]$$

$$\overline{x^2} = \frac{\sum x_i^2 \cdot f_i}{\sum f_i} \rightarrow \overline{x^2} = \frac{28798.759}{77} \rightarrow \overline{x^2} = 374.0098571$$

$$\sigma_x^2 = \overline{x^2} - (\overline{x})^2 \rightarrow 374.0098571 - 18.91792208^2 \rightarrow \sigma_x^2 = 16.12208132$$

$$\sigma_x = 4.015231166 \ [\sigma_x = 4.130748697]$$

The Student is advised to look closely at the results obtained on the base of the grouped data, and compare them with the above given results for the raw statistics (partially repeated here in the brackets). The mean differs by 0.067%, but the variance differs by 5.51% (dispersion by 2.8%). For the grouped data the mean became greater but the variance smaller than appropriate raw data averages. It seems that the sign of the differences for both means can be either positive or negative. The Student is recommended to observe these changes in the following examples. Unfortunately typical examples for this Unit according to common practice deal with the data which has been already grouped. We will have a look at this matter below. Before that we consider how to group the discrete raw statistics.

Problem 2.5 (see [2], Prob.5.10) – The Discrete Case

The sports programming coordinator of a major network has requested her administrative assistant to compile the length of the last 50 televised professional football games. The data collected show the following lengths (in minutes):

103	107	095	110	115	123	096	107	115	097	090	125	123	127	095	102	107
115	093	108	110	111	115	139	116	105	097	144	098	104	114	119	129	133
122	121	111	127	118	115	123	118	094	132	093	106	114	115	111	112	

Group the data into a frequency table and represent it on a histogram. Determine the mean, variance, standard deviation.

Solution

SD procedure gives the following raw statistics results:

$$\sum x_i = 5619 \quad \sum x_i^2 = 639143 \quad x_{ave}^2 = 12782.86 \quad \overline{x} = 112.38$$

$$\sigma_x^2 = 153.5956 \quad \sigma_x = 12.39336919$$

The grouping procedure for the discrete data has some specific features but the algorithm of the procedure is the same as described above. The first two points of the Algorithm can be proceeded in either direction. Let us determine the range first, i.e. (144 – 90) = 54, then the greatest and the smallest terms must be combined to give the answer. Then let us suggest the number of the classes as $n = 5$. To determine the common interval we calculate $\Delta x = 54/5 \rightarrow 10.8 \approx 11$ which has to be also a discrete value – so, we should always round up.

Table 2P.2 Direct method to determine averages - the discrete case

Class limits	*Midpoints x*	f	$f x$	x^2	$f\ x^2$
134 – 144	139	02	278	19321	38642
123 – 133	128	09	1152	16384	147456
112 – 122	117	15	1755	13689	205335
101 – 111	106	14	1484	11236	157304
90 - 100	95	10	950	9025	90250
		N= 50	5619		638987

Determining the interval limits and the midpoints due to the specific differences with respect to the continuous case requires some comments. Let us look at the lowest class. Its left band is determined by the smallest value of the grouped data. Each class has 11 terms and each class contains terms appearing at the class bands. These requirements allow to determine the right band as equal to 100. The immediately following higher class has the lowest term 101 as the next following after the term 100. To determine the right band for this class we have to proceed exactly in the same way as for the lowest class. In the end we have to check whether the highest class contains the greatest term 144. To determine midpoints we commence again from the lowest class. Among the 11 discrete terms the midpoint is the sixth term counting from each side. For an even case of the number of terms, the two middle terms have to be selected and their average determined. This average will serve as the midpoint.

Each subsequent midpoint of the case under consideration is greater by 11 than the previous one. Generally the midpoints are distanced by the value of the interval – as it is the case also for the continuous data.

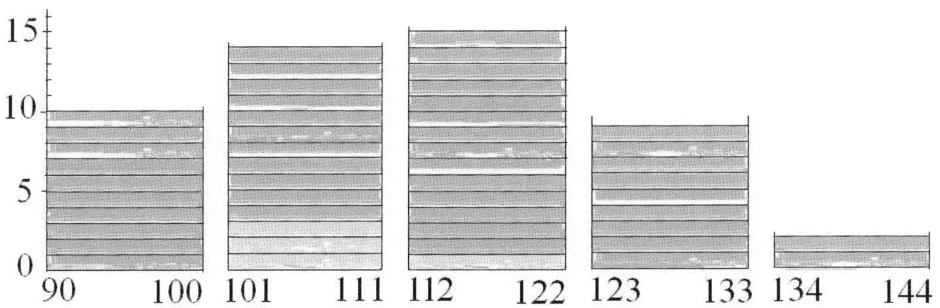

Fig. 2P.2 Frequency bar histogram – the discrete case (see Table 2P.2).

The next stage makes use of the calculations given in Table 2P.2. First to derive the mean – we see that the grouped data gives exactly the same value as that obtained for the raw statistics. Also there is no significant difference regarding the measures of variability. The variance of the grouped data is 2% smaller than the variance obtained for the raw statistics. Below in squared brackets are repeated raw statistics results (indicating the variance data as different for both cases).

$$\overline{x}^2 = \frac{\sum x_i^2 \cdot f_i}{\sum f_i} \quad \rightarrow \quad \overline{x}^2 = \frac{638987}{50} \quad \rightarrow \quad \overline{x}^2 = 12779.74\,[x_{ave}^2 = 12782.86]$$

$$\sigma_x^2 = \overline{x}^2 - (\overline{x})^2 \quad \rightarrow \quad 12779.74 - 112.38^2 \quad \rightarrow \quad \sigma_x^2 = 150.4756 \quad \sigma_x = 12.26684964$$

$$[\sigma_x^2 = 153.5956],\ [\sigma_x = 12.39336919]$$

Any doubts that the obtained results could be wrong may be removed if remember in relation to the mean, that the results are identical – such a rare coincidence cannot be incidental. The remaining doubts regarding variability may be resolved thus: the results in the last column of Table 2P.2 are calculated once more by multiplying the results of the second column with the results of the fourth column to get the same values as obtained from multiplying the third and fifth columns.

Problem 2.6 (see: [4], pp.32-33)

The final marks in mathematics of 80 students at State University are recorded in the accompanying table.

68	84	75	82	68	90	62	88	76	93
73	79	88	73	60	93	71	59	85	75
61	65	75	87	74	62	95	78	63	72
66	78	82	75	94	77	69	74	68	60
96	78	89	61	75	95	60	79	83	71
79	62	67	97	78	85	76	65	71	75
65	80	73	57	88	78	62	76	53	74
86	67	73	81	72	63	76	75	85	77

By making frequency bar histogram determine the all major averages for the grouped data.

Solution

In the book [4] Spiegel proposed $n = 10$ i.e. ten classes - while grouping the above given $N = 80$ grades – called here *raw statistics* or *descriptive statistics*. According to our guidelines, the highest number of classes would be $n = 6$ - as this number satisfies the recommended rule of the thumb. It is seen that with

$n = 6$ we have an interval $\Delta x = 44/6 \rightarrow 7.3 \approx 8$. But we recommend $n = 5$ - which gives the interval $\Delta x = 44/5 \rightarrow 8.8 \approx 9$. Let us discus these proposals.

Apparently Spiegel's proposal seems to offer a worse proposal. He disregards the true range of the initial statistics and proposes ten classes: (50-54), (55-59), (60-64), (65-69), … , (90-94), (95-99). Nevertheless this proposal also has an evident advantage: the grading system with grades in the range of (50-99) is commonly used at universities. Therefore, all such results will fall into the array proposed by Spiegel. However, for the particular case under consideration our proposal suits it better. Taking $n = 6$ we get the following classes: (53-60), 61-68), (69-76), (77-84), (85-92), and (93-100). Only the highest class has the upper limit greater than the greatest value of the initial statics. From this point of view the proposal $n = 5$ is the ideal one with the following limits: (53-61), 62-70), (71-79), (80-88), and (89-97). Moreover, the widely used "traditional marks" suit this scale – offering respectively: 3.0, 3.5, 4.0, 4.5 and 5.0 for the above class limits. So, the Student can see a number of valuable proposals taken into account here and consider the arguments for each particular one. If we end our discussion at this point it is not because we exhausted the possible remarks on the matter but rather because of limitations of the length of this passage.

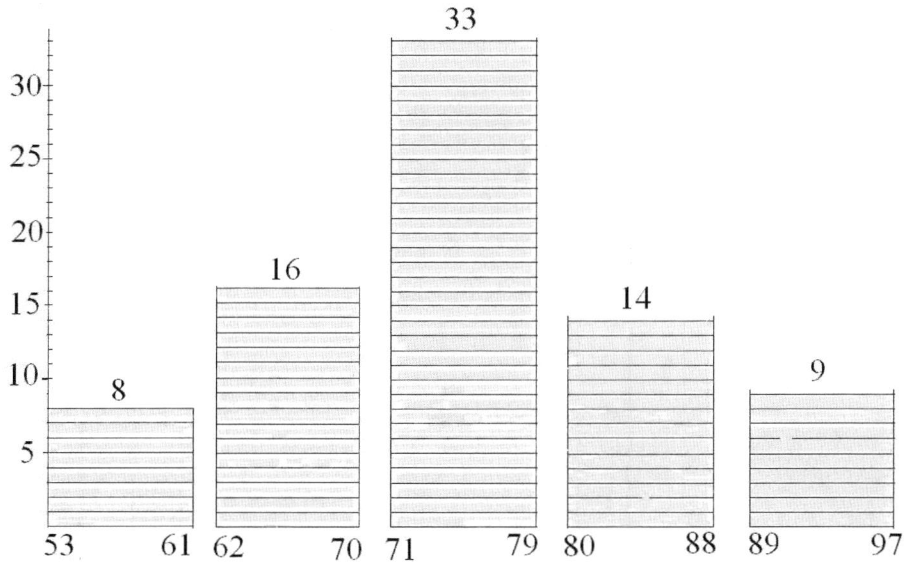

Fig. 2P.3 Frequency bar histogram – the discrete case

The Student is also recommended to make a histogram according to class limits given by Spiegel. It will provide a good opportunity to understand the advantages of accepting our rule of the thumb regarding the choice of the number of classes: the *zig-zag* multimodal histogram obtained by Spiegel serves as the best recommendation of this thumb rule. But there is a second comment which points to the histogram shown in Fig. 2P.3 – its unimodal shape is shifted towards the higher grades but the normal law of grade distribution suggests that the maximum should be closer to the medium grades, say 3.5 in this scale – but not 4.0. It follows that the passing grade for these examination was set too low. Then we turn to numerical calculations.

Table 2P.3 Direct method to determine averages – the discrete case

Class limits	*x midpoints*	*f*	*f x*	*f x²*
89 – 97	93	09	837	77841
80 – 88	84	14	1176	98784
71 – 79	75	33	2475	185625
62 – 70	66	16	1056	69696
53 - 61	57	08	456	25992
		N=80	6000	457938

$\sum x_i = 6020 \quad \sum x_i^2 = 461508 \quad x_{ave}^2 = 5768.85 \quad \overline{x} = 75.25 \quad \sigma_x^2 = 106.2875$

$$\overline{x} = \frac{\sum x_i \cdot f_i}{\sum f_i} \rightarrow \overline{x} = \frac{6000}{80} \rightarrow \overline{x} = 75 \quad [75.25]$$

$$\overline{x^2} = \frac{\sum x_i^2 \cdot f_i}{N} \rightarrow \frac{457938}{80} = 5724.225 \quad [5768.85]$$

$$\sigma_x^2 = \overline{x^2} - (\overline{x})^2 \rightarrow 5724.225 - 5625 = 99.225 \quad [106.2875]$$

$\sigma_x \cong 9.96117463$ [10.3095829] in brackets – results obtained for the original raw statistics (obtained using the SD procedure).

***

Before presenting the next problem we present Table IX from [1], (see p.95 in Polish translation or p.114 in the original edition) which will be used below as a reference to at least one problem solved here.

Table IX – Weights for Adult Men [1]

Weight in lbs.	Number of Men within given Limits of Weight. Place of Birth—				Total.
	England.	Scotland.	Wales.	Ireland.	
90–	2	—	—	—	2
100–	26	1	2	5	34
110–	133	8	10	1	152
120–	338	22	23	7	390
130–	694	63	68	42	867
140–	1240	173	153	57	1623
150–	1075	255	178	51	1559
160–	881	275	134	36	1326
170–	492	168	102	25	787
180–	304	125	34	13	476
190–	174	67	14	8	263
200–	75	24	7	1	107
210–	62	14	8	1	85
220–	33	7	1	—	41
230–	10	4	2	—	16
240–	9	2	—	—	11
250–	3	4	1	—	8
260–	1	—	—	—	1
270–	—	—	—	—	—
280–	—	—	1	—	1
Total	5552	1212	738	247	7749

***

Problem 2.7

Taking into account the grouped data of Table IX regarding Men born in Wales, draw both histograms and find P_{40} and $PR_{X=143}$; also find the main averages.

<u>Solution</u> – we commence by calculating the mean values following the results from Table 2P.4.

$$\overline{x} = \frac{\sum x_i \cdot f_i}{\sum f_i} \rightarrow \overline{x} = \frac{116650}{738} \rightarrow \overline{x} = 158.0623306233062330623306233062330623$$

$$\overline{x^2} = \frac{\sum x_i^2 \cdot f_i}{\sum f_i} \rightarrow \overline{x^2} = \frac{18694400}{738} \rightarrow \overline{x^2} = 25331.16531165311653116531165311653117$$

$$\sigma_x^2 = \overline{x^2} - \left(\overline{x}\right)^2 \rightarrow 25331.16531 - 158.0623306^2$$

$$\sigma_x^2 = 347.4649495817451399446243784931 1$$

$$\sigma_x = 18.64041173316043306181402789633 38$$

Table 2P.4 Direct method to determine averages – the continuous case

Class limits	*x*	*f*	*x f*	*x x f*	Class limits	*x*	*f*	*x f*	*x x f*
190 – 200	195	14	2730	532350	---				
180 – 190	185	34	6290	1163650	280 - 290	285	01	285	81225
170 – 180	175	102	17850	3123750	270 - 280	275	00	---	---
160 – 170	165	134	22110	3648150	260 - 270	265	00	---	---
150 – 160	155	178	27580	4274900	250 - 260	255	01	255	65025
140 – 150	145	153	22185	3216825	240 – 250	245	00	---	---
130 – 140	135	68	9180	1239300	230 – 240	235	02	470	110450
120 – 130	125	23	2875	359375	220 – 230	225	01	225	50625
110 – 120	115	10	1150	132250	210 - 220	215	08	1720	369800
100 – 110	105	02	310	32550	200 – 210	205	07	1435	294175
----	---		---	----			**738**	**116650**	**18694400**

An interesting numerical experiment is shown below: eliminating the last record given in Table IX leads to the following numerical results.

Table 2P.5 Both methods to determine averages – the continuous case - modified

Class	*X*	*F*	*F-*	*F*X*	*X*X*	*F*X*X*	*U*	*F*U*	*U*U*	*F*U*U*
250 - 260	255	01	737	255	65025	65025	8	08	64	64
240 – 250	245	00	736	--	60025	---------	7	0	49	0
230 – 240	235	02	736	470	55225	110450	6	12	36	72
220 – 230	225	01	734	225	50625	50625	5	05	25	25
210 - 220	215	08	733	1720	46225	369800	4	32	16	128
200 – 210	205	07	725	1435	42025	294175	3	21	09	63
190 – 200	195	14	718	2730	38025	532350	2	28	04	56
180 – 190	185	34	704	6290	34225	1163650	1	34	01	34
170 – 180	175	102	670	17850	30625	3123750	0	0	0	0
160 – 170	165	134	568	22110	27225	3648150	-1	-134	01	134
150 – 160	155	178	434	27590	24025	4276450	-2	-356	04	712
140 – 150	145	153	256	22185	21025	3216825	-3	-459	09	1377
130 – 140	135	68	103	9180	18225	1239300	-4	-272	16	1088
120 – 130	125	23	35	2875	15625	359375	-5	-115	25	575
110 – 120	115	10	12	1150	13225	132250	-6	-60	36	360
100 – 110	105	02	02	210	11025	22050	-7	-14	49	98
-----------	----	**737**		**116275**	------	**18604225**	----	**-1270**	----	**4786**

Direct method results:
Mean value: 157.76797829036635006784260515604 lb
Variance: 352.44684435231023861818329101992
Standard deviation: 18.773567704416500668614996734586 lb

Coded method results:

Mean: $$-\frac{1270}{737}\cdot 10+175\cong 157.767978290366500678426 \text{ lb}$$

Variance: $\sigma_U^2 = \dfrac{4786}{737} - \left(\dfrac{1270}{737}\right)^2 \cong 3.52446844352310238618183291$

The simple but meaningful conclusion is left for the Student.

The next part of the solution is devoted to derivation of the indicated percentiles. In the procedure which leads to their determination we have to draw a graphic frequency histogram. It is shown in Fig.2P.4.

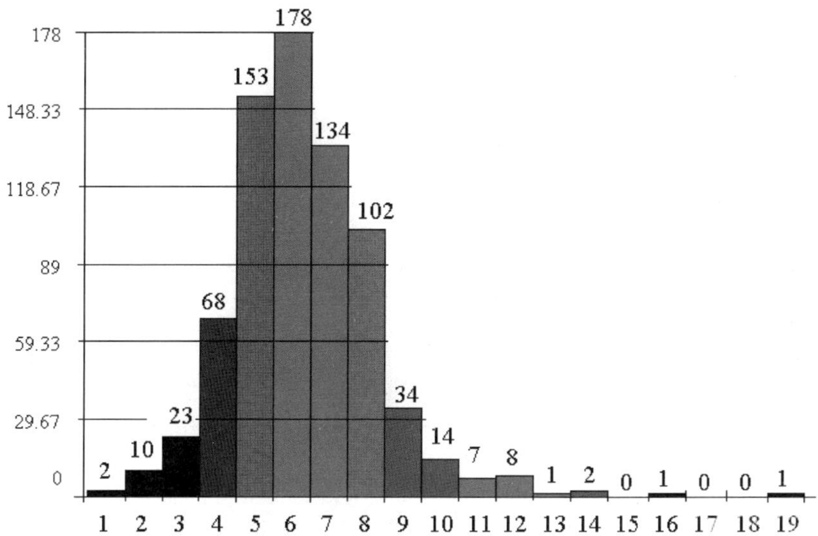

Fig. 2P.4 Frequency histogram – the continuous case

Before coming to the next step we propose a brief comment which is related to the rule of the thumb recommended in this book. According to this rule the proposal with respect to the number of classes seen in the statistics under consideration $n = 19$ would be in a harmony with $N = 524288$ individuals to be considered. Keeping in mind that we have in fact only the total number of individuals $N = 737$ - the right tail of the histogram shown in Fig.2P.4 presents a serious deficiency. On the other hand, from the point of view of the procedure collecting such statistics, the presented case seems to have been conducted correctly and the interval $\Delta x = 10 lb$ evidently cannot be broadened if we are to maintain the accuracy of measuring adult's weight. Now let us proceed to solve the problem.

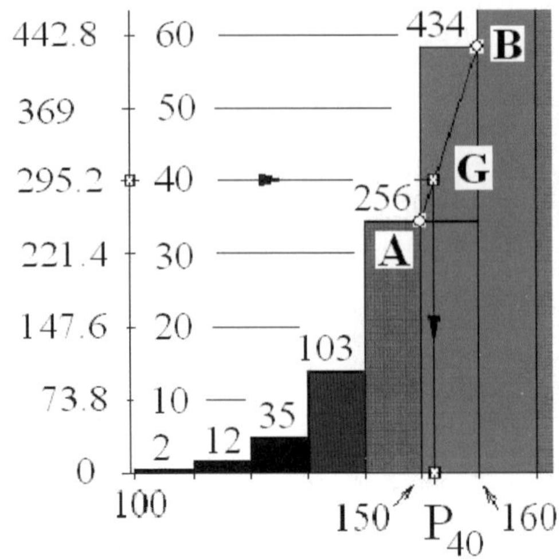

Fig. 2P.5 Cumulated frequency histogram, left tail – the continuous case

In order to determine the 40th percentile, we propose first to follow a rough graphic solution which is depicted in Fig.2P.5. It contains an initial part of the cumulated frequency histogram. This initial part uses values from Table 2P.5, although it does not have the greatest term which has to be added in the last step, leading to the final value of the cumulative frequency 738.

The vertical scale in Fig.2P.5 has two numerical values: on the left side we see the scale of cumulative frequencies – on the right side the corresponding values of the percentiles are indicated. Therefore the cumulative frequency 73.8 corresponds to the percentile "10", the value 147.6 – to the percentile "20" and so on. Subsequently it is seen that the percentile "40" – corresponds to the cumulated frequency "295.2". Looking for the weight of an individual, which corresponds to this cumulated frequency, we draw a horizontal line through the value "295.2" – until the line reaches the frequency bar corresponding to the sixth class of the weight – i.e. a value between 150 and 160 pounds. To determine roughly the appropriate weight, which is called the "*position of the percentile forty*" – there is a segment denoted as A B, the intersection between this segment and the horizontal line denoted as G - has to be projected vertically to the horizontal axes, the projection point determines the *position of the percentile forty* P_{40}. The last step has to give an approximate number corresponding to this point, for instance $P_{40} \approx 151$ lb. Now let us explain the meaning of this weight: individuals with a weight of less than 151 lb belong to the subset of 40% of the entire set of the investigated people, while those with weight bigger than 151 lb belong to the complement subset. To determine a more accurate value of this percentile we will present an appropriate procedure (see also Chapter 2) using an important part of Fig.2P.5 – denoted as Fig.2P.6.

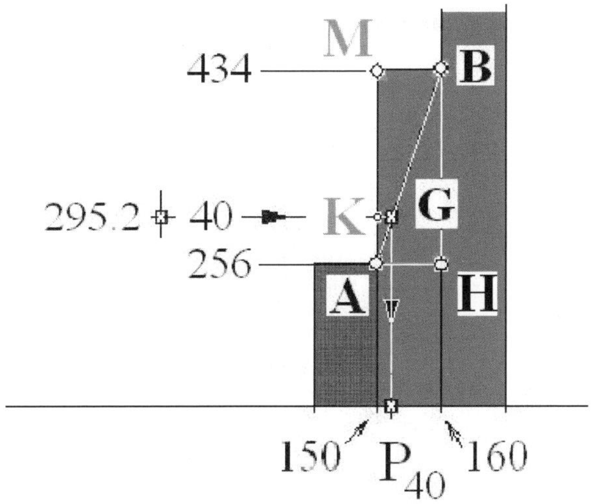

Fig. 2P.6 To explain the procedure of determining percentiles

$$P_{40} = 150 + \frac{295.2 - 256}{434 - 256}(160 - 150) = 152.2022471910112359550561797752 8$$

The above result hardly needs any comments. To see *Thales' Theorem,* the Student has to consider the triangle AMB with its sides subdivided by points K and G - and that is all. There is also one important point to mention: the presented method implicitly assumes that every class is filled by a continuum of numbers which are supposed to represent a continuum of individuals. In practice we always deal with a finite number of objects/individuals. So, the mentioned assumption cannot be literally carried out in practice as it presents an idealization of the statistical reality. The same remark goes for the second concept – the ranking percentile of a given term. We are going to present the solution to the problem stated above showing how to determine the ranking percentile for an individual weighing 177 lb. Its numerical part is presented below, and contains two figures giving a rough solution in Fig. 2P7, and then in Fig. 2P.8.

$$P_{X=177} = \frac{568 + 0.7 \cdot 102}{738} 100\ \% = 86.63956(63956)\ \% \approx 87\ \%$$

Before we turn the Student's attention to the drawings, we look at the numerical details seen in the above result: there is interesting cycle "639566" which it is possible to see using the Windows calculator but impossible with the 10 digit scientific calculators available on the market, which gives us the number of "86.6395664".

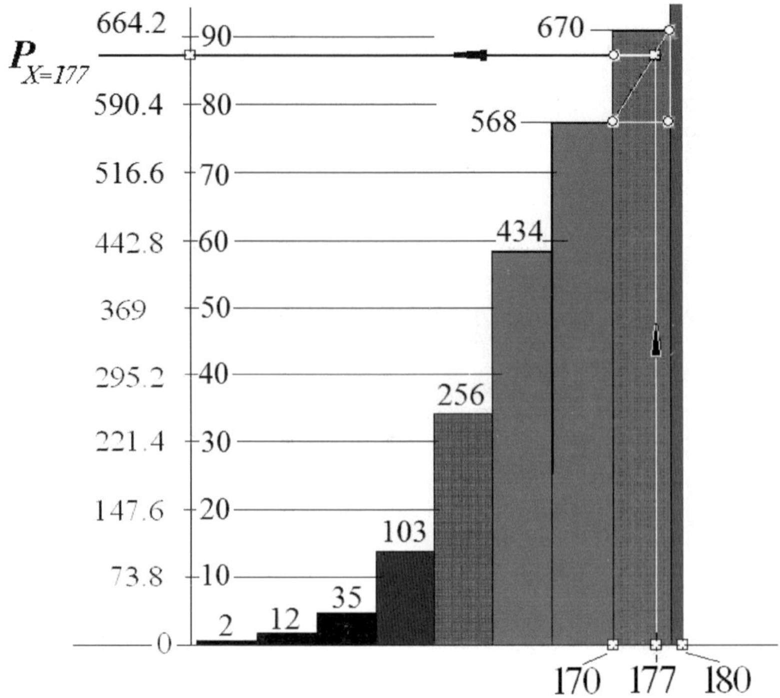

Fig. 2P.7 Cumulated frequency histogram explaining how to determine the ranking percentile

To see more details regarding the possibility of applying once more *Thales' theorem* we resort to Fig.2P.8 which vividly tells us that the procedure to determine the ranking percentile for the individual whose weight is 177 pounds is exactly the reversal of the above procedure determining here the position of the forty percentile.

To finish explanations regarding the coefficient "0.7" from the above numerical formula, we add a short comment: the number "0.7" indicates the relative position of number "177" in the class of individuals whose weight ranges from 170 lb to 180 lb - excluding this real number. In fact it would be determined in the following way:

$$(177-170)/(180-170)=7/10$$

In a subsequent step, this ratio from the MB side of the AMB triangle according to *Thales' theorem* is "exported" to the AM side where the ratio AK: AM must give the same 0.7. In the scale of the cumulated frequency histogram, point A marks the number "568", while the position of point M is higher than A for "102" points, therefore to determine the position of point K in this scale we have to add 0.7 * 102 = 71.4 to 568 and then determine the position

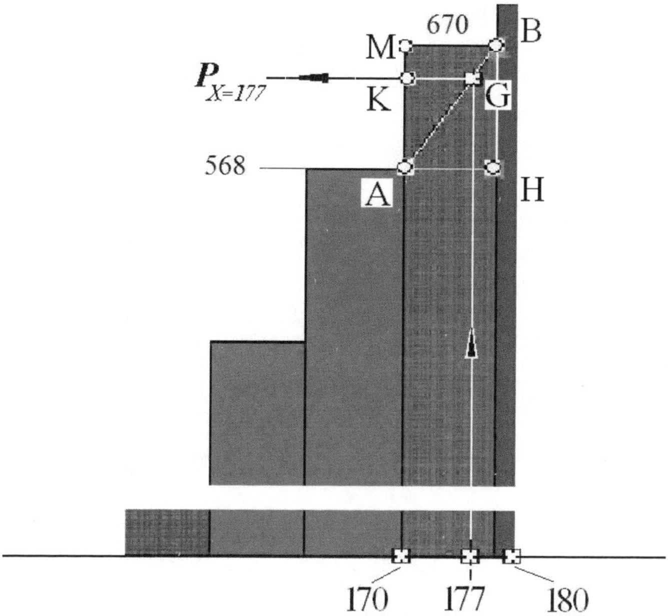

Fig. 2P.8 To explain the procedure of determining the ranking percentiles

of number 71.4 + 568 = 639.4 in the scale showing percentiles. In the end this will give us number 86.63956 (63956) – which rounded to the nearest integer will give the final answer "87".

Problem 2.8 (see [2], Prob.5.21)

The following Table 2P.6 shows the number of turkey in each weight class delivered to one of its outlets in metropolitan area by a large turkey distributor. Determine (1) the mean, (2) variance, (3) draw both frequency and cumulative histograms, (4) determine the 30th percentile, (5) determine the ranking percentile for the turkey weight 20.3 lb.

Table 2P.6 Grouped Data to the Problem 2.8

Class limits lb	Class frequency	Class limits lb	Class frequency
9.5-11.5	3	17.5-19.5	70
11.5-13.5	7	19.5-21.5	60
13.5-15.5	10	21.5-23.5	25
15.5-17.5	60	23.5-25.5	15

Solution

X	F	XF	XXF	U	UF	UUF	FFF
24.5	15	367.5	9003.75	-3	-45	135	250
22.5	25	562.5	12656.25	-2	-50	100	235
20.5	60	1230.0	25215.00	-1	-60	60	210
18.5	70	1295.0	23957.50	0	0	0	150
16.5	60	990.0	16335.00	1	60	60	80
14.5	10	145.0	2102.50	2	20	40	20
12.5	7	87.5	1093.75	3	21	63	10
10.5	3	31.5	330.75	4	12	48	3
	250	4709.0	90694.50	--	-42	506	----

Using the above numerical values the following results have been obtained:

μ = 18.836 lb $\quad \bar{U} = -0.168 \quad$ i = 2 $\quad$ R = 18.5 $\quad$ a check: $\quad \mu$ = 18.836

$\sigma_x^2 = 7.983104$ a check $\quad \sigma_U^2 = 1.995776 \quad \sigma_x^2 = i^2 \cdot \sigma_U^2 \quad \sigma_x^2 = 7.983104$

$$P_{30} = 17.3(3) \qquad PR_{20.3} = 69.6\%$$

Comment 1:

(i) note the bin interval of 2 lb; (ii) to determine P_{30} the fourth bin of the cumulative histogram which corresponds to the mark 0.3 * 250 = 75 is to be used; (iii) the fourth bin shows an increase of 60 marks in the class frequency (with respect to the third bin), while the mark 20 corresponds to the top of the third bin; (iv) further calculations, therefore, are as follows:

$$(75 - 20) / 60 * 2 + 15.5 = 17.3\ (3)$$

Comment 2:

to determine the ranking percentile corresponding to the object 20.3 lb it has to be pointed out that this value belongs to the sixth class with the limits (19.5, 21.5). The mark showing cumulated frequency corresponding to the top of the bin spanning the sixth class is 210 (see the above Table). The sixth bin has a 60 pint bigger class frequency than the previous one. The mark corresponding to the top of the fifth bin indicates 150. All of this is inserted in the following calculations:

$$[\ (\ [(20.3 - 19.5) / 2\ ] * 60 + 150\ ) / 250\ ] * 100\ \% = 69.6\ \% \rightarrow 70\ \%$$

The Student is requested to draw a cumulative histogram to follow directly the above given comments.

Problem 2.9 (see [2], Review I, Prob.1.13)

There is often a considerable time span between the discovery of a drug and its introduction to the market. Testing, government regulations, and production-marketing activities account for much of this time. A study of 5000 drugs available today, some by prescription and some over-the-counter, shows the time between development and introduction for each drug. A summary of the results follows.

Table 2P.6 Grouped Data to the problem 2.9

Class limits lb	Class frequency	Class limits lb	Class frequency
0.5-3.5	137	12.5-15.5	84
3.5-6.5	85	15.5-18.5	52
6.5-9.5	44	18.5-21.5	70
9.5-12.5	17	21.5-24.5	11

Determine (1) the mean, (2) variance, (3) draw both frequency and cumulative histograms, (4) determine the 28th percentile, (5) determine the ranking percentile for the drug tested 13.8 years.

Solutions

X	F	FFF	XF	XXF	U	UF	UUF
23	11	500	253	5819	4	44	176
20	70	489	1400	28000	3	210	630
17	52	419	884	15028	2	104	208
14	84	367	1176	16464	1	84	84
11	17	283	187	2057	0	---	-----
8	44	266	352	2816	-1	-44	44
5	85	222	425	2125	-2	-170	340
2	137	137	274	548	-3	-411	1233
---	500	------	4951	72857	----	-183	2715

To determine the mean

Direct method $\overline{x} = 9.902$

Coded method $\bar{U} = -0.366$ i = 3 R = 11 $\overline{x} = U * i + R$ $\rightarrow$ $\overline{x} = 9.902$

To determine the variance and standard deviation

Direct method

$$\sigma_x^2 = 72857/500 - 9.902 * 9.902 = 145.714 - 98.049604 = 47.664396$$

$$\sigma_x = 6.903940614$$

Coded method

$$\sigma_U^2 = 2715/500 - 0.366 * 0.366 = 5.43 - 0.133956 = 5.296044$$

$$\sigma_x^2 = i^2\ \sigma_U^2 = 9 * 5.296044 = 47.664396$$

Comment 1: note that the bin interval equals 3; determining P_{28} the second bin of the cumulative histogram is to be used as the appropriate one for the mark 0.28 * 500 = 140 ; the second bin has limits (3.5 – 6.5), its top indicates marks 222 and the bin frequency indicates an increase of 87 marks in comparison to the first bin, which top corresponds to the mark 137, therefore further calculations are as follows:

$$P_{28} = \frac{[(0.28 \cdot 500 - 137)]}{85} \cdot 3 + 3.5 = 3.605882353$$

Comment 2: to determine the ranking percentile corresponding to the term 13.8 years, it has to be noted that this term belongs to the fifth class with the boundaries of (12.5, 15.5). The mark showing cumulated frequency corresponding to the top of the bin spanning the fifth class is 367 (see the above Table). The fifth bin has class frequency 84 points bigger than the previous one. The cumulated frequency marking the top of the fourth bin shows 283. All of this is inserted in the following calculations:

$$PR_{13.8} = 100\left(\frac{(13.8 - 12.5)}{3} 84 + 283\right) / 500 = 63.88\ \% \rightarrow 64\%$$

Problem 2.10 (see [1], p.96)

For the data of Table X [1] showing the frequency distribution of fecundity (i.e. the ratio of the number of yearling foals produced to the number of coverings for brood-mare race-horse, covered at least eight times) after 2000 observations, in 1899, Karl Pearson determined all the statistical measures i.e. means, percentile 35, and ranking percentile for fertility 12/30 and also drew frequency histograms.

Answers:

Mean: given as a real number 0.632808333 or as convenient assignment 18.98/30 $\approx$ 19/30

Standard deviation: 0.156452431 or as 4.69/30

$P_{35} = 19.187730159/30$ $P_{12/30} = 7.4875$

Table X – Horse races fecundity [1]

Fecundity.	Number of Mares with Fecundity between the Given Limits.	Fecundity.	Number of Mares with Fecundity between the Given Limits.
1/30– 3/30	2	17/30–19/30	315
3/30– 5/30	7·5	19/30–21/30	337
5/30– 7/30	11·5	21/30–23/30	293·5
7/30– 9/30	21·5	23/30–25/30	204
9/30–11/30	55	25/30–27/30	127
11/30–13/30	104·5	27/30–29/30	49
13/30–15/30	182	29/30–1	19
15/30–17/30	271·5		
		Total	2000·0

Numerical solutions:

Table 2P.6 Direct method determining averages

X	F	FX	X²	F X²
29.5/30	19.0	18.683(3)	0.96694(4)	18.37194(4)
28/30	49.0	45.73(3)	0.871(1)	42.684(4)
26/30	127.0	110.06(6)	0.751(1)	95.391(1)
24/30	204.0	163.2	0.64	130.56
22/30	293.5	215.23(3)	0.537(7)	157.837(7)
20/30	337.0	224.6(6)	0.4(4)	149.7(7)
18/30	315.0	189.0	0.36	113.4
16/30	271.5	144.79(9)	0.284(4)	77.226(6)
14/30	182.0	84.93(3)	0.217(7)	39.635(5)
12/30	104.5	41.8	0.16	16.72
10/30	55.0	18.3(3)	0.1(1)	6.1(1)
8/30	21.5	5.73(3)	0.071(1)	1.528(8)
6/30	11.5	2.3	0.04	0.46
4/30	7.5	0.9(9)	0.017(7)	0.13(3)
2/30	2.0	0.13(3)	0.004(4)	0.08(8)
	2000	**1265.616(6)**		**849.84749(9)**

Direct method to derive averages – numerical results:

Mean: $\overline{x} = \dfrac{\sum FX}{\sum F} = \dfrac{1265.616(6)}{2000} = 0.632800332 \qquad \overline{x} = \dfrac{18.984249(9)}{30} \approx 19/30$

Variance: $\sigma_x^2 = \dfrac{\sum FX^2}{\sum F} - \left(\dfrac{\sum FX}{\sum F}\right)^2 = \dfrac{849.84749(9)}{2000} - (0.6328083(3))^2$

$\sigma_x^2 = 0.424923749 - 0.400446386 = 0.024477363$

Standard deviation: $\sigma_x = 0.156452432 \qquad \sigma_x = 4.69357296/30 \approx 4.69/30$

Table 2P.7 Coded method determining averages

X	F	U	FU	U²	F U²
29.5/30	19.0	6.75	128.25	45.5625	865.6875
28/30	49.0	6	294.0	36	1764
26/30	127.0	5	635.0	25	3175
24/30	204.0	4	816.0	16	3264
22/30	293.5	3	880.5	9	2641.5
20/30	337.0	2	674.0	4	1348
18/30	315.0	1	315.0	1	315
16/30	271.5	0	0	0	0
14/30	182.0	-1	-182	1	182
12/30	104.5	-2	-209	4	418
10/30	55.0	-3	-165	9	495
8/30	21.5	-4	-86	16	344
6/30	11.5	-5	-57.5	25	287.5
4/30	7.5	-6	-45	36	270
2/30	2.0	-7	-14	49	98
	2000		**2984.25**		**15467.6875**

Coded method to derive averages:

Mean: $\overline{U} = \dfrac{\sum FU}{\sum F} = \dfrac{2984.25}{2000} = 1.492125 \qquad i = 2 \quad \text{and} \quad R = 16$

$\overline{x} = \overline{U} \cdot i + R = 1.492125 \cdot 2 + 16 = 18.98425$

Variance: $\sigma_U^2 = \dfrac{\sum FU^2}{\sum F} - \left(\dfrac{\sum FU}{\sum F}\right)^2 = \dfrac{15467.6875}{2000} - 1.492125^2$

$\sigma_U^2 = 7.733841375 - 2.226437015625 = 5.507406734375$

$\sigma_U = 2.3467864697017 \qquad \sigma_x = \sigma_U \cdot i = 4.693572939403$

$\sigma_x = 4.69357296/30 \approx 4.69/30$

Note: a necessary comment regarding two numerical procedures to determine the desired averages is left for the Student. In the end it has to be noted that instead of our own drawing of the frequency histogram we reproduce here the figure from [1], p.94 and encourage our Student to draw this histogram and next also a cumulative frequency histogram following the examples presented here.

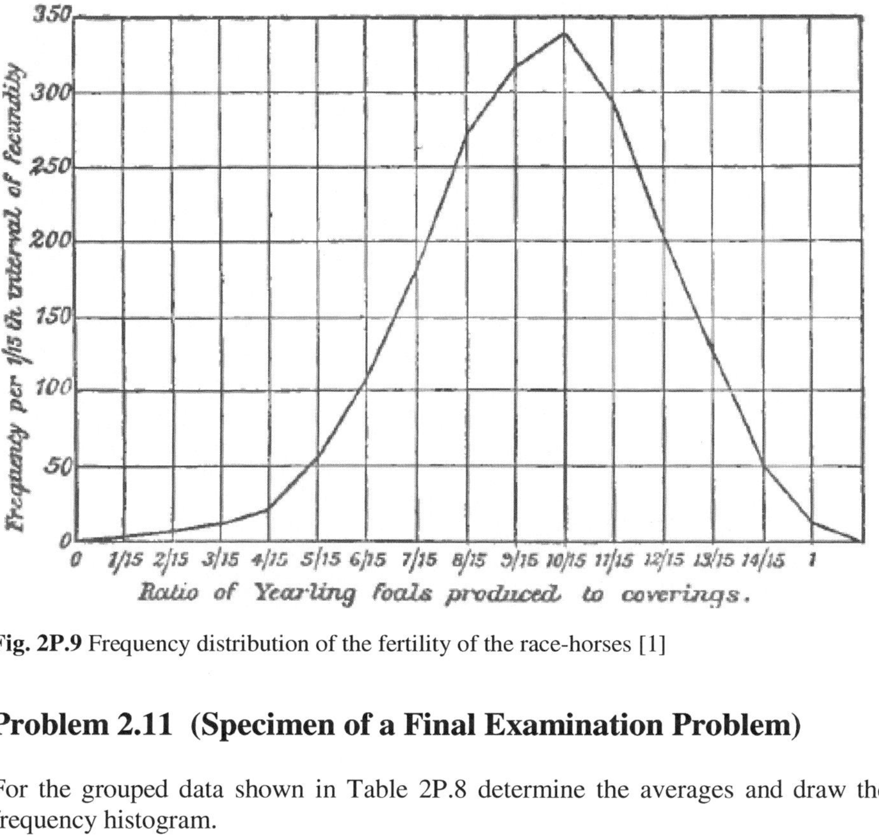

Fig. 2P.9 Frequency distribution of the fertility of the race-horses [1]

Problem 2.11 (Specimen of a Final Examination Problem)

For the grouped data shown in Table 2P.8 determine the averages and draw the frequency histogram.

Table 2P.8 Grouped Data to Problem 2.11

Age classes	Class frequency thousands	Age classes	Class frequency thousands
0 - 10	6840	40 -50	2890
10 -20	4980	50 -60	3120
20 -30	4560	60 -70	1950
30 -40	4200	70 -80	460

Solution

X	F	FF	XF	XXF	U	UF	UUF
75	460	29000	34500	2587500	4	1840	7360
65	1950	28540	126750	8238750	3	5850	17550
55	3120	26590	171600	9438000	2	6240	12480
45	2890	23470	130050	5852250	1	2890	2890
35	4200	20580	147000	5145000	0	0	0
25	4560	16380	114000	2850000	- 1	-4560	4560
15	4980	11820	74700	1120500	- 2	-9960	19920
5	6840	6840	34200	171000	- 3	-20520	61560
	29000		832800	35403000		-18220	126320

Both methods of determining averages:

$$\mu = \frac{832800}{29000} = 28.71724138 \qquad E\left(x^2\right) = \frac{35403000}{29000} = 1220.793103$$

$$\bar{U} = -\frac{18220}{29000} = -0.628275862$$

$$\sigma_x^2 = E\left(x^2\right) - \bar{x}^2 \quad \rightarrow \quad 1220.793103 - (28.71724138)^2 = 396.113151$$

$$i = 10,\ \mathrm{R} = 35 \quad \sigma_x = 19.90259156$$

$$E\left(U^2\right) = \frac{126320}{29000} = 4.355862069 \qquad \sigma_U^2 = E\left(U^2\right) - \bar{U}^2 \quad \rightarrow \quad \sigma_U^2 = 3.96113151$$

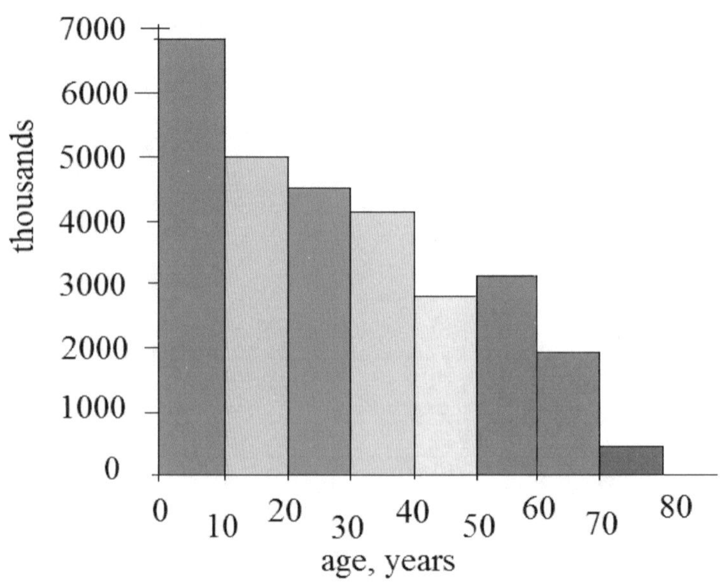

Fig. 2P.10 Frequency distribution of the population age, Poland, year 1970

Problem 2.12 (see [1], p.104)

Making use data of Table XV (see: [1] p.103), originally published by Karl Pearson [6] we suggest examining the frequencies of estimated intervals of cloudiness registered in the decade of 1876-85 in a city which has since that time changed its name from Breslau to Wrocław. The point is that the shape of the histogram belongs to the category of bimodal distributions.

Table XV – Cloudiness at Breslau (now Wrocław) [1]

Cloudiness.	Frequency.	Cloudiness.	Frequency.
0	751	6	21
1	179	7	71
2	107	8	194
3	69	9	117
4	46	10	2089
5	9		
		Total	3653

Answers and comments:

$\mu = 6.821516562$, $\sigma_x = 4.29004631$

$\sum F = 3653$; $\sum FX = 24919$; $\sum FX^2 = 237217$.

Comment: this form of distribution, as in Fig.2P.11 is a rare case in the original comments of F. Galton, see also [1]. On our side we note that the case can be

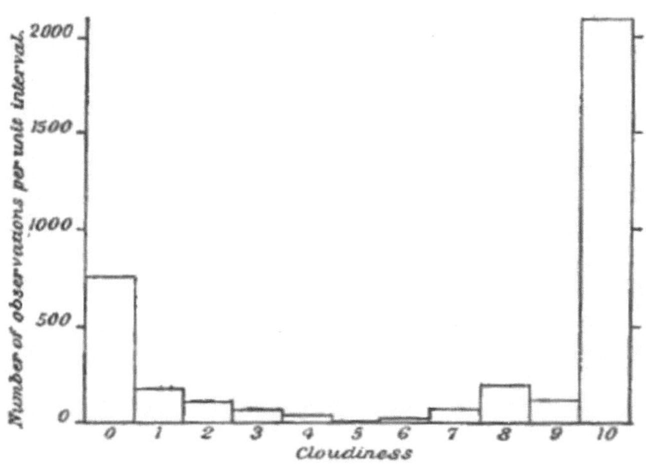

Fig. 2P.11 Frequency distribution of degrees of cloudiness at Breslau/Wrocław

considered as a discrete case. But the presented figure suggests that Karl Pearson who originally investigated these data claimed that they were a "novel case of frequency" considered as continuous data (see [6]). In fact this distribution shows that the mean value is meaningless for such cases. Francis Galton in his *Natural Inheritance* associated this case with "consumptivity" amongst the offspring of consumptives.

A very brief summary of the histogram shown in Fig.2P.12 is as follows: "sky completely or almost completely covered with clouds" such a forecast at the time of observation is the most common, then "practically clear sky" comes next and intermediates are much more rare.

Problem 2.13 (Another Example of a Final Examination Problem)

In a town the taxi cab club collects 60 cabs. Monthly the number of courses they provide for the public service is approximated in the following table. Draw to scale a frequency histogram for the grouped data assuming five classes and determine the basic averages. To get the full mark use both methods of determining averages: direct and coded.

738	729	743	740	736	741	735	731	726	737
728	737	736	735	724	733	742	736	739	735
745	736	742	740	728	738	725	733	734	732
733	730	732	730	739	734	738	739	727	735
735	732	735	727	734	732	736	741	736	744
732	737	731	746	735	735	729	734	730	740

Solution

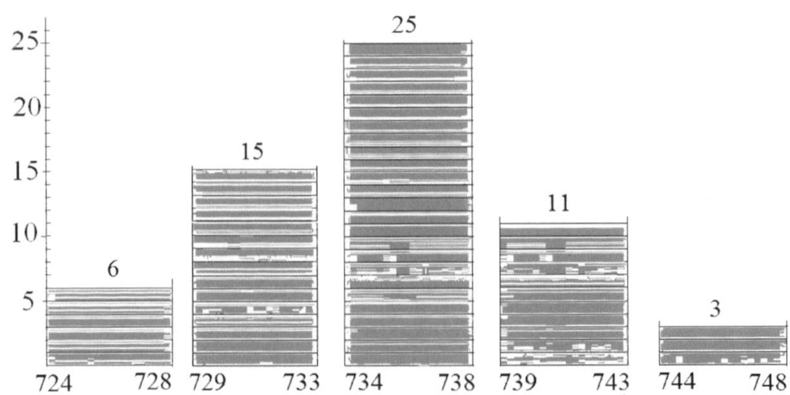

Fig. 2P. 12 Frequency bar histogram following data from informal table of Problem 2.13.

X	F	FF	XF	XXF	U	UF	UUF
746	3	60	2238	1669548	2	6	12
741	11	57	8151	6039891	1	11	11
736	25	46	18400	13542400	0	0	0
731	15	21	10965	8015415	-1	-15	15
726	6	6	4356	3162456	-2	-12	24
	60		44110	32429710		-10	62

$$\bar{x} = 735.16(6) \quad i = 5 \quad R = 736 \quad \bar{U} = -\frac{10}{60} = -0.16(6) \rightarrow \bar{x} = 735.16(6)$$

$$\sigma_x^2 = 25{,}138(8) \quad \sigma_u^2 = 1.005(5) \rightarrow \sigma_x^2 = 25{,}138(8)$$

Comment: the coded method shows a huge numerical simplification of the numerical procedure, however, the numerical results are identical with the direct method. The specificity of this problem makes it, so to speak, a typical examination problem. Regarding the data compare Prob.2.27 of [4], p.43.

References

[1] Yule, G.U.: An Introduction to the Theory of Statistics. Charles Griffin and Co., London (1911); 2nd Edition translated into Polish by Z. Limanowski: Wstęp do Teorji Statystyki, Gebethner i Wolff, Warszawa 1921; pp. 1–446. Vi-th Edition of 1922 accessible by Internet, pp.1–415. 14-th edition, co-author M.G. Kendall, 1950 translated into Polish as Wstęp do Teorii Statystyki. PWN, Warszawa (1966)

[2] Weinberg, G.H., Schumaker, J.A., Oltman, D.: Statistics – An Intuitive Approach, 4th edn., pp. 1–447. Brooks/Cole, Monterey (1981)

[3] Laudański, L.M.: Statystyka nie tylko dla Licencjatów (in Polish: Statistics not only for undergraduates), 2nd edn., vol. 1. Publishing House of the Rzeszow TU, Rzeszów (2009)

[4] Spiegel, M.R.: Schaum's Outline of Theory and Problems of Statistics, pp. 1–359. McGraw-Hill, New York (1972), 870 solved problems

[5] Reichmann, W.J.: Use and Abuse of Statistics. Penguin Books, Middlesex (1961), p. 345 (1976); polish translation by Robert Bartoszyński (1933-1998) Drogi i bezdroża statystyki, pp. 1–395. PWN, Warszawa (1968)

[6] Pearson, K.: Cloudiness: Note on a Novel Case of Frequency. Proceedings of the Royal Society of London, Vol.62 (1897-1898), pp. 287-290

Unit 3
Regression vs. Correlation

The subject of this illustrative material following the division of Chapter 3 is presented in two major parts. First we deal with the derivation of the two (always two!) regression lines with respect to the data which we call in this book two descriptive statistics combined into a set of paired values taken from each boundary statistics. In the second step, where some problems can be incorporated at the stage of deriving the regression lines, we determine the coefficient of correlation. Then we present a few examples of dealing with the subject regarding the grouped data, which require a much more developed tool, in view of what we call *great array of correlation*. The Student can note, that for instance Udny Yule [2], skipped the first stage dealing with descriptive statistics, moreover he entitled Chapter IX "*Correlation*", and Yule's examples offered to the reader are exclusively related to the grouped data. Paying a tribute to Sir Francis Galton in the opening part of our examples we strove to present problems from Galton's publications even if we sometimes cannot indicate which of Galton's papers exactly they come from.

Problem 3.1 (see [3])

Considering data in the below Table 3P.1 examine the regression line approach to fit the data describing the relation between MRI sizing of the brain and the resulting IQ value for men's data . Data was obtained in a special medical examination to 40 carefully selected students of both genders.

The Student has to decode the way which is applied to present the major part of the solution provided by Table 3P.2. Here instead of an additional row at the bottom of Table 3P.2 to save space for the appropriate sums, we present them separately, below.

L.M. Laudański: Between Certainty and Uncertainty, ISRL 31, pp. 217–243.
springerlink.com

Table 3P.1 Two dimensional statistics MRI vs. QI – data

Men				Women			
MRI	IQ	MRI	IQ	MRI	IQ	MRI	IQ
1,001,121	140	1,038,437	139	816,932	133	951,545	137
965,353	133	904,858	89	928,799	99	991,305	138
955,466	133	1,079,549	141	854,258	92	833,868	132
924,059	135	945,088	100	856,472	140	878,897	96
889,083	80	892,420	83	865,363	83	852,244	132
905,940	97	955,003	139	808,020	101	790,619	135
935,494	141	1,062,462	103	831,772	91	798,612	85
949,589	144	997,925	103	793,549	77	866,662	130
879,987	90	949,395	140	857,782	133	834,344	83
930,016	81	935,863	89	948,066	133	893,983	88

Table 3P.2 MRI vs. IQ – Men's data

X	Y	X	Y	X X	X X	Y Y	Y Y	X Y	X Y
1001121	140	1038437	139	1002243256641	1078351402969	19600	19321	140156940	144342743
965353	133	904858	89	931906414609	818768000164	17689	7921	128391949	80532362
955466	133	1079549	141	912915277156	1165426043401	17689	19881	127076978	152216409
924059	135	945088	100	853885035481	893191327744	18225	10000	124747965	94508800
889083	80	892420	83	790468580889	796413456400	6400	6889	71126640	74070860
905940	97	955003	139	820727283600	912030730009	9409	19321	87876180	132745417
935494	141	1062462	103	875149024036	1128825501444	19881	10609	131904654	109433586
949589	144	997925	103	901719268921	995854305625	20736	10609	136740816	102786275
879987	90	949395	140	774377120169	901350866025	8100	19600	79198830	132915300
930016	81	935863	89	864929760256	875839554769	6561	7921	75331296	83291807

$$\sum X = 19097108, \quad \bar{X} = 954855.4 \quad \sum Y = 2300 \quad \bar{Y} = 115 \quad \sum Y^2 = 276362 \tag{3P.1}$$

$$\sum X^2 = 18294372210308 \qquad \sum X\,Y = 2209395807 \tag{3P.2}$$

To facilitate the subsequent substitutions, we provide all major formulae necessary in further calculations.

$$A^* = \frac{\sum x^2 \cdot \sum y - \sum x \cdot \sum x \cdot y}{N \sum x^2 - \left(\sum x\right)^2} \tag{3P.3}$$

$$B^* = \frac{N \cdot \sum x \cdot y - \sum x \cdot \sum y}{N \sum x^2 - \left(\sum x\right)^2} \tag{3P.4}$$

Next, numerical results (3P.1) and (3P.2) were substituted into formulae (3P.3) and (3P.4) for both coefficients of the first regression line and the obtained results were shown.

$$A^* = \frac{18294372210308 \cdot 2300 - 19097108 \cdot 2209395807}{20 \cdot 18294372210308 - 19097108^2} \tag{3P.5}$$

$$A^* = -97.6624774898736929696601170368 97 \tag{3P.6}$$

$$B^* = \frac{20 \cdot 2209395807 - 19097108 \cdot 2300}{20 \cdot 18294372210308 - 19097108^2} \tag{3P.7}$$

$$B^* = 2.22716944879689315229992014536 33\text{e-}4 \tag{3P.8}$$

A similar approach concerns the second regression line:

$$A_* = \frac{\sum y^2 \cdot \sum x - \sum y \cdot \sum x \cdot y}{N \sum y^2 - \left(\sum y\right)^2} \tag{3P.9}$$

$$B_* = \frac{N \cdot \sum x \cdot y - \sum x \cdot \sum y}{N \sum y^2 - \left(\sum y\right)^2} \tag{3P.10}$$

$$A_* = \frac{276362 \cdot 19097108 - 2300 \cdot 2209395807}{237240} \tag{3P.11}$$

$$A_* = 826608.518782667341089192379025 46 \tag{3P.12}$$

$$B_* = \frac{264567749}{20 \cdot 276362 - 2300^2} \tag{3P.13}$$

$$B_* = 1115.19030939133367054459618951 27 \tag{3P.14}$$

Here we calculate the correlation coefficient which gives the first numerical value in this succession which possesses unambiguous meaning and preserves the range (-1, +1), therefore we can see for ourselves whether the obtained results are correct.

$$r = \sqrt{B^* \cdot B_*} \quad r = 0.498369118893891071726217273709 1 \tag{3P.15}$$

Of course, we do not pretend to be professional investigators of the subject (see [6]), but we use this opportunity to draw our Student's attention to the meaning of the correlation coefficient obtained now: it confirms that the brain size/weight has a positive correlation to the intellect of the examined person. We now draw the regression lines on the x,y plane with all the 20 pairs of data shown in Table 3P.1. Moreover, we emphasize that there is a division of the data in Fig.3P.1 – corresponding with the division into two types of people which we briefly mentioned in the opening pages of Chapter 1 – we mean a division according to specific terminology of Witold Gombrowicz into "bright" and "dumb/dim" people. Fig.3P.1 very clearly also shows that the first regression line best fits the data seen vertically, while the second regression line is related to the horizontal distances between the points and the regression line.

To determine the position of the first regression line $y = A^* + B^* x$ a point was chosen whose horizontal coordinate equals 1050000 – taking (3P.6) and (3P.8) its vertical coordinate was determined as:

$$136.19031463380008802183149766311$$

Then the first regression line was drawn (blue line in Fig.3P.1).

In a similar way the second regression line $x = A_* + B_* y$ was determined. This time a point was chosen whose vertical coordinate equals 85 and using (3P.12) and (3P.14) its horizontal coordinate was found to be:

$$921399.69508093070308548305513392$$

The second regression line was drawn through this point and an arbitrary point, it is shown as the green line in Fig.3P.1.

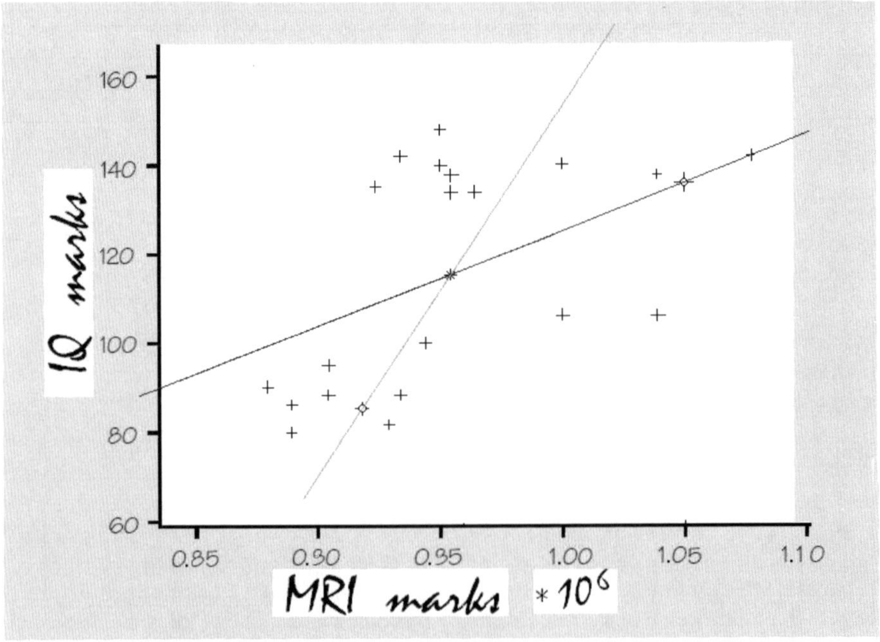

Fig. 3P.1 Two regression lines, data of Table 3P.1

Problem 3.2

Professor Moore swims 2000 yards regularly in a vain attempt to undo middle age. Here are his times (in minutes) and his pulse rate after swimming (in beats per minute) for 23 sessions in the pool.

Table 3P.3 Two dimensional statistics Time vs. Pulse – data

Time:	34.12	35.72	34.72	34.05	34.13	35.72	36.17	35.57	35.37
Pulse:	152	124	140	152	146	128	136	144	148
Time:	35.57	35.43	36.05	34.85	34.70	34.75	33.93	34.60	34.00
Pulse:	144	136	124	148	144	140	156	136	148
Time:	34.35	35.62	35.68	35.28	35.97				
Pulse:	148	132	124	132	139				

Provide a routine examination of Professor Moore's health activities.

Table 3P.4 Professor Moore's health activities

X	Y	X	Y	X X	X X	Y Y	Y Y	X Y	X Y
34.12	152	34.85	148	1164.1744	1214.5225	23104	21904	5186.24	5157.80
35.72	124	34.70	144	1275.9184	1204.09	15376	20736	4429.28	4996.8
34.72	140	34.75	156	1205.4784	1207.5625	19600	24336	4860.8	5421
34.05	152	33.93	140	1159.4025	1151.2449	24649	19600	5175.6	4750.2
34.13	146	34.60	136	1164.8569	1197.16	21316	18496	4982.98	4705.6
35.72	128	34.00	148	1275.9184	1156	16384	21904	4572.16	5032
36.17	136	34.35	148	1308.2689	1179.9225	18496	21904	4919.12	5083.8
35.57	144	35.62	132	1265.2249	1268.7844	20736	17424	5122.08	4701.84
35.37	148	35.68	124	1251.0369	1273.0624	21904	15376	5234.76	4424..32
35.57	144	35.28	132	1265.2249	1244.6784	20736	17424	5122.08	4656..96
35.43	136	35.97	139	1255.2849	1293.8409	18496	19321	4818.48	4999.83
36.05	124	-----	----	1299.6025	-------	15376	---	4470.2	----
-----	----	806.35	3221	-----	28281.26	---	453053	----	112823.93

$\sum x = 806.35 \qquad \bar{x} = 35.05869565 \qquad \sum x^2 = 28281.2605$

$\sigma_x = 0.712657636 \qquad \sigma_x^2 = 0.507880907$

$\sum y = 3221 \quad \bar{y} = 140.0434783 \quad \sum y^2 = 453053$

$\sigma_y = 9.261788075 \quad \sigma_y^2 = 85.78071834$

$\sum x\,y = 112823.93$

Note: Coordinates $\bar{x}$ and $\bar{y}$ determine an arbitrary point K shown in Fig.3P.2.

From this place onwards we proceed in a routine way substituting first the above numerical results into formulae (3P.3) and (3P.4) to derive both coefficients of the first regression line.

$$A^* = \frac{28281.2605 \cdot 3221 - 806.35 \cdot 112823.93}{23 \cdot 28281.2605 - 806.35^2}$$

$$A^* = \frac{118364.115}{268.669} = 440.557395903509522870148770420 11$$

$$B^* = \frac{23 \cdot 112823.93 - 806.35 \cdot 3221}{268.669} = \frac{-2302.96}{268.669}$$

$$B^* = -8.5717369700263149079350427477677$$

Then we apply formulae (3P.9) and (3P.10) to derive the coefficients determining the second regression line and the appropriate coefficient of linear correlation.

$$A_* = \frac{453053 \cdot 806.35 - 3221 \cdot 112823.93}{23 \cdot 453053 - 3221^2}$$

$$A_* = \frac{1913408.02}{45378} = 42.165983956983560315571422275111$$

$$B_* = \frac{-2302.96}{45378} = -0.050750583983428092908457843007625$$

$$r = \sqrt{B^* \cdot B_*} \qquad r = -0.65746595827549178906927043739 32$$

To determine the position of the first regression line $y = A^* + B^* x$ point L was chosen whose horizontal coordinate equals $x_L = 36.2$ – while its vertical coordinate was calculated to be:

130. 26051758855692320290022295092

Then, eventually the first regression line (blue line in Fig.3P.2) was drawn.
In a similar way the second regression line $x = A_* + B_* \ y$ was determined. This time point M was chosen whose vertical coordinate was selected as $y_M = 160$, and using (3P.12) and (3P.14) its horizontal coordinate was found to be:

34.045890078892855568777821851999

Then the second regression line (green line in Fig.3P.2) was drawn.

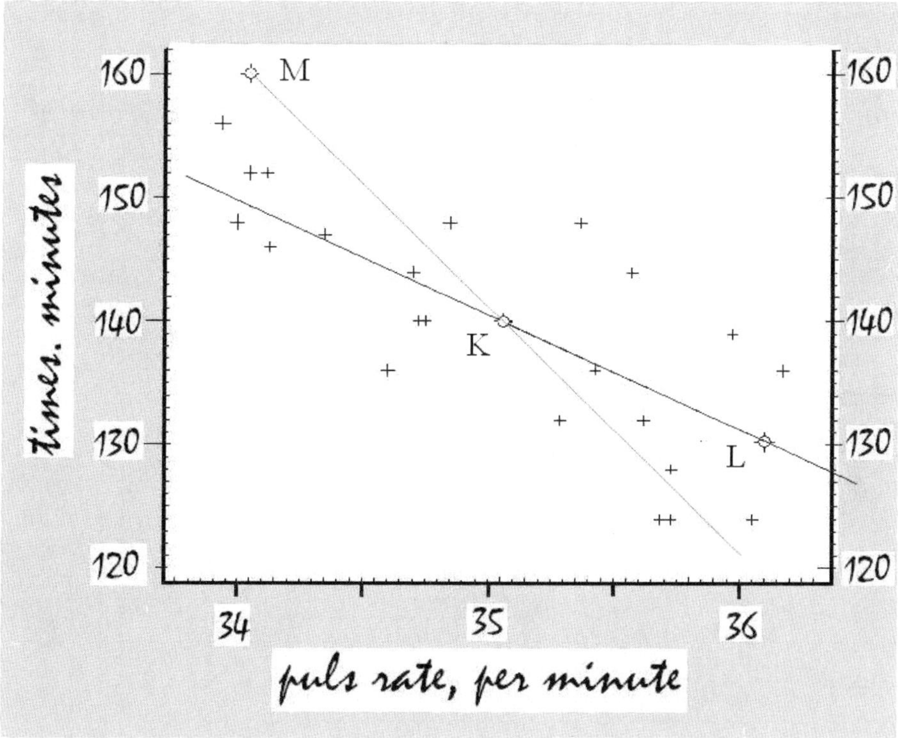

Fig. 3P.2 Two regression lines, data of Table 3P.4

Comment: with respect to professor Moore's attempts to improve his health– negative correlation shows that the shorter time for swimming 2000 yards (by the way: 1 yard = 0.914 m) goes with a higher pulse rate measured just after swimming this distance. The Student should also notice the relatively high correlation between those two variables.

Problem 3.3 [7]

Examine the provided data (see Table 3P.5) presenting relations between the distances and velocities of 24 nebulae in Table 1 of paper [7] by Edwin Powell Hubble (1889-1953) published in 1929.

$\sum r = 21.873$ $\quad \bar{r} = 0.911375$ $\quad \sum r^2 = 29.517795$ $\quad \sigma_r = 0.631904846$
$\sigma_r^2 = 0.399303734$

$\sum \upsilon = 8955$ $\quad \bar{\upsilon} = 373.125$ $\quad \sum \upsilon^2 = 6511425$ $\quad \sigma_\upsilon = 363.4379031$
$\sigma_\upsilon^2 = 132087.1094$ $\quad \sum r\,\upsilon = 12513.695$

Table 3P.5 Edwin Hubble's paper [7] – Table 1 p.169

r	v	r	v	$r\,r$	$r\,r$	$v\,v$	$v\,v$	$r\,v$	$r\,v$
0.032	170	0.9	650	1164.1744	1214.5225	23104	21904	5.44	585
0.034	290	0.9	150	1275.9184	1204.09	15376	20736	9.86	135
0.214	**130**	0.9	500	1205.4784	1207.5625	19600	24336	**27.82**	450
0.263	**70**	1.0	920	1159.4025	1151.2449	24649	19600	**18.41**	920
0.275	**185**	1.1	450	1164.8569	1197.16	21316	18496	**50.875**	495
0.275	**220**	1.1	500	1275.9184	1156	16384	21904	**60.5**	550
0.45	200	1.4	500	1308.2689	1179.9225	18496	21904	90	700
0.5	290	1.7	960	1265.2249	1268.7844	20736	17424	145	1632
0.5	270	2.0	500	1251.0369	1273.0624	21904	15376	135	1000
0.63	200	2.0	850	1265.2249	1244.6784	20736	17424	126	1700
0.8	300	2.0	800	1255.2849	1293.8409	18496	19321	240	1600
0.9	**30**	2.0	1090	1299.6025	-------	15376	---	**27**	2180
-----	----	21.873	8955	-----	29.517795	---	6511425	----	12513.695

Note 1: coordinates $\overline{r}$ and $\overline{v}$ determine the arbitrary point K seen in Fig.3P.3 and with this data obtained mainly through the SD procedure using a scientific calculator available on the market we shall commence a further routine examination of Problem 3.3, whose final result is shown in Fig. 3P.3. To derive final numerical results we also apply the Word Calculator which shows 24 digits.

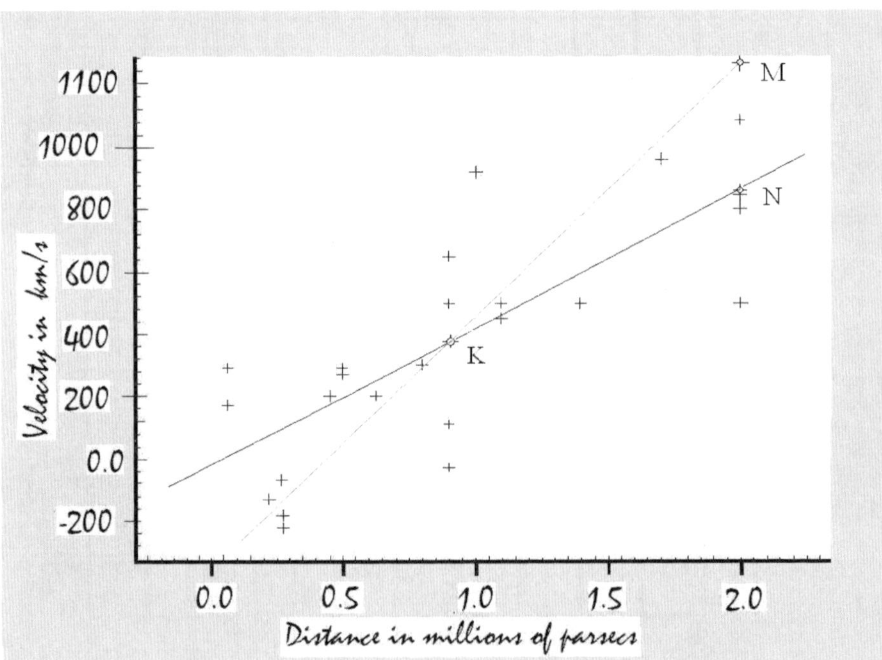

Fig. 3P.3 Hubble's paper, graphic representation of the data in Table 3P.5

Note 2: The picture in Fig.3P.3 presents not only an essential set of data from Hubble's paper but also both regression lines derived below in a routine manner. First the coefficients of the blue regression line were determined. To draw this line, coordinates of point N (2, 868) were determined. A similar approach produced the green regression line with the point M (2, 1166). Although this book is not about Cosmology and the data examined here is considered simply illustrative material for the regression and correlation, as a side note, we may mention temporary doubts regarding Hubble's discovery on whether our Universe is really expanding (see [8]-[9]).

Now returning to the point, we will derive the parameters of the first regression line, and then draw it.

$$A^* = \frac{-9380.19651}{229.998951} = -40.783649095860441554796482528305$$

$$B^* = \frac{104455.965}{229.998951} = 454.15844092262838190075049516204$$

Substitution of the coordinates of the arbitrary point into the equation of the first linear regression line confirms that the above derived coefficients are exact as they have to be. If we choose a new point with horizontal coordinate of 2.0, then we obtain its vertical coordinate equal to 867.5332327493963222467045077961 which allows to draw the first regression line (blue line in Fig.3P.3). Similarly we derive the coefficients of the second regression line and draw the second regression line (the green line in Fig.3P.3).

$$A_* = \frac{30364260.3}{76082175} = 0.39909821584359279949607118881657$$

$$B_* = \frac{104455.965}{76082175} = 0.0013729361049417948422215847535904$$

Again, it is easy to check that the arbitrary point lies in this second regression line. Moreover, the position of the second regression line is shown in Fig.3P.3 whose vertical coordinate of the point M gives 1166.0424533917234884575524246255.

But the most concise result concerns the coefficient of correlation, and thus according to (3P.15) we obtain:

$$r = 0.78963948793531827355547883319608$$

This result says that there is quite a significant linear relation between the position of the nebulae and their speed.

Below we enclose four problems to be solved in a routine way by the Student.

Problem 3.4

Manatees, sirenian mammals, are large sea creatures that live along the Florida coast (but not only). Many manatees are killed or injured by power boats. Statistics shows X data on powerboats registrations (in tons) and Y the number of manatees killed by boats in Florida during the period 1977-90.

Table 3P.6 Two dimensional statistics X vs. Y – data

Year	Powerboat registrations (1000)	Manatees killed	Year	Powerboat registrations (1000)	Manatees killed
1977	447	13	1984	559	34
1978	460	21	1985	585	33
1979	481	24	1986	614	33
1980	498	16	1987	645	39
1981	513	24	1988	675	43
1982	512	20	1989	711	50
1983	526	15	1990	719	47

Problem 3.5

Analyze the possibility of predicting the collapse of Franklin National Bank, position 19 in the list showing assets in billions of dollars and income in millions of dollars, based on a general trend showing a weakening of US banks in 1973 (see [9]).

Table 3P.7 Two dimensional statistics of US Banks

Bank:	1	2	3	4	5	6	7	8	9	10
Assets:	49.0	42.3	36.3	16.4	14.9	14.2	13.5	13.4	13.2	11.8
Income:	218.8	265.6	170.9	85.9	88.1	63.6	96.9	60.9	144.2	53.6
Bank:	11	12	13	14	15	16	17	18	19	20
Assets:	11.6	9.5	9.4	7.5	7.2	6.7	6.0	4.6	3.8	3.4
Income:	42.9	32.4	68.3	48.6	32.2	42.7	28.9	40.7	13.8	22.2

Problem 3.6

Alcohol consumption seems to be connected with heart diseases. The below data is to be examined from this point of view (see [11]).

Table 3P.8 Two dimensional statistics wine vs heart disease

Country	Alcohol from wine (liters/year)	Heart disease death rate (per 100,000)	Country	Alcohol from wine (liters/year)	Heart disease death rate (per 100,000)
Australia	2.5	211	Netherlands	1.8	167
Austria	3.9	167	New Zealand	1.9	266
Belgium/Lux.	2.9	131	Norway	0.8	227
Canada	2.4	191	Spain	6.5	86
Denmark	2.9	220	Sweden	1.6	207
Finland	0.8	297	Switzerland	5.8	115
France	9.1	71	United Kingdom	1.3	285
Iceland	0.8	211	United States	1.2	199
Ireland	0.7	300	West Germany	2.7	172
Italy	7.9	107			

Problem 3.7

Examine the provided set showing the age of first word spelled vs. the Gesell score (see [10]).

Table 3P.9 Two dimensional statistics Gesell score – age first word spelled

Child	Age	Score	Child	Age	Score	Child	Age	Score
1	15	95	8	11	100	15	11	102
2	26	71	9	8	104	16	10	100
3	10	83	10	20	94	17	12	105
4	9	91	11	7	113	18	42	57
5	15	102	12	9	96	19	17	121
6	20	87	13	10	83	20	11	86
7	18	93	14	11	84	21	10	100

***

From this place onwards we will draw the Student's attention towards the grouped data examination. The first problem which we would like to present here is taken from the famous paper by F. Galton [1], which will be solved as Prob. 3.8 afterwards.

Problem 3.8

Examine the grouped data of Francis Galton [1], presented in Table 3P.10.

Table 3P.10 Two dimensional grouped data of F. Galton

Heights of the Mid-parents in inches.	Heights of the Adult Children.														Total Number of		Medians.
	Below	62·2	63·2	64·2	65·2	66·2	67·2	68·2	69·2	70·2	71·2	72·2	73·2	Above	Adult Children.	Mid-parents.	
Above ..	..	..	..	..	..	..	..	..	..	..	..	1	3	..	4	5	..
72·5	..	..	..	..	..	..	..	1	2	1	2	7	2	4	19	6	72·2
71·5	..	..	..	..	1	3	4	3	5	10	4	9	2	2	43	11	69·9
70·5	1	..	1	..	1	1	3	12	18	14	7	4	3	3	68	22	69·5
69·5	..	..	1	16	4	17	27	20	33	25	20	11	4	5	183	41	68·9
68·5	1	..	7	11	16	25	31	34	48	21	18	4	3	..	219	49	68·2
67·5	..	3	5	14	15	36	38	28	38	19	11	4	..	..	211	33	67·6
66·5	..	3	3	5	2	17	17	14	13	4	..	..	..	..	78	20	67·2
65·5	1	..	9	5	7	11	11	7	7	5	2	1	..	..	66	12	66·7
64·5	1	1	4	4	1	5	5	..	2	..	..	..	..	..	23	5	65·8
Below ..	1	..	2	4	1	2	2	1	1	..	..	..	..	..	14	1	..
Totals ..	5	7	32	59	48	117	138	120	167	99	64	41	17	14	928	205	..
Medians ..	..	..	66·3	67·8	67·9	67·7	67·9	68·3	68·5	69·0	69·0	70·0	..	..	..	..	..

To solve Problem 3.8 we apply Table 3P.11 whose structure and notations follow exactly the content of Table 3.2 from Chapter 3. The Student should note however, that the boundaries given by Galton were not used (rows and columns denoted as "*below*" and "*above*"), so we skip them reducing slightly the totals in both groups – parents and children (from 928 to 892). The first results given below concern the total number of objects, denoted *x* for parents, and *y* for children while their coded symbols are denoted by *u* and *w*, respectively:

$$n = \sum_{i=1}^{N} n_{i\mu} = \sum_{j=1}^{M} n_{\mu j} \qquad N = 9,\ M = 12 \qquad n = 892 \qquad i = 1 \qquad R^{*} = 67.5$$

$$R_{*} = 67.2$$

In the second step the main averages were derived:

$$\sum_{i=1}^{N} u_i \, n_{i\mu} = 731 \quad \bar{u} = \frac{731}{892} \Rightarrow 0.819506726 \rightarrow \mu_u = 68.3190673 \text{ parents}$$

$$\sum_{j=1}^{M} w_j \, n_{\mu j} = 759 \quad \bar{w} = \frac{759}{892} \Rightarrow 0.850896861 \rightarrow \mu_w = 68.05089686 \text{ children}$$

$$Var\,u = \overline{u}^{2} - (\bar{\mathrm{u}})^2 \qquad \overline{u}^{2} = \frac{3073}{892} \quad \rightarrow \quad \sigma_u^2 = 2.773475991$$

$$Var\,w = \overline{w}^{2} - (\bar{\mathrm{w}})^2 \qquad \overline{w}^{2} = \frac{1638}{270} \quad \rightarrow \quad \sigma_w^2 = 5.604449868$$

Table 3P.11 Francis Galton's grouped statistics [1]

u_i	x_i \ y_j w_j	-5	-4	-3	-2	-1	0	1	2	3	4	5	6	$n_{i\mu}$	$u_i \cdot n_{i\mu}$	$u_i^2 \cdot n_{i\mu}$	$\sum_j w_j \cdot n_{ij}$	$u_i \cdot \sum_j w_j \cdot n_{ij}$
		62.2	63.2	64.2	65.2	66.2	67.2	68.2	69.2	70.2	71.2	72.2	73.2					
-3	64.5	1	4	4	1	5	5	-	2	-	-	-	-	22	**66**	198	**38**	114
-2	65.5	-	9	5	7	11	11	7	7	5	2	1	-	65	**130**	260	**27**	54
-1	66.5	3	3	5	2	17	17	14	13	4	-	-	-	78	**78**	78	**11**	11
0	67.5	3	5	14	15	36	38	28	38	19	11	4	-	211	0	0	**82**	0
1	68.5	-	7	11	16	25	31	34	48	21	18	4	3	218	218	218	85	185
2	69.5	-	1	16	4	17	27	20	32	25	20	11	4	178	356	712	243	486
3	70.5	-	1	-	1	1	3	12	18	14	7	4	3	64	192	576	149	447
4	71.5	-	-	-	1	3	4	3	5	10	4	9	2	41	164	656	111	444
5	72.5	-	-	-	-	-	-	1	2	1	2	7	2	15	75	375	63	315
$n_{\mu j}$		7	30	55	47	115	136	119	166	99	64	40	14	892	731	3073		2056
$w_j \cdot n_{\mu j}$		**35**	**120**	**165**	**94**	**1115**	0	119	332	297	256	200	84	759				
$w_j^2 \cdot n_{\mu j}$		165	480	495	188	115	0	119	664	891	1024	1000	504	5645				
$\sum_i u_i \cdot n_{ij}$		**6**	**21**	16	12	20	56	99	165	144	101	107	38					
$w_j \cdot \sum_i u_i \cdot n_{ij}$		30	84	**48**	**24**	**20**	0	99	330	432	404	535	228	2050				

Note: **Bold** fonts denote negative numbers

Next, both regression lines were determined. Let us start this step by determining the directional coefficients:

$$B^* = \frac{E(uw) - \bar{u}\,\bar{w}}{Var\,u} \qquad B_* = \frac{E(uw) - \bar{u}\,\bar{w}}{Var\,w}$$ keeping in mind that:

$$E(u\,w) = \frac{1}{n}\sum_{j=1}^{M} w_j \sum_{i=1}^{N} u_i\, n_{ij} = \frac{1}{n}\sum_{i=1}^{N} u_i \sum_{j=1}^{M} w_j\, n_{ij}$$

Here we face a disturbing fact which cannot be explained by the Author of this book: instead of a common value we get, for each of the above formulae, two slightly different values: 2050, and 2056. Although they differ by practically a negligible amount nevertheless we do not know why. And this, somewhat disturbing question we have to leave unanswered. After making the above point we proceed further bringing into calculations this strange non dimensional moment value equal to 2050.

$$E(u\,w) = \frac{2050}{892} = 2.298206278$$

$$B^* = \frac{E(uw) - \bar{u}\,\bar{w}}{Var\,u} \quad \rightarrow \quad B^* = \frac{1.600890577}{2.77347599} = 0.577214507$$

$$B_* = \frac{E(uw) - \bar{u}\,\bar{w}}{Var\,w} \quad \rightarrow \quad B_* = \frac{1.600890577}{5.604449868} = 0.285646337$$ they lead to the correlation coefficient:

$$r = \sqrt{B_1 \cdot B_2} \quad \rightarrow \quad r \cong 0.406053209$$

Therefore the equations of both regression lines take the following form:

$$y = 28.6159049 + 0.577214507\ x$$

$$x = 48.88057788 + 0.285646337\ y$$

We get them taking into account that these lines have to contain the same arbitrary point denoted by K whose coordinates (according to the symbols used in Table 3P.4) we denote as (68.319, 68.051). Then we determined the coordinates of point L (73, 70.75256391) and point M (67.16194345, 64). In the last step we drew a graph of these two lines, Fig. 3P.4 where horizontal coordinate x presents the data of the stature of parents, and the vertical coordinate y denotes stature of children. In the end we would like to note that the paper [1] by Galton in comparison with the above (and below) given results probably shows some differences. But because this famous paper was written with more improvisation than the provided mathematical details, it is not easy to compare its results with any other results.

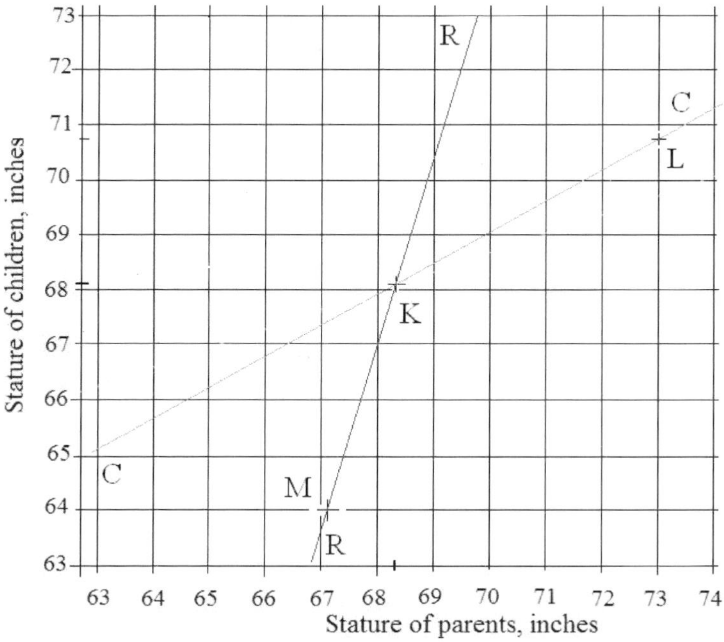

Fig. 3P.4 Galton's paper [1], graphic representation of the data given in Table 3P.11

The opportunity given by Fig.3P.4 encourages us to present the equations of both regression lines inserting imaginary coordinate axes at the arbitrary point: x horizontally, and y vertically, then skipping values A^* and A_*, both regression lines are presented in the following way

line CC $y = r\dfrac{\sigma_w}{\sigma_u}x$ line RR $x = r\dfrac{\sigma_u}{\sigma_w}y$ which applies

$$B^* = \frac{\sigma_w}{\sigma_u}r \qquad B_* = \frac{\sigma_u}{\sigma_w}r$$

The above convention was in general used in the book by Yule [2].

Problem 3.9

Our next example, Problem 3.9, examines a similar problem, this time taken from another well known paper, by Karl Pearson [13]. Initial data have already been presented in Part One (see end of Chapter 3). We are going to examine the grouped data considering the relation between stature of fathers and sons given by Karl Pearson and copied from the book by Udny Yule [2] – see Table 3P.12.

Table 3P.12 Karl Pearson's approach to the problem of stature of fathers and sons

	-9	-8	-7	-6	-5	-4	-3	-2	-1	0	1	2	3	4	5	6	7						
	59	60	61	62	63	64	65	66	67	68	69	70	71	72	73	74	75						
-9	-	-	-	-	.5	.5	1	-	-	-	-	-	-	-	-	-	-	2	60	-18	162	**7.5**	67.5
-8	-	-	-	-	.5	-	-	-	1	-	-	-	-	-	-	-	-	1.5	61	-12	96	**3.5**	28
-7	-	.25	.25	-	.5	1	.25	.25	.5	.5	-	-	-	-	-	-	-	3.5	62	-24.5	171.5	**12**	84
-6	-	.25	.25	2.25	2.25	2	4	5	2.75	1.25	-	.25	.25	-	-	-	-	20.5	63	-123	738	**60**	360
-5	1	-	1.5	3.75	3	4.25	8	9.25	3	1.25	1.5	.75	1.25	-	-	-	-	38.5	64	-192.5	962.5	**110.75**	553.75
-4	2	1	.5	2	3.25	9.5	13.5	10.75	7.5	5.5	3.5	2.5	-	-	-	-	-	61.5	65	-246	984	**157.75**	631
-3	-	.5	1	2.25	5.25	9.5	10	16.75	17.5	16	5.25	2	2.5	1	-	-	-	89.5	66	-268.5	805.5	**150**	450
-2	-	1.5	2	4.75	3.5	13.75	19.75	26.5	25.75	19.5	12.5	13.75	3.25	.5	1	-	-	148	67	-296	592	**208.25**	416.5
-1	-		1.5	2	7.5	10	10.25	24.25	31.5	23.5	29.5	13.25	8.5	9.5	2.25	-	-	173.5	68	-173.5	173.5	**80**	80
0	-		1	-	5.25	5	12.75	18.25	16	24	29	21.5	10	3.5	2.25	-	1	149.5	69	0	0	**9.75**	0
1	-	-	-	-	1	2.5	5.75	18.75	11.75	19.5	22.5	19.5	14.5	6.25	3.5	1.5	1	128	70	128	128	82	82
2	-	-	-	-	-	3.25	5	8.75	10.75	19	14.75	20.75	10.75	8	5	1	1	108	71	216	432	102.25	204.5
3	-	-	-	-	-	.25	3	1.25	7	7.75	10.75	11.25	10	8.5	2.75	.5	-	63	72	189	567	94.5	283.5
4	-	-	-	-	-	-	.75	.75	2.5	7.5	6 .5	6	7.5	6.25	3.25	.5	.5	42	73	168	672	82.5	330
5	-	-	-	-	1	-	1.5	1.5	-	5.25	2.25	2.5	6.5	3.25	3.25	-	2	29	74	145	725	57.5	287.5
6	-	-	-	-	-	-	-	-	-	1	2	-	2.5	.75	1.75	.5	-	8.5	75	51	306	24.25	145.5
7	-	-	-	-	-	-	-	-	-	1.25	.25	-	.5	1	1	-	-	4	76	28	196	10.75	75.25
8	-	-	-	-	-	-	-	-	-	1.25	.25	1	-	-	1.5	-	-	4	77	32	256	9.75	78
9	-	-	-	-	-	-	-	-	-	-	1	1	-	.25	.75	-	-	3	78	27	243	7.75	69.75
10	-	-	-	-	-	-	-	-	-	-	-	-	-	.25	.25	-	-	.5	79	50	500	2.25	22.5
	3	3.5	8	17	33.5	61.5	95.5	142	137.5	154	141.5	116	78	49	28.5	4	5.5	1078		**320**	8710	**----**	4249.25
	27	**28**	**56**	**102**	**167.5**	**246**	**286.5**	**284**	**137.5**	---	141.5	232	234	196	142.5	24	38.5	**326**					
	243	224	392	612	837.5	984	859.5	568	137.5	----	141.5	464	702	784	712.5	144	269.5	8075					
	13	**11.75**	**21.25**	**58.5**	**82.75**	**139**	**173.25**	**198**	**144.25**	12	54.5	86.25	116.75	91.75	85.5	9.5	15	-------					
	117	94	148.75	351	413.75	556	519.75	396	144.25	0	54.5	172.5	350.25	367	427.5	57	105	4249.25					

$$n = \sum_{i=1}^{N} n_{i\mu} = \sum_{j=1}^{M} n_{\mu j} \qquad N = 20,\ M = 17 \qquad n = 1078 \qquad i = 1 \qquad R^* = 68.0 \text{ (Fathers)} \quad R_{*} = 69.0 \text{ (Sons)}$$

$$\sum_{i=1}^{N} u_i n_{i\mu} = -320 \quad \bar{u} = \frac{-320}{1078} = -0.296846011 \rightarrow \mu_u = 68.0 - 0.296846011 = 67.70315399 \quad Var\,u = \overline{u^2} - (\bar{u})^2 \quad \overline{u^2} = \frac{8710}{1078} \rightarrow \sigma_u^2 = 7.991659811$$

$$\sum_{j=1}^{M} w_j n_{\mu j} = -326 \quad \bar{w} = \frac{-326}{1078} = -0.302411873 \rightarrow \mu_w = 69.0 - 0.302411873 = 68.69758813 \quad Var\,w = \overline{w^2} - (\bar{w})^2 \quad \overline{w^2} = \frac{8075}{1078} \rightarrow \sigma_w^2 = 7.399270621$$

Initiating calculations based on the results of Table 3P.12 we start by deriving the correlation coefficient. We choose the procedure which starts with the directional coefficients of the regression lines.

$$E(uw) = \frac{1}{n}\sum_{j=1}^{M} w_j \sum_{i=1}^{N} u_i n_{ij} = \frac{1}{n}\sum_{i=1}^{N} u_i \sum_{j=1}^{M} w_j n_{ij}$$

$$B^* = \frac{E(uw) - \bar{u}\,\bar{w}}{Var\,u} \qquad B_* = \frac{E(uw) - \bar{u}\,\bar{w}}{Var\,w}$$

$$E(uw) = \frac{4249.25}{1078} = 3.941790353,$$ therefore:

$$B^* = \frac{E(uw) - \bar{u}\,\bar{w}}{Var\,u} \quad \rightarrow \quad B^* = \frac{3.852020594}{7.991659811} = 0.482005076$$

$$B_* = \frac{E(uw) - \bar{u}\,\bar{w}}{Var\,w} \quad \rightarrow \quad B_* = \frac{3.852020594}{7.399270621} = 0.520594635$$ they lead to correlation coefficient:

$$r = \sqrt{B_1 \cdot B_2} \quad \rightarrow \quad r \cong 0.500928395$$

As a side note we suggest that the Student consider a point which is evident at first glance at the *Great Correlation Array,* given as Table 3P.12 in view of the fractional frequencies shown there. Udny Yule commented this absolutely unusual situation in the following way (see [2], p.162):

The difficulty as to the intermediate observations – if the value of one variable alone is intermediate, the unit of frequency being divided between two adjacent compartments. If both values of the pair are intermediates, the observation must be divided among four *adjacent compartments, and thus quarters as well as hales may occur in the table. In this case* (of the above Table) *the stature of fathers and sons were measured to the nearest quarter-inch and subsequently grouped by 1-inch intervals: a pair in which the recorded stature of the father is 60.5 in. and that of the son 62.5 in. is accordingly entered as 0.25 to each of four compartments under the columns 59.5-60.5, 60.5-61.5, and the rows 61.5-62.5, 62.5-63.5.*

In the above quotation the Student will see a trivial fact known in the theory of probability as the *left band continuity*. We explained this concept first in Chapter 2 of Part One and then in Unit 2 of Part Two. Taking the above case by U.Yule – its contemporary solution is to regard the result of the father's recorded stature such as 60.5 as one to be inserted in the column 60.5-61.5, and of the son's such as 62.5 in the row 62.5-63.5 as a single reading each. Also it is difficult to guess what the accuracy of measurements can add to this matter. Unfortunately *ex post* it is impossible to change in this way the data shown by K. Pearson (and Alice Lee), therefore we have to accept them as such – although they are in an apparent contradiction to common sense.

Now let us prolong the routine procedure of examining the results obtained in the *Great Correlation Table.* We come to the point which allows to draw both regression lines shown in Fig.3P.5.

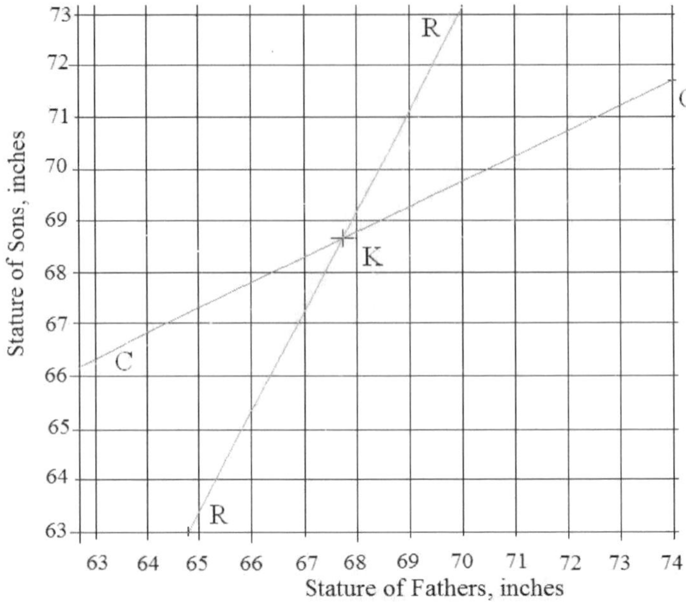

Fig. 3P.5 Pearson's paper [13], and his approach to the problem of stature, see Table 3P.12

The Student can check that both regression lines shown in Fig.3P.5 are given by the following equations:

$$y = 36.06432425 + 0.0.482005076\ x$$

$$x = 31.93955817 + 0.520594635\ y$$

Scant description given by Udny Yule regarding Problem 3.9 considered here does not allow for a close comparison of the results obtained here with the results shown by Yule. Particularly his Fig. 37 in comparison with our Fig. 3P.5 – shows a different convention in solving this problem. Therefore, we decided to include and solve in detail Problem 3.10 which will be in the same spirit as the solution shown in Fig. 37 of [2]. Nevertheless, we have to note that the coefficient of correlation obtained here with comparatively high accuracy shows value r = 0.50 – while U. Yule's result is close to 0.51. We think this is due to low accuracy in considering the same data by Pearson. Also the mean values which can be read from Fig.37 are close to the values obtained here. We also note that the results obtained in Problem 3.8 exploring Galton's data [1] differ from the ones derived presently according to Pearson's statistics [13]. It is easy to see that Pearson's statistics shows a much wider range of measured statures of fathers and sons which generally slightly changed the mean, and more essentially resulted in an increase of the variances.

Problem 3.10

Exploring data by Pearson [13] determine the so called column and row averages and compare them with the values shown by U. Yule in Fig. 37 of [2].

Solution

We commence from the column averages. To solve the problem we first have to copy two selected rows from the *Great Correlation Array 3P.12* (at the bottom). The first row in Table 3P.13 shows class frequencies of the stature of fathers denoted by $n_{\mu j}$ (symbols follow those from Table 3P.12).

Table 3P.13 Selected cells from Table 3P.12

3.5	8	17	33.5	61.5	95.5	142	137.5	154	141.5	116	78	49	28.5
11.75	**21.25**	**58.5**	**82.75**	**139**	**173.25**	**198**	**144.25**	12	54.5	86.25	116.75	91.75	85.5
60	61	62	63	64	65	66	67	68	69	70	71	72	73

From Table 3P.11 we left out the boundaries leaving fourteen essential columns, and we added appropriate midpoint values of the fathers' stature in the third row. Then for the sake of convenience of further calculations Table 3P.13 was transformed into Table 3P.14.

Table 3P.14 Transformed Table 3P.13

3.5	**11.75**	65.64286	60	137.5	**144.25**	67.95091	67
8	**21.25**	66.34375	61	154	12	69.07792	68
17	**58.5**	65.55882	62	141.5	54.5	69.38516	69
33.5	**82.75**	66.52985	63	116	86.25	69.74353	70
61.5	**139**	66.73984	64	78	116.75	70.49679	71
95.5	**173.25**	67.18586	65	49	91.75	70.87245	72
142	**198**	67.60563	66	28.5	85.5	72.	73

The point is that we have to determine the average for each of all the fourteen selected columns. These values are inserted into a new column in Table 3P.14. Beside this new, essential column, the values of corresponding midpoints indicating the stature of fathers were also inserted. Then all such pairs of the stature have to be presented in a way similar to that show in Fig.37 of [2]. Below we provide the first example of the calculations leading to the first column average, that is the stature of the "average son":

$$\frac{-13}{3}+69=64.6(6) \text{ corresponding to stature of father equal to 59 in.}$$

Calculations to derive the all row averages are similar. In order to present them we have to copy two appropriate columns from Table 3P.5. And for the first pair of values below is an appropriate example of calculations to establish the stature of the average father.

$$\frac{-7.5}{2}+68=64.25 \text{ corresponds to the son whose midpoint indicates 60 in.}$$

The first and the fourth column from the right end of Table 3P.12 was copied and then re-arranged, this time without skipping any entries, into Table 3P.15. According to the pattern of calculations shown above, the values of the new column were added, repeating the midpoints of the stature of sons.

Table 3P.15 Selected cells (columns) from Table 3P.12

2	**7.5**	64.25	60	128	82	68.64063	70
1.5	**3.5**	65.6(6)	61	108	102.25	68.94676	71
3.5	**2**	64.57143	62	63	94.5	69.02381	72
20.5	**60**	65.07317	63	42	82.5	69.96429	73
38.5	**10.75**	65.12338	64	29	57.5	69.98276	74
61.5	**157.75**	65.43496	65	8.5	24.25	70.85294	75
89.5	**150**	66.32402	66	4	10.75	70.6875	76
148	**208.25**	66.59291	67	4	9.75	70.375	77
173.5	**80**	67.53757	68	3	7.75	70.5833	78
149.5	**9.75**	67.93478	69	.5	2.25	68.45	79

In the next step the pairs of values from Table 3P.15 were inserted into Fig.3P.6 along the line denoted as $R - R$. To allow comparison with the mentioned above Fig. 37 from U. Yule's book we provide a copy of this figure with some small geometrical re-arrangements for the sake of convenience.

The first impression is that there is something wrong: possibly a reverse case, i.e. the positions of data were interchanged. But it is an illusion caused by an insignificant difference connected with how the coordinates in both cases are designated. The Student has to be warned: the right comparison requires some effort! Unintentionally we interchanged the significance of *circles* and *crosses* in our Fig.3P.6 as compared to *circles* and *crosses* in Fig. 3P.7 presenting Udny Yule results. Apart from this difference, the remaining symbolic conventions are the same in this book and in the book by U. Yule [2] – and the examined results in both figures correspond to one another.

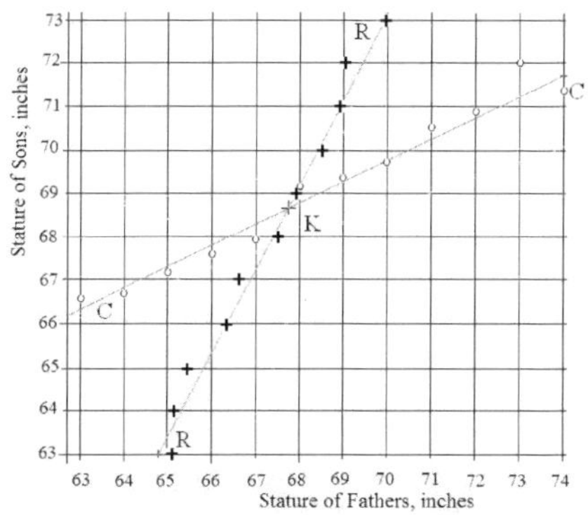

Fig. 3P.6 Pearson's data [13] to show the row and column averages

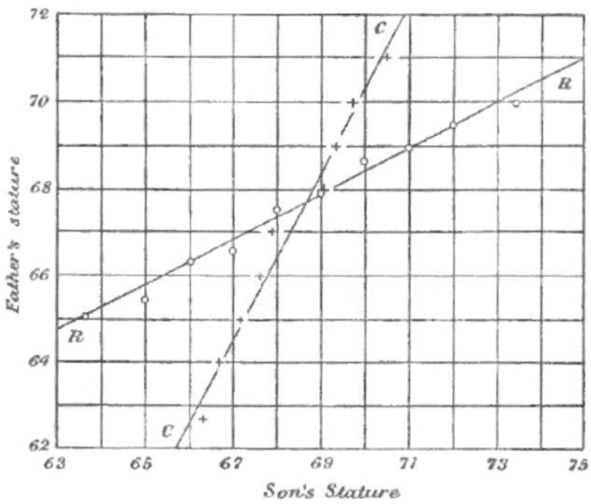

Fig. 3P.7 Udny Yule's [2] way of presenting Pearson's data [13]

***

Coming to the end of this unit below we propose two problems to the Student. The first problem – Problem 3.11 - again offers a chance to investigate the descriptive case. The second problem – Problem 3.12 - investigates once more the grouped data, this time taking into consideration the discrete case data taken also from U. Yule's book [2].

Problem 3.11 (see [4], Problem.16.11)

Table 3P.16 presents the pairs of measurements, apparently related each to another: the chest girth X in inches, and the lung capacity Y in cubic inches of college 15 freshmen.

Table 3P.16 The chest girth vs. the lung capacity

X	30.8	31.5	30.0	30.3	31.3	35.0	38.9	33.7	37.6	34.5	32.6	37.5	34.3	34.4	37.2
Y	305	238	269	210	330	305	311	219	226	278	310	275	220	219	265

Solution uses the SD procedure twice and produces the following results:

$$\sum x_i = 509.6 \quad \overline{x} = 33.973(3) \quad \sum x_i^2 = 17429.28 \quad \sigma_x = 2.786507173$$

$$\sum y_i = 3980 \quad \overline{y} = 265.3(3) \quad \sum y_i^2 = 1079408 \quad \sigma_y = 39.48107845$$

Therefore, the only term to be determined is the mixed term $\sum x\,y$ and its value was obtained according to the results given in Table 3P.17.

The desired sum is as follows: $\sum x\,y = 135250.9$. The most intriguing factor in this problem shows the correlation coefficient whose value obtained here shows a lack of correlation, because $r = +0.022441525$.

Note, that this value calculated using Pearson's correlation formula (see (3.37) of Chapter 3) results mainly from distraction of two big numbers whose values are close to one another (see the results below). If calculations are not provided with sufficient accuracy, the obtained result may differ significantly (the answer to this problem in [4] claims $r = 0.04$).

Table 3P.17 The chest girth vs. the lung capacity; mixed term values

X	Y	X Y	X	Y	X Y	X	Y	X Y
30.8	305	9394	35.0	305	10675	32.6	310	10106
31.5	238	7497	38.9	311	12097.9	37.5	275	10312.5
30.0	269	8070	33.7	219	7380.3	34.3	220	7546
30.3	210	6363	37.6	226	8497.6	34.4	219	7533.6
31.3	330	10329	34.5	278	9591	37.2	265	9858

2028763.5 – 2028208 = 555.5
41.797607587037801654370062726372
592.21617674629591293284971866399
24753.21936233749777577489427921; $r \approx 555.5 / 24753.21936233749778$
$r \approx 0.02244152535751$

Expecting that the above given lines will be rightly deciphered by the Student we present the last problem in this Unit.

Problem 3.12

Discuss the case of the interrelation between the number of children in two successive generations called conventionally mothers and daughters, using data shown in Table 3P.18.

Table 3P.18 Mothers and daughters

Number of her Daughter's Children	Number of Mother's Children																Total
	1	2	3	4	5	6	7	8	9	10	11	12	13	14	15	16	
0	5	9	11	18	21	15	8	9	6	3	2	3	—	—	—	—	110
1	12	5	14	15	10	13	9	8	5	3	2	2	—	—	—	—	98
2	9	9	10	15	18	15	9	3	2	4	2	—	—	—	—	1	97
3	5	10	16	11	9	14	13	10	4	8	2	3	—	—	—	—	105
4	5	5	19	17	21	15	18	10	14	2	1	5	1	—	—	—	133
5	7	6	7	17	23	9	12	13	14	8	3	2	2	—	—	—	123
6	4	5	8	11	15	12	15	14	7	5	3	3	1	—	—	—	103
7	5	4	3	8	4	13	9	8	5	10	2	1	1	—	—	—	73
8	1	2	4	12	9	9	8	5	12	3	4	1	2	1	—	—	73
9	—	—	4	3	3	4	7	5	3	2	2	1	—	—	—	—	34
10	—	—	1	2	1	3	4	6	3	2	—	1	—	1	—	—	24
11	—	—	2	1	1	1	—	—	1	2	—	—	—	—	—	—	8
12	—	2	1	2	3	—	1	1	—	—	1	—	1	—	1	—	13
13	—	—	—	—	2	1	—	—	—	—	1	—	2	—	—	—	6
Total	53	57	100	132	140	124	113	92	76	52	25	22	10	2	1	1	1.000

Table 3P.19 Mothers and daughters

u_i \ w_j	x_i \ y_j	-5	-4	-3	-2	-1	0	1	2	3	4	5	6	7	8	9	10	$n_{i\mu}$	$u_i \cdot n_{i\mu}$	$u_i^2 \cdot n_{i\mu}$	$\sum_j w_j \cdot n_{ij}$	$u_i \cdot \sum_j w_j \cdot n_{ij}$
		1	2	3	4	5	6	7	8	9	10	11	12	13	14	15	16					
-4	0	5	9	11	18	21	15	8	9	6	3	2	3	-	-	-	-	110	**440**	1760	**67**	268
-3	1	12	5	14	15	10	13	9	8	5	3	2	2	-	-	-	-	98	**294**	882	**88**	264
-2	2	9	9	10	15	18	15	9	3	2	4	2	-	-	-	-	1	97	**194**	388	**102**	204
-1	3	5	10	16	11	9	14	13	10	4	8	2	3	-	-	-	-	105	**105**	105	**39**	39
0	4	5	5	19	17	21	15	18	10	14	2	1	5	1	-	-	-	133	0	0	**27**	0
1	5	7	6	7	17	23	9	12	13	14	8	3	2	2	-	-	-	123	123	123	16	16
2	6	4	5	8	11	15	12	15	14	7	5	3	3	1	-	-	-	103	206	412	23	46
3	7	5	4	3	8	4	13	9	8	5	10	2	1	1	-	-	-	73	219	657	33	99
4	8	1	2	4	12	9	9	8	5	12	3	4	1	2	1	-	-	73	292	1168	56	224
5	9	-	-	4	3	3	4	7	5	3	2	2	1	-	-	-	-	34	170	850	29	145
6	10	-	-	1	2	1	3	4	6	3	2	-	1	-	1	-	-	24	144	864	39	234
7	11	-	-	2	1	1	1	-	-	1	2	-	-	-	-	-	-	8	56	392	2	14
8	12	-	2	1	2	3	-	1	1	-	-	1	-	1	-	1	-	13	104	832	6	48
9	13	-	-	-	-	2	1	-	-	-	-	1	-	2	-	-	-	6	54	486	17	153
$n_{\mu j}$		53	57	100	132	140	124	113	92	76	52	25	22	10	2	1	1	1000	335	8919		1754
$w_j \cdot n_{\mu j}$		**265**	**228**	**300**	**264**	**140**	0	113	184	228	208	125	132	70	16	9	10	**102**				
$w_j^2 \cdot n_{\mu j}$		1325	912	900	528	140	0	113	368	684	832	625	792	490	128	81	100	8018				
$\sum_i u_i \cdot n_{ij}$		**45**	**27**	**26**	03	12	19	78	78	84	59	38	05	41	10	08	**02**					
$w_j \cdot \sum_i u_i \cdot n_{ij}$		225	108	78	**06**	**12**	0	78	156	252	236	190	30	287	80	72	**20**	1754				

The majority of the necessary results derived in Table 3P.19 will be now used to present the desired components of the solution. First the results necessary to determine the main averages are derived: the mean and the variance for both variables.

$$\overline{u} = \frac{335}{1000} = 0.335 \qquad \overline{w} = \frac{-102}{1000} = -0.102 \qquad \overline{u \cdot w} = \frac{1754}{1000} = 1.754$$

$$\overline{u^2} = \frac{8919}{1000} = 8.919 \qquad \overline{w^2} = \frac{8018}{1000} = 8.018$$

With the help of the above derived values the directional coefficients of both regression lines can be derived and then the correlation coefficient.

$$B^* = \frac{1.754 + 0.335 \cdot 0.102}{8.806775} \cong 0.203044814 \; B_* = \frac{1.754 + 0.335 \cdot 0.102}{8.007596} \cong 0.223309217$$

$$r = \sqrt{B^* \cdot B_*} \cong \sqrt{0.203044814 \cdot 0.223309217} \cong 0.21293609$$

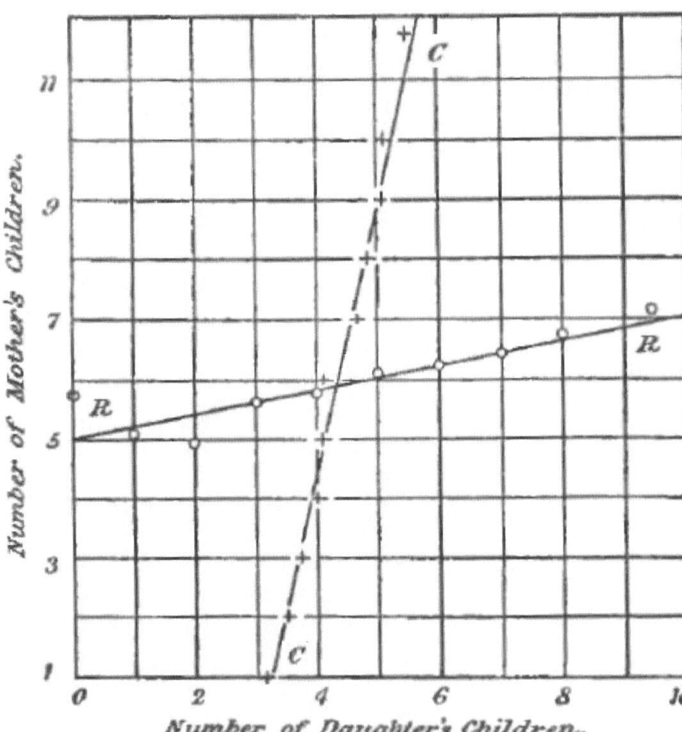

Fig. 3P.8 Udny Yule's [2] way of presenting solution of mother-daughter regression

In the final step the averages of both variables can be determined – expressing the mean number of children born by daughters and by mothers, together with the corresponding values of the variances and standard deviations.

$$\overline{x} = \overline{u} + R^{*} \cong 0.335 + 4 \cong 4.335 \quad \textit{daughters}$$

$$\overline{y} = \overline{w} + R_{*} \cong -0.102 + 6 \cong 5.898 \quad \textit{mothers}$$

$$\sigma_u^2 = \sigma_x^2 = \overline{u^2} - (\overline{u})^2 \cong 8.919 - 0.335^2 \cong 8.806775 \qquad \sigma_x \cong 2.967621101$$

$$\sigma_w^2 = \sigma_y^2 = \overline{w^2} - (\overline{w})^2 \cong 8.018 - 0.102^2 \cong 8.007596 \qquad \sigma_y \cong 2.829769602$$

Results of the solution given by Udny Yule [2] p.175 show satisfactory correspondence with the above results. In particular the values of the correlation coefficients are the same, also the point of intersection lines *R–R* and *C-C* agree with the values of both averages indicated above. The Student can easily continue our example.

Problem – an **extra** case for consideration - presents IQ and school grades statistics intending to examine whether students with higher IQ test scores tend to do better in school. Data sheet presents Fig.3P.9 displaying the pairs of results for 76 students examined from the above stated point of view.

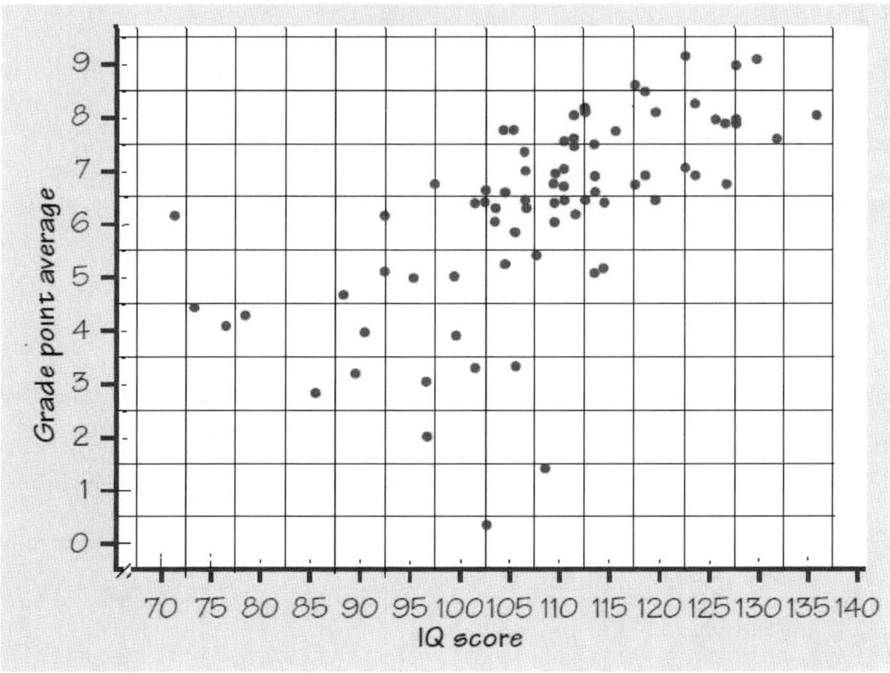

Fig. 3P.9 Scatter plot of school grade point average versus IQ test score for seventh-grade students.

The first task we concerns the grouping. We would like to recall the "*trick*" which used with respect to so called "*Houbolt's cloud*", see Chapter 3. We have the two dimensional statistics in the form of a graph depicted in Fig.3P.9. That is we have no numbers. The "*trick*" offers a procedure which does not need numbers for grouping the data shown this way: it is enough to count the dots corresponding to the appropriate squares. For the dots lying exactly on the boundaries we propose to use the rule of the *right hand continuity* which here is generalized as the rule of the *lower band continuity*. There is also a second remark: the net shown in Fig.3P.9 can be chosen in a way which reduces the cases of the dots on the mesh lines. Once the data has been grouped as in Tab.3P.20, further procedure takes us back to the procedure well documented in the Unit.

Table 3P.20 School grades vs. IQ points

w_j y_j / u_i x_i		IQ scores of the school students														$n_{i\mu}$
	w_j	-7	-6	-5	-4	-3	-2	-1	0	1	2	3	4	5	6	
School grades averages	y_j	70	75	80	85	90	95	100	105	110	115	120	125	130	135	
-4	0	-	-	-	-	-	-	1	-	-	-	-	-	-	-	1
-3	1	-	-	-	-	-	-	-	-	1	-	-	-	-	-	1
-2	2	-	-	-	-	-	1	-	-	-	-	-	-	-	-	1
-1	3	-	-	-	1	1	1	1	1	-	-	-	-	-	-	5
0	4	-	2	1	-	1	-	1	-	-	-	-	-	-	-	5
1	5	-	-	-	-	2	1	1	2	-	2	-	-	-	-	8
2	6	1	-	-	-	1	-	2	5	5	1	1	-	-	-	16
3	7	-	-	-	-	-	1	1	3	5	4	2	2	-	-	18
4	8	-	-	-	-	-	-	-	2	5	1	2	5	1	1	17
5	9	-	-	-	-	-	-	-	-	-	1	1	1	1	-	4
$n_{\mu j}$		1	2	1	1	5	4	7	13	16	9	6	8	2	1	76

References

[1] Galton, F.: Regression towards mediocrity in hereditary stature. Journal of the Anthropological Institute of Great Britain and Ireland 15, 246–263 (1886)

[2] Yule, G.U.: An Introduction to the Theory of Statistics. Charles Griffin and Co., London (1911); 2-nd Edition translated into Polish by Z. Limanowski: Wstęp do Teorji Statystyki, Gebethner i Wolff, Warszawa 1921; pp. 1–446. Vi-th Edition of 1922 accessible by Internet, pp.1–415. 14-th edition, co-author M.G. Kendall, 1950 translated into Polish as Wstęp do Teorii Statystyki, PWN, Warszawa (1966)

[3] Willerman, L., Schultz, R., Rutledge, J.N., Bigler, E.: In vivo brain size and intelligence. Intelligence 15, 223–228 (1991)

[4] Weinberg, G.H., Schumaker, J.A., Oltman, D.: Statistics – An Intuitive Approach, 4th edn., pp. 1–447. Brooks/Cole, Monterey (1981)

[5] Laudański, L.M.: Statystyka nie tylko dla Licencjatów (in Polish: Statistics not only for undergraduates), 2nd edn., vol. 1. Publishing House of the Rzeszow TU, Rzeszów (2009)

[6] Spiegel, M.R.: Schaum's Outline of Theory and Problems of Statistics, pp. 1–359. McGraw-Hill, New York (1972), 870 solved problems

[7] Hubble, E.P.: A relation between distance and radial velocity among extra-galactic nebulae. Proceedings of the National Academy of Sciences 15, 168–173 (1929)
[8] von Brzeski, J.G.: Application of Lobatchevsky's Formula on the Angle of Parallelism to Geometry of Space and to the Cosmological Redshift. Russian Journal of Mathematical Physics 14, 366–369 (2007)
[9] von Brzeski, J.G.: Expansion of the Universe – Mistake of Edwin Hubble? Cosmological Redshift and Related Electromagnetic Phenomena in Static Lobatchevskian (Hyperbolic) Universe. Acta Physica Polonica 39, 1501 (2007)
[10] Booth, D.E.: Regression Methods and Problem Banks. COMAP, Inc. (1986)
[11] Draper, N.R., John, J.A.: Influential Observations and Outliers in Regression. Technometrics 23, 21–26 (1981)
[12] Criqui, M.H.: University of California, San Diego, reported in the New York Times, December 28 (1994)
[13] Pearson, K., assisted by Lee, A.: On the Laws of Inheritance on Man. Biometrica 2(4), 357–462 (1903).

Unit 4
Binomial Distribution

We consider the subject of the binomial distribution as the subject opening the theoretical background of Statistics. All the problems in this part of Statistics have some reference to the Theory of Probability which is sometimes bigger and sometimes smaller, but always important. The binomial distribution [we know that *binomial* can be *positive* or *negative*] exemplifies this reference in a special way. Moreover, *binomial* distribution represents a topic, which we briefly presented in Chapter 4, and which has exceptionally rich history. The subject of this illustrative part, mirroring the divisions from the Chapter 4, is presented through problems partly expanding the practice, and partly supporting the theory. It will be seen that the Unit presents a deep dependence between binomial and Poisson distributions. This time we cannot recommend to the Student a single textbook neither from the Author's own books, nor from any other sources. The Student can note, that for instance Udny Yule [1] investigated binomial distribution making it the subject of one of the final chapters, [but here it is rather of no use] while in Weinberg's book [2] it is only episodically mentioned in the context of the limiting behavior of the normal distribution. In this Unit we have decided to follow the rule: it will not be said in advance which particular part of the theory we are going to illustrate by the given problem. Therefore the Student is advised after reading the problem to close the book and try to solve the problem on his/her own, and only then consult the solution in this book. We are not always able to indicate which book served as the source of the considered problem, but more frequently than not we do provide this information. For instance, we owe the first two problems to the book by Emanuel Parzen [12]. However, the answer given by Parzen to the first problem is unfortunately wrong. The second problem has no answers. Then we would like to note that in this, and the following Unit we try to use the *MathCad* package available on the Internet, making frequent suggestions commenting the results derived in this way.

Problem 4.1 (see [12], Problem 3.1, p.256, p.453)

The incidence of polio during the years 1949-54 was approximately 25 per 100,000 population. (i) In a city of 40,000 what is the probability of having 5 or fewer cases? (ii) In a city of 100,000 what is the probability of having 5 or fewer cases? State your assumptions.

L.M. Laudański: Between Certainty and Uncertainty, ISRL 31, pp. 245–274.
springerlink.com

Solution. For the above circumstances, the so called "*small probability*" corresponds to 25 per 100,000 therefore $p = 0.00025$. Using the same notation as in Chapter 4, in the first city we have $n = 40,000$. So, let us apply the Poisson distribution determined by $\lambda = n\,p$ here equal to 10. The Student should decipher numbers obtained in this way in Table 4P.1.

Table 4P.1 Poisson distribution, $\lambda = 10$

k	$P(k)$
0	4.539992976248485153559151556055 1e-5
1	4.5399929762484851535591515560551e-4
2	0.0022699964881242425767795757780 3
3	0.0075666549604141419225985859267 6
4	0.0189166374010353548064964648169
5	0.0378332748020707096129929296337 9

Question 1: $\sum P(k) = 0.0670859628790317822857590628266 5$ (the answer given by Parzen is a number ten times greater, 0.671).

Question 2: for $n = 100{,}000$ and respectively for $\lambda = 25$, therefore looking for probabilities corresponding to $k \le 5$ we obtain negligible values, the greatest of them is

1.1573286554136683828884803121739e-8

This answer may give a misleading impression: if you want to avoid polio – settle down in a big city! The right explanation indicates that the average number of people with polio in this big town is 25 (while in the smaller town the average is only 10).

The Student is advised to compare the *results* obtained by applying the Poisson distribution – considering them as *approximate* results in comparison to the *exact* solution given by the binomial distribution. Therefore, starting with this first problem, we are faced with an important question of what circumstances allow to expect successes in such a procedure. There are some general indicators which have to be taken into account. We know, that the Poisson distribution presents a limit of the procedure corresponding to the sufficiently high n, and sufficiently small p. From a practical point of view we recommend using *MathCad* but this is not always possible due to numerical restrictions. For instance the case under consideration rules out such a possibility. While with the help of the Windows 7 calculator it is possible to derive the above six probabilities from binomial distribution, and such results have been inserted in Table 4P.2. Therefore, an assessment of the results from Table 4P.1 may be conducted using the results from Table 4P.2.

Table 4P.2 Binomial distribution, $n = 40\,000$, $p = 0.00025$

k	$b(k)$
0	4.53432058559290889469310284476e-5
1	4.535454449205210197242413448122e-4
2	0.00226823759081969495723304945626
3	0.00756230450592699499036488390026
4	0.018909070600372719344187563225714
5	0.03782381534046047926432236953117
40000	6.2230152778611417071440640537801e-361

Therefore, the exact answer to Question 1 based on binomial distribution indicates:

0.06706231668835633866477903848666

Comments. Rounding both final results to a practical number of digits, we express the Poisson result as 0.067086, and the binomial as 0.067062. They differ by 0.0358%. Also the difference between the presented probabilities can be analyzed in the same fashion showing an interesting variability of errors. The last result shown in Table 4P.2 gives the probability of all the inhabitants of the town getting polio. With *MathCad* one can also get a drawing of the Poisson distribution. It is interesting to note that the diagram shown in Fig.4P.1 apparently presents a continuous distribution, however, it is the discrete distribution, though infinite.

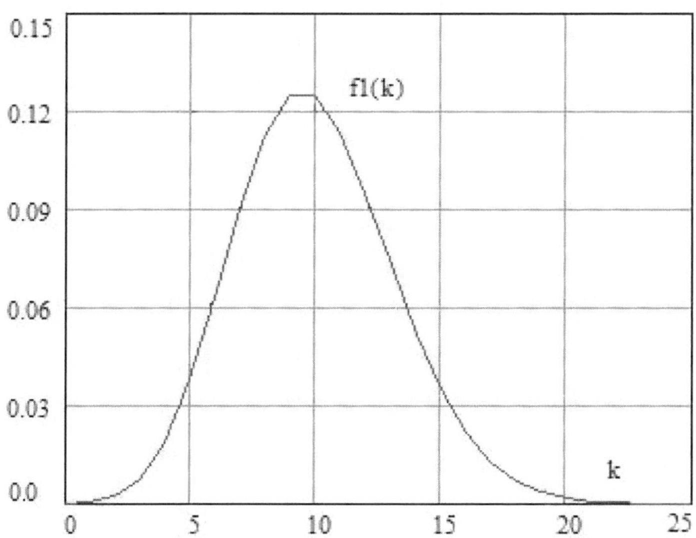

Fig. 4P.1 Diagram of Poisson distribution $\lambda = 10$

The Student is encouraged to draw a diagram of Poisson distribution on his/her own. To support such effort we provide in Table 4P.3 a set which covers a practical range of values for this distribution.

Table 4P.3 Poisson distribution, $\lambda=10$

k	P(k)	k	P(k)
0	0.0000453999	12	0.0947803301
1	0.0004539993	13	0.072907946
2	0.0022699965	14	0.052077104
3	0.0075666550	15	0.034718069
4	0.0189166374	16	0.021698793
5	0.0378332748	17	0.012763996
6	0.0630554580	18	0.007091108
7	0.0900792257	19	0.003732162
8	0.1125990321	20	0.001866081
9	0.1251100357	21	0.00088861
10	0.1251100357	22	0.000403905
11	0.1137363961	23	0.000175611

Problem 4.2 (see [12], Problem 2.12, p.251)

Suppose that among 10,000 students of a certain college 100 are red-haired. What is the probability, that a sample of 100 students, selected with replacement, will contain at least one red-haired student?

Solution

To solve the above problem, let us first solve the following problem "*what is the probability, that a sample of 100 students, selected with replacement, will contain no red-haired student*"?

We will present simultaneously both solutions, on the base of binomial and Poisson distribution. The parameters of the distributions are as follows $p = 0.01$, $q = 0.99$, $n = 100$, $\lambda=1$. The case of "no red-haired students" corresponds to $k = 0$. Calculations make use of the formulas

$b(k;n{=}100, p{=}0.01) = \dfrac{n!}{k!(n-k)!}\, p^k q^{n-k}$ $\quad P(k;\lambda) = e^{-\lambda}\dfrac{\lambda^k}{k!}$ leading to

$0.99^{100} = 0.36603234127322950493061602657252$ and

$e^{-1} = 0.36787944117144232159552377016146$

Then we return to the first question looking for the probability of the *complement event* and we get

0.63396765872677049506938397342748 and 0.63212055882855767840447622983854

These two results provide an answer to Problem 4.2

To complement the case we present a diagram showing both distributions for an important range of the argument. Here, for the first time we draw the Student's attention to a very simple formal condition: if the mean and the variance of the binomial are close to each other – it guarantees the applicability of the Poisson distribution where these two means are identical. In this respect Table 4P.4 also offers the first opportunity of a formal comparison between even four related distributions.

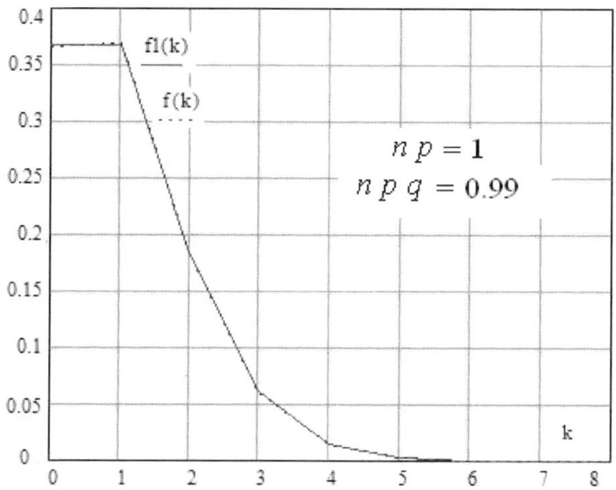

Fig. 4P.2 Binomial and Poisson coupled distributions, n = 100, p = 0.01; $\lambda = 1$

Problem 4.3

If it is assumed that a newly printed a book of 500 pages usually contains 50 printing errors, determine the probability that a randomly chosen page has: (1) exactly 3 errors, (2) at least 3 errors (3) not less than 3 errors?

Solution

The formal statement using standard notation of this book (see also Table 4P.2) determines the mean number of errors per page as $n\,p$ = 50/500 = 0.1. Therefore applying the binomial distribution to solve the problem we have n = 500, p = 0.0002 and solving the problem by applying the Poisson distribution we have $\lambda = 0.1$.

The first solution requires the use of the Word calculator whereas if we apply the Poisson approximation a scientific calculator available on the market is enough. Below, in order to answer the first question we will use both distributions, then to answer the remaining two problems we will limit ourselves to the Poisson distribution.

Table 4P.4 Basic averages of the four distributions

	Binomial distribution	Normal distribution	Poisson distribution	Negative binomial
Mean	$\mu = n\,p$	μ	$\mu = \lambda$	$\mu = k/q$
Variance	$\sigma^2 = n\,p\,q$	σ^2	$\sigma^2 = \lambda$	$\sigma^2 = k\,p/q^2 \quad \sigma^2 = \mu + \mu^2/k$
Standard deviation	$\sigma = \sqrt{n\,p\,q}$	σ	$\sigma = \sqrt{\lambda}$	$\sigma = \sqrt{k\,p}/q$
Moment coefficient of skewness	$a_3 = \dfrac{q-p}{\sqrt{n\,p\,q}}$	$a_3 = 0$ $a_4 = 3$	$a_3 = 1/\sqrt{\lambda}$	$a_3 = (1+p)/\sqrt{k\,p}$
Moment coefficient of curtosis	$a_4 = 3 + \dfrac{1-6pq}{n\,p\,q}$	Mean deviation $\sigma\sqrt{2/\lambda}$	$a_4 = 3 + 1/\lambda$	$a_4 = 6/k + (1+p)^2/k\,p$

First question "exactly 3 errors" – a scientific calculator available on the market used to derive Poisson approximation gives us:

$$p\,(k=3;\lambda=0.1) = e^{-0.1}\,\frac{0.1^3}{3!} = 0.904837418\cdot 0.00016(6) = 0.00015080623$$

while the Word 7 calculator gives us the following result:

0. 90483741803595957316424905944644 * 0.00016(6) =
= 1. 5080623633932659552737484324107e-4

The Word 7 calculator used to derive the binomial '*exact*' solution gives:

$$\frac{500!}{3!\cdot 497!}0.0002^3\cdot 0.9998^{498} \quad\rightarrow\quad 1.4996108458811648041671345841231\text{e-}4$$

The Poisson distribution approximates the result slightly greater than the '*exact*' solution (it exceeds the exact value by 0.54%)

Second question "at least 3 errors", requires that we calculate :

$$\sum_{k=0}^{3} b(k) = b\,(0) + b\,(1) + b\,(2) + b\,(3)$$

$$b\,(0) = 0.904837418 \qquad b\,(1) = 0.0904837418$$

$$b\,(2) = 0.004524187 \qquad b\,(3) = 0.00015080623$$

$$\sum_{k=0}^{3} b(k) = 0.999996383$$

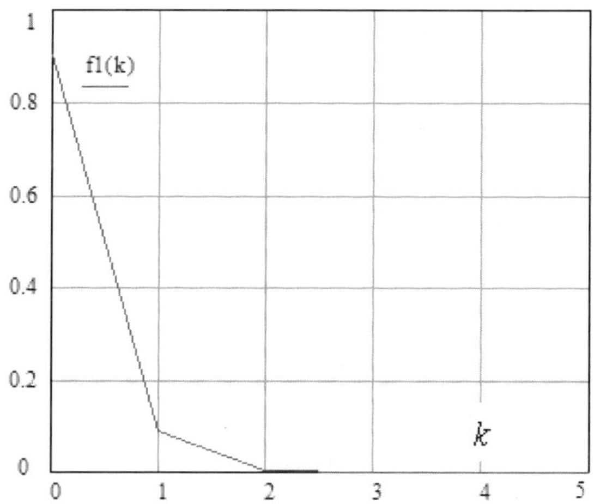

Fig. 4P.3 Poisson distribution, $\lambda = 0.1$

Third question "not less than 3 errors" requires us to calculate :

$$1 - [b(0) + b(1) + b(2)] \quad \rightarrow \quad 0.00015442291$$

This quite unusual distribution is shown in Fig.4P.3. When we say "quite unusual" we mean that the distribution has practically only two (non zero) values.

Problem 4.4

Bernoulli trials were repeated $n = 20$ times, with $p = 0.4$; determine the number of successes corresponding to the maximum probability and investigate its closest neighborhood.

Solution

It is known, that the maximum of the binomial distribution is associated with the mean. Therefore because $np = 8$ the probability $b(k{=}8)$ has to be determined. Auxiliary steps are as follows: $\frac{20!}{8! \cdot 12!} = 125970$, then $0.4^8 = 0.00065536$ and $0.6^{12} = 0.002176782$. Combining them will give the desired result of 0.179705787.

To investigate the neighboring values, we have to determine probabilities $k = 7$, and $k = 9$. The results shown below precisely present the maximum position.

$b(k = 7) = 77520 * 0.0016384 * 0.0013060694 = 0.165882265$
$b(k = 9) = 167960 * 0.000262144 * 0.0036279705 = 0.159738478$

Problem 4.5

An *unfair* coin $p = 0.4$ has been tossed $n = 11$ times. Investigate the case by drawing the diagram of the binomial and Poisson distributions. Provide all the numerical values for the binomial distribution.

Solution

The complete solution was obtained using the *MathCad* package. Fig. 4P.4 shows a diagram of both distributions and Table 4P.5 contains numerical values. Discuss and comment the results.

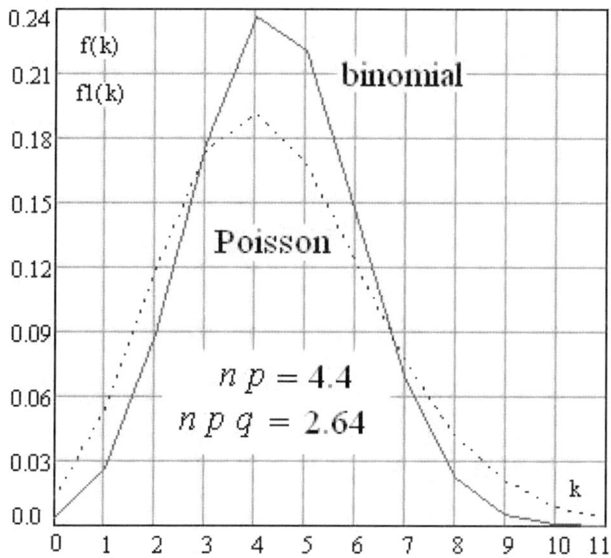

Fig. 4P.4 Binomial and Poisson distributions, $n = 11,\ p = 0.4;\ \lambda=4.4$

Table 4P.5 Binomial distribution $n = 11,\ p = 0.4$

f(0)=.0036279706	f(4)=.2364899328	f(8)=.0233570304
f(1)=.0266051174	f(5)=.2207239373	f(9)=.0051904512
f(2)=.0886837248	f(6)=.1471492915	f(10)=.0006920602
f(3)=.1773674496	f(7)=.0700710912	f(11)=.000041943

Problem 4.6

A Corporation employs 90 young managers. Assuming that with a probability of 0.1 each of them will need a secretary when starting his/her morning duties, determine *r*, i.e. how many secretaries should be employed to assure their services in at least 0.95 chances. [Hint: use De Moivre-Laplace theorem].

Solution

To remind the Student once more about both distributions applied in this context, binomial and Poisson:

$$b(k;n,p)=\frac{n!}{k!(n-k)!}\,p^k\,q^{n-k} \qquad P(k;\lambda)=e^{-\lambda}\frac{\lambda^k}{k!}$$

In the considered problem $n = 90$, $p = 0.1$ $\rightarrow$ $n\,p = 9$ and $n\,p\,q = 8.1$ and the formal requirement has the following direct formulation:

$$\sum_{k=0}^{r} b(k) = 0.95$$

and the following indirect formulation brought forward with respect to the complement event

$$\sum_{k=n-r}^{n} b(k) = 0.05$$

A priori it is impossible to make a rational guess regarding the value of r therefore, we have no idea which way to recommend to obtain an answer most conveniently. Thus, it is seems reasonable to make use of the hint and resort to the theorem of de Moivre-Laplace, but this is beyond the present scope and to obtain a solution based on de Moivre-Laplace we have to go to Unit 5. In this situation we decided to solve the problem by direct approach - obtaining the complete results by using the Word 7 calculator. The problem can also be solved using *MathCad.* The numerical method is presented in Table 4P.6 and Fig.4P.5 shows the graphic results.

Table 4P.6 Binomial and Poisson distributions $n = 90,\ p = 0.1;\ \lambda=9$

k	$P(k)$	$b(k)$
$k = 0$	0.00012340980408667954949763669073	7.6177348045866392339289727720616e-5
$k = 1$	0.00111068823678011594547873021657	7.6177348045866392339289727720616e-4
$k = 2$	0.00499809706551052175465428597457	0.00376654665337894939899821431507
$k = 3$	0.0149942911965315652639628579237	0.01227615205545731655969788369358
$k = 4$	0.03373715519219602184391643032832	0.02966736746735518168593655225948
$k = 5$	0.06072687934595283931904957459098	0.05669763560427879166645652209589
$k = 6$	0.09109031901892925897857436188647	0.08924627826599439428979267366946
$k = 7$	0.11711612445290904725816703671118	0.11899503768799252571972356489261
$k = 8$	0.13175564000952267816543791630008	0.13717483511254693937134799841787
$k = 9$	0.13175564000952267816543791630008	0.13886835159541788924013007247242
$k = 10$	0,11858007600857041034889412467007	0.12498151643587610031611706522517
$k = 11$	0.09702006218883033574000428382096	0.10099516479666755581100368907085
$k = 12$	0.07276504664162275180500321286572	0.07387609276793274915804899478331
$k = 13$	0.05037580152112344355730991659935	0.0492507285119551661053659965222
$k = 14$	0.03238444383500792800112780352815	0.03009766742397260150883477565246
$k = 15$	0.01943066630100475680067668211689	0.01694387203127346455312179962657
$k = 16$	0.01092974979431517570038063369075	0.00882493334962159612141760397217
$k = 17$	0.00578633812640215184137798254216	0.00426826841746403995414969081007
$k = 18$	0.00289316906320107592068899127108	0.00192335552144984516452424338972
$k = 19$	0.00137044850362156227822110112841	8.0983390376835585874704984830434e-4
$k = 20$	0.00555031643966732722679545957005	1.628500647801375348717212096967e-4

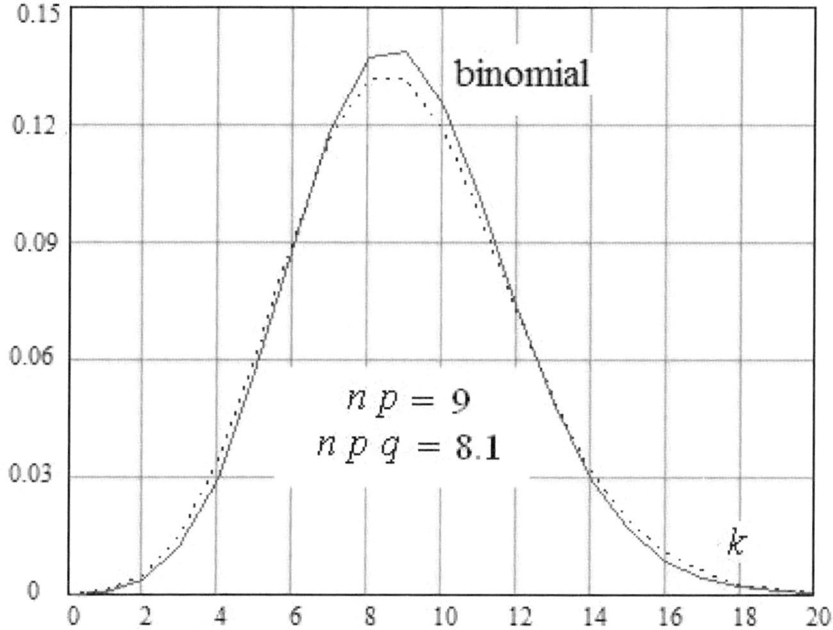

Fig. 4P.5 Binomial and Poisson coupled distributions, n = 90, p = 0.1; λ=9

Final numerical results:

Taking the exact – binomial – solution we find out that
employing 13th secretary leads to: 0.93663365778335808963835141442284
while employing 14th secretary we get: 0.96673132520733069114718619007 53

Taking the Poisson approximation we find out that
employing 13th secretary leads to: 0.92614923069208834769538828487878
employing 14th secretary we get: 0.95853367452709627569651608840693

By using *MathCad* for solutions requiring 17 digits, the appropriate results are as follows. For the binomial distribution we get:

$F(0,13) = 0.9366336577833609$

$F(0,14) = 0.9667313252073336$

While for the Poisson distribution the appropriate results are as follows:

$$F1(0,13) = 0.9261492306920884$$

$$F1(0,14) = 0.9585336745270964$$

The results obtained using both approaches coincide which eliminates the possibility of errors. Similarity between the mean and the variance of the binomial distribution documents the applicability of the Poisson approximation. Therefore in the end we arrive at the same conclusion as obtained using the binomial distribution. However, the Poisson distribution gives a conservative account of both probabilities with respect to the exact binomial probabilities.

At the beginning of Unit 5 we return to this Problem deriving the best approximation to this question using the Second Theorem of De Moivre-Laplace, proving, that this third solution also determines $r = 14$.

Problem 4.7 (see: [13], p.125)

The management of Premiere theater knows from past experience that 20 percent of the complimentary tickets sent to critics are not used. This percentage is apparently independent of the type of play and any other unidentifiable circumstances. The theater reserves 10 seats for critics. If 12 complimentary tickets are sent out, what is the probability of accommodating all those critics who actually go to the theatre?

Solution 1

It is obvious that the ambiguous wording of the problem does not express the question clearly. In the first solution we assume that *success* corresponds to the use of complimentary tickets, therefore from the formal point of view it is the case of $p = 0.8$. Moreover, to be more precise, we assume that the question concerns the probability that with $n = 12$, and $p = 0.8$, the number of successes is $k = 10$. Therefore, the answer is obtained in a single calculation

$$b(k=10) = \frac{12!}{10!\cdot 2!}\ 0.8^{10}\cdot 0.2^{2} = 0.283467841536$$

An attempt to derive an approximate solution by applying the Poisson distribution with $\lambda = 12\cdot 0.8 = 9.6$ gives the following result

$$e^{-9.6}\,\frac{9.6^{10}}{10!} = 0.124085853212048464625664788732 64$$

As we see this "approximation" completely failed – the reason is that the circumstances do not justify the application of the Poisson distribution: value $n\,p$ is 9.6 and value $n\,p\,q$ is 1.92 – so they are exactly the opposite of what is required: *they have to be close to each other.*

To complete this solution we derived the full set of values of the binomial distribution.

Table 4P.7 Binomial, n = 12, $p = 0.8$

k	$P(k)$	6	0.015502147584
0	0.000000004096	7	0.053150220288
1	0.000000196608	8	0.13287555072
2	0.000004325376	9	0.23622320128
3	0.00005767168	10	0.283467841536
4	0.00051904512	11	0.206158430208
5	0.003321888768	12	0.068719476736

Solution 2 together with some comments

In the book [13], Hawkins & Weber, present a solution based upon binomial distribution defining *success* as corresponding to probability $p = 0.2$ i.e. they consider the empty seats as success and moreover claim that the answer to the problem is given by the formula $b(k<2;\ n=12,\ p=0.2)$ giving result 0.2749. Accidentally this answer is numerically close to the answer found above, but that is just by luck. We suggest discussing this point more closely. First, let us note, that this case gives an opportunity to make use of the complimentary event, therefore

$$b(k<2;\ n=12,\ p=0.2) = b(k=0)+b(k=1)$$

Then, both required probabilities can be found in Table 4P.3. This is the result of the fact that by reversing the meaning of *success* we can make use of the fact that for instance

$$b(k=0;\ n=12,\ p=0.2)=b(k=12;\ n=12,\ p=0.8)$$

Therefore $b(k=0)$= 0.068719476736 and $b(k=1)$= 0.206158430208, their sum 0.274877906944. Moreover, this case of the binomial distribution can be well approximated by the Poisson distribution. Let us first find the result corresponding to k = 2 and $\lambda = 12 \cdot 0.2 = 2.4$ applying the Poisson distribution, the result is

$$e^{-2.4}\frac{2.4^2}{2!} = 0.26126770547350800972049599382951$$

The accuracy of this particular Poisson result with respect to the above obtained result $b(k=10;\ n=12, p=0.8)=0.283467841536$ remains within a range of 4.95 %.

Moreover, if we look at Fig. 4P.6, it will be evident that both these distributions are so close to each other that the calculations of any probability shown in Table 4P.7 can be done based on the Poisson distribution.

Note: consulting [13] (see p.125) the Student has to be ready to correct the errors which in the solution to this problem were provided by Hawkins and Weber.

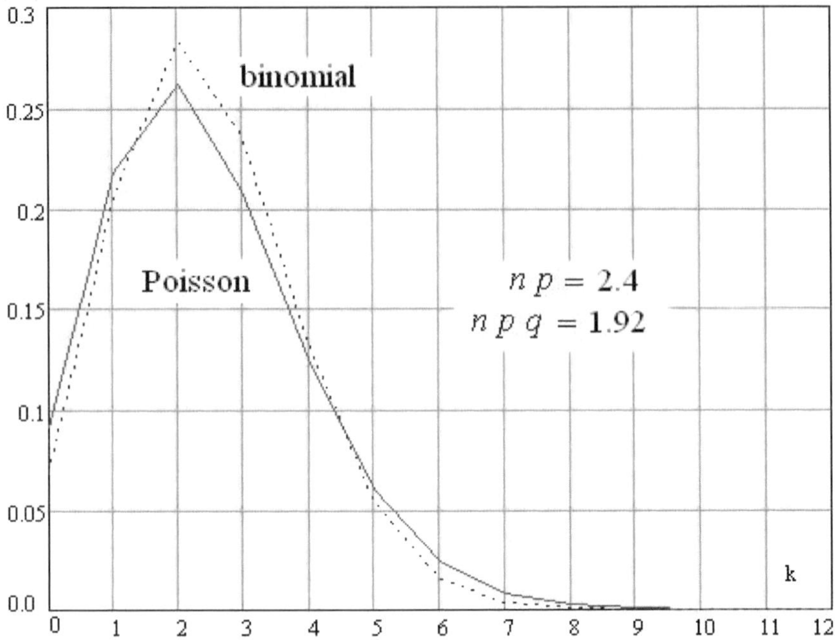

Fig. 4P.6 Coupled distributions, binomial $n = 12, p = 0.2$, and Poisson $\lambda = 2.4$

To illustrate how the binomial distribution $n = 12$, $p = 0.2$ and the Poisson distribution $\lambda = 2.4$ contrast with the binomial distribution $n = 12$, $p = 0.8$ and the Poisson distributions $\lambda = 9.6$ compare Fig. 4P.6 and Fig. 4P.7. The Student should notice that the source of this discrepancy is the difference in the values of the mean and the variance for both pairs of binominal distributions. In the first case it is 2.4 and 1.92 and in the second, 9.6 and 1.92.

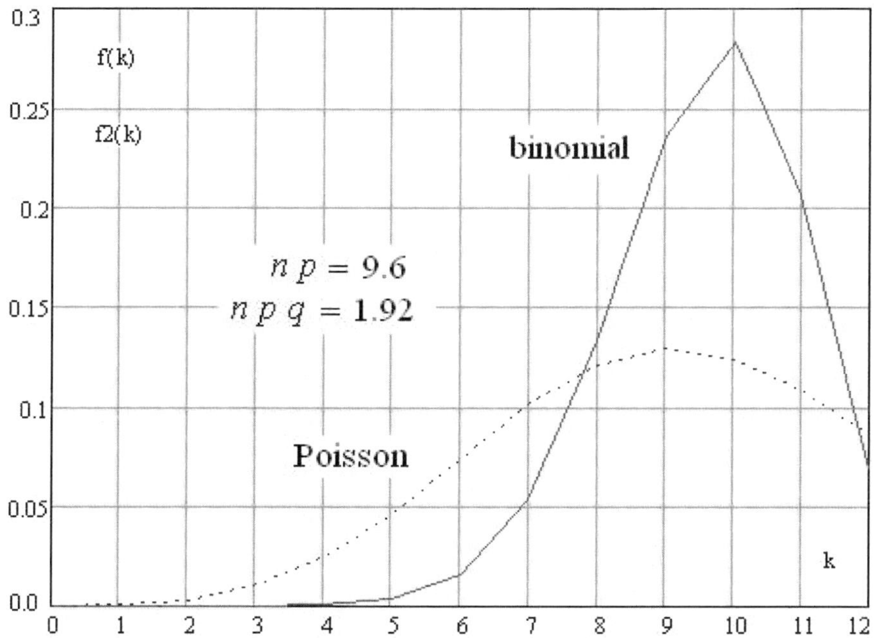

Fig. 4P.7 Two distributions, binomial $n = 12$, $p = 0.8$, and Poisson $\lambda = 9.6$

Problem 4.8

The Department of Municipal Services in a town monitors the number of failures of the sewage system. The provided sample was obtained over a monitoring period of one month:

Table 4P.8 Sewage system failures 1

Number of daily failures	0	1	2	3	4
Frequencies (days)	22	30	22	16	10

Examine whether the failures can be approximated by the Poisson distribution, restore this distribution, and then study the empirical and theoretical results.

Solution

Table 4P.9 Sewage system failures 2

Number of daily failures	0	1	2	3	4
Empirical Probability	0. 22	0.30	0.22	0.16	0.10

Estimating the mean and the variance:

$$\widehat{\lambda_1} = \overline{k} = \sum k\,P(k) = 0.3 + 2*0.22 + 3*0.16 + 4*0.1 = 1.62$$

$$\overline{k^2} = \sum k^2\,P(k) = 0.3 + 4*0.22 + 9*0.16 + 16*0.1 = 4.22$$

$$\widehat{\lambda_2} = \overline{k^2} - \left(\overline{k}\right)^2 \rightarrow \quad 4.22 - 1.62^2 = 1.5956 \approx 1.6$$

The similarity of both estimators of a single Poisson distribution parameter λ indicates the applicability of the Poisson distribution to the empirical data. On the other hand - restoring both parameters of the associated binomial distribution as $n = 100$ and $p = 0.0162$ - which gives the mean $n\,p = 1.62$, and the variance equal to $n\,p\,q = 1.59$ - so similar to the estimators based on the empirical data – puts both distributions very close to each other.

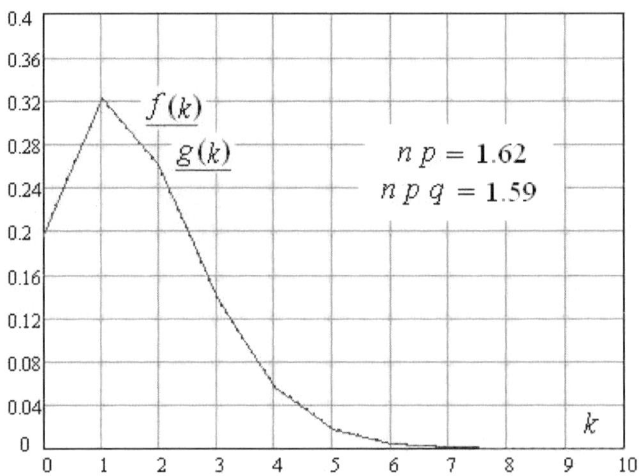

Fig.4P.8 Binomial and Poisson distributions, $n = 100$, $p = 0.0162$; $\lambda = 1.62$

It is seen that with the scale of Fig.4P.8 both curves are identical. Small numerical differences are presented in Table 4P10 .

Table 4P.10 Binomial and Poisson distributions, $n = 100$, $p = 0.0162$ $\lambda = 1.62$

Poisson distribution	Binomial distr.	Poisson distribution	Binomial distr.
f1(0) = 0.1978986991	f2(0) = 0.195290817	f1(1) = 0.3205958925	f2(1) = 0.32158073
f1(2) = 0.2596826729	f2(2) = 0.26212196	f1(3) = 0.1402286434	f2(3) = 0.140999129
f1(4) = 0.0567926006	f2(4) = 0.056303627	f1(5) = 0.0184008026	f2(5) = 0.017801057

Problem 4.9

A leading insurance company assumes that each year 1% of their male customers die due in accidences. What is the probability that within one year the company will pay claims more than three times if the number of the insured is 100 persons?

Solution

Binomial with $n = 100$ and $p = 0.01$ is the mother distribution while the Poisson approximation is determined by the mean $\lambda = 1.0$. Instead of discussing the details of such distributions we remind to our Student that such a case has already been considered - therefore we only add that Fig.4P.2 presents both these distributions and there is no need to repeat it again. But we suggest complementing Problem 4.2 with Table 4P.11 which presents an analogy to Table 4P.10 – showing coupled numerical values of both distributions.

Table 4P.11 Binomial and Poisson distributions, $n = 100$, $p = 0.01$; $\lambda = 1.0$

Poisson distribution	Binomial distr.	Poisson distribution	Binomial distr.
f3(0) = 0.3678794412	f1(0) = 0.3660323413	f3(1) = 0.3678794412	f1(1) = 0.3697296376
f3(2) = 0.1839397206	f1(2) = 0.1848648188	f3(3) = 0.0613132402	f1(3) = 0.0609991658
f3(4) = 0.01532831	f1(4) = 0.0149417149	f3(5) = 0.003065662	f1(5) = 0.0028977871
f3(6) = 0.0005109437	f1(6) = 0.0004634508	f3(7) = 0.000072992	f1(7) = 0.0000628635

To give a numerical answer to the question in Problem 4.9 we need to use *MathCad* and copy the results. In the notation used there the symbol "*F*" stands for Poisson, and "*F1*" for the binomial distributions. They show:

Table 4P.12 Probability of more than three claims

F(0, 3) = 0.9816259636	F1(0, 3) = 0.9810118431
1 – F(0, 3) = 0.0183740364	1 – F1(0, 3) = 0.0189881569

To be more specific we also provide symbolic notations used to obtain results given in Table 4P.12:

$$F(k_1, k_2) = \sum_{i=k_1}^{k_2} f_i(k) \qquad F_1(k_1, k_2) = \sum_{i=k_1}^{k_2} g_i(k)$$

Where in turn, symbol $g_i(k)$ denotes binomial distribution, and symbol $f_i(k)$ denotes Poisson distribution.

Problem 4.10 (see [2], Problem 9.15, p.178)

Historically, 65% of Senator Pete Low's constituents have approved of his decisions. In a recent random sample of voters, in his district, more than 65 of 90 people approved of his action on the state banking amendment. What was the probability of this occurring – based on his historical approval rate?

Answers. Weinberg: 0.0606, LML: 0.0587 due to binomial

Solution

Here we recognize the binomial with the following parameters: $n = 90$, $p = 0.65$ and understand the question as the probability of $k > 65$ taking place. Therefore we look for the answer by solving:

$$f(k) = \frac{90!}{k! \cdot (90-k)!} 0.65^k \cdot (1-0.65)^{(90-k)}; \quad F(k_1, k_2) = \sum_{i=k_1}^{k_2} f_i(k)$$

MathCad solution:

$F(66,90) = 0.05871866696485527$ [see Unit 5 for the normal solution]

Fig.4P.9 shows "unmatched" binomial and Poisson distributions. In other words we cannot use the Poisson distribution with $\lambda = 58.5$ as a valuable approximation to the binominal distribution $n\,p = 58.5$ and $n\,p\,q \approx 20.5$.

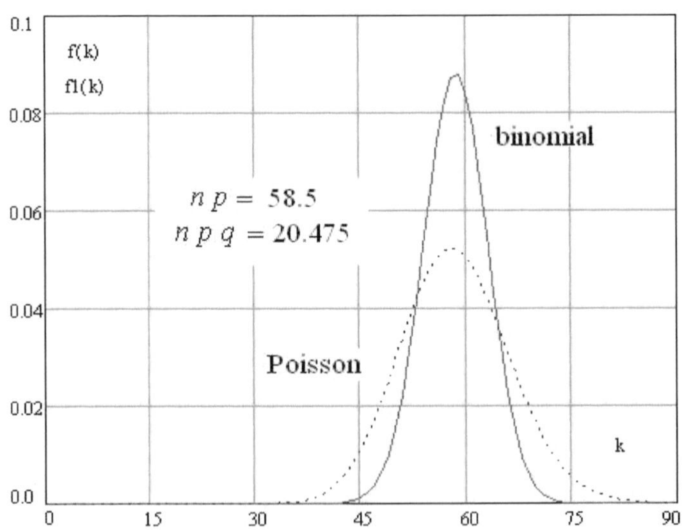

Fig. 4P.9 'Unmatched' binomial and Poisson distributions $n = 90$, $p = 0.65$; $\lambda = 58.5$

To be certain that it “does not work” we use *MathCad* to derive the appropriate answer and get the following numerical result:

$$F_1(66,90) = 0.178938474670244$$

***

<u>Short remainder.</u> From this place onwards we will concentrate on the third distribution examined in this Unit. This third distribution is the *negative binomial distribution*. This distribution is completely different from binominal distribution, especially in the most advanced version presenting continuous distributions. So far we recall two main versions, both closely related and both presenting the discrete distribution but with infinite terms. We commence by listing the two main descriptions of this distribution.

As the first version let us take the case where we count the n number of Bernoulli trials where the *r-th* success occurs. Here:

$$P(X_1 = n \mid p, r) = \binom{n-1}{r-1} \cdot p^r \cdot q^{n-r}$$

In this formula there are two parameters: real p satisfying the following condition: $0 < p < 1$, and r a positive integer ; integer $n \geq 0$ is independent variable and infinite.

The second version counts k i.e. the number of failures before the *r-th* success and here,

$$P(X_2 = k \mid p, r) = \binom{r+k-1}{k} \cdot p^k \cdot q^{k-r}, \qquad X_2 = X_1 - r$$

More equivalences are listed in Chapter 4, therefore the Student is advised to revise this material before starting the applications below. And we have to add that further on we also propose an extension of the results given in Chapter 4.

Problem 4.11 (see [13], p.128)

Market studies have established that 20 percent of the housewives in Oatsville use Happiness detergent. Find the probability that in a random sample of Oatsville housewives, the 25th person interviewed is the 10th user of Happiness detergent.

<u>Solution</u>

Let us recall formula (4.96) from Book One

$f(n) = \binom{n-1}{r-1} \cdot p^r \cdot (1-p)^{n-r}$ assuming $n = 25$, $r = 10$, and $p = 0.2$ we get

$$f(25) = \binom{24}{9} \cdot 0.2^{10} \cdot (1-0.2)^{15}$$; and finally

$f(\ 25\ ) = 0.0047107796217483464015872$

Considering "n" as an independent variable, we can generalize the answer to the form shown in Fig.4P.10.

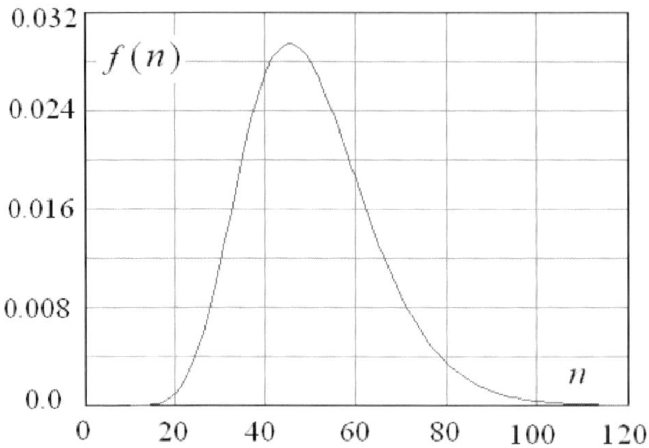

Fig. 4P.10 Negative binomial distribution $r = 10,\ p = 0.2$

Problem 4.12 (see [5], p.165, p.63)

If $f(k) = \binom{k+r-1}{k} \cdot p^r \cdot (1-p)^k$ then, prove that

$$f(k) = (-1)^k \cdot \binom{-r}{k} \cdot p^r \cdot (1-p)^k$$

by using an intermediate result stating that for any $r > 0$

$$\binom{-r}{k} = (-1)^k \binom{k+r-1}{k}$$

Solution

Expanding the left hand side we will find out that

$$\binom{-r}{k} = \frac{(-r)\cdot(-r-1)\cdot\ldots\cdot(-r-k+1)}{k!},$$ then from the right side of the above

tautology the term (-1) occurring k-times can be excluded which will give us:

$$\frac{(-r)\cdot(-r-1)\cdot\ldots\cdot(-r-k+1)}{k!} = (-1)^k \cdot \frac{r\cdot(r+1)\cdot\ldots\cdot(r+k-1)}{k!}$$; then we apply

$$\frac{r\cdot(r+1)\cdot\ldots\cdot(r+k-1)}{k!} = \binom{k+r-1}{k}$$. With this result we come closer to the desired result. Now, let us finally prove that:

$$\binom{k+r-1}{k} = (-1)^k \cdot \binom{-r}{k}$$ which can be established in the following way:

$$\frac{(r)\cdot(r+1)\cdot\ldots\cdot(r+k-1)}{k!} = (-1)^k \frac{(-r)\cdot(-r-1)\cdot\ldots\cdot(-r-k+1)}{k!} = (-1)^k \binom{-r}{k}$$

Problem 4.13 (Source: Internet)

Pat has to sell candy bars to earn his pocket money. There are forty houses in the neighborhood, and Pat cannot return home until he sells eight candy bars. So he goes from door to door, selling them. At each house, there is a 0.3 probability of selling one candy bar and a 0.7 probability of selling nothing. What is the probability of selling the last candy bar at the nth house?

Solution

The answer to such a general question presented in this problem can either take the form of a table or of a diagram (or both forms). We have chosen the second option and provided Fig. 4P.11.

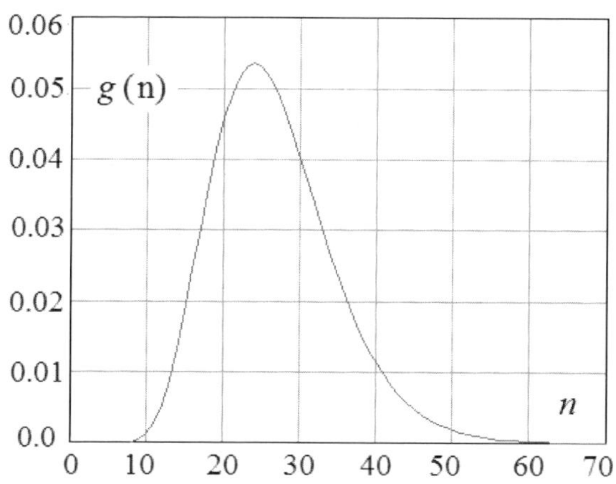

Fig. 4P.11 Negative binomial, $r = 8,\ p = 0.3$

It can be noted here what a significant change takes place if probability is changed, e.g. to the compliment value of $p = 0.7$. The result is seen in Fig. 4P.12. Both diagrams were obtained using the *MathCad* tools.

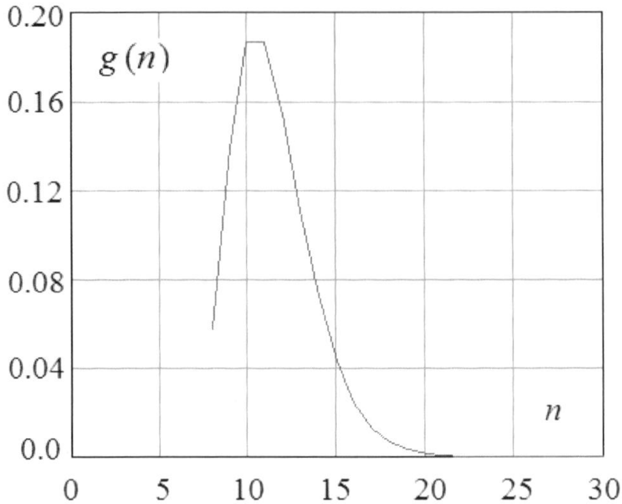

Fig. 4P.12 Negative binomial, $r = 8,\ p = 0.7$

Maybe it is also a good opportunity to underline an important fact which is in a way overshadowed by all our diagrams: the binomial distribution belongs to the family of the *finite discrete distributions*, and on the other hand, the negative binomial distributions belong to the family of *continuous infinite distributions*.

Problem 4.14

Investigate the probability ranges for the above given three binomial distributions of $r = 10, p = 0.2$; $r = 8, p = 0.3$ and $r = 8, p = 0.7$ supplementing them with $r = 5,\ p = 0.4$.

Solution

We suggest presenting the solution using Table 4P.13. The results shown there were obtained with *MathCad* . Unfortunately, the necessity of integration rules out the use of scientific calculators available on the market. By rounding the figures in Table 4P.13 we wanted to save some space.

Table 4P.13 Negative Binomial distributions, four cases

$r = 10,\ p = 0.2$		$r = 8,\ p = 0.3$		$r = 8,\ p = 0.7$		$r = 5,\ p = 0.4$	
10 - 20	0.026	8 – 10	0.0016	8 -10	0.383	5 – 10	0.367
21 - 30	0.058	11 – 20	0.226	11 –15	0.567	11 – 15	0.416
31 - 40	0.207	21 – 30	0.491	16 - 20	0.049	16 – 20	0.166
41 - 50	0.288	31 – 40	0.226	8 - 20	0.999	21 - 30	0.049
51 - 60	0.231	41 – 50	0.048			5 - 30	0.998
61 –70	0.129	8 - 60	0.999				
71 - 80	0.056						
81 - 90	0.02						
10-100	0.998						

***

Important Generalization. Before giving the Student the next problem we suggest generalizing the negative binomial distribution formula to the form which recalls the Gamma function – as given in Chapter 2 by formula (2.16). So, it is:

$$\binom{k+r-1}{k} = \frac{(r)\cdot(r+1)\cdot\ldots\cdot(r+k-1)}{k!} = \frac{\Gamma(k+r)}{k!\cdot\Gamma(r)}$$

In short, it states that:

$$\binom{k+r-1}{k} = \frac{\Gamma(k+r)}{\Gamma(k+1)\cdot\Gamma(r)}$$

Finally, the promised generalization takes the form:

$$f(k) = \binom{k+r-1}{k}\cdot p^r\cdot(1-p)^k \quad\rightarrow\quad f(k) = \frac{\Gamma(k+r)}{\Gamma(k+1)\cdot\Gamma(r)}\cdot p^r\cdot(1-p)^k \qquad \text{(A)}$$

Important Remarks. In the above formula, numbers k remain *positive integer*, playing the role of the independent variable, although r is now a *positive real* number. Moreover, this parameter loses its previous meaning as a kind of a *restraint*. Also the range of the validity for the variable k includes now the value *zero*. Moreover this distribution belongs to the class of continuous distributions. Interestingly enough, the mentioned features are hardly to be seen in the publications which we recommend in Chapter 4 and to which we add here a new paper [14]. The generalization described here is useful in applications, the paper [14] serves just as an example. Other examples can be found in references [42]-[43] from Literature to Chapter 4, which we repeat here as references [15]-[16], complementing them with two short papers by Fisher [17]-[18]. To discuss this topic we propose the following example.

Problem 4.15 (see [14])

Investigate the provided statistics in Table 4P. 14 which presents *family size distribution* in Jordan presented in Table 1 of paper [14]. This data are based on the general census in Jordan in 1994.

Table 4P.14 Jordanian family size statistics

classes	0	1	2	3	4	5	6	7	8	9	10	11	12	13+	*total*
class frequencies	59979	64047	78838	82384	77575	68431	57795	46127	35983	25766	18410	8303	5315	7843	636796

Check in particular the mean and the variance with respect to the values given in [14], that is $\mu = 4.32$ and $\sigma^2 = 9.19$. Then derive both parameters of the approximation assuming that the negative binomial distribution fits the case. Provide calculations based on generalized formula (A) . Recommended way suggests using the following formulae to determine values of both parameters:

$$p = \frac{\mu}{\sigma^2} \qquad r = \frac{\mu^2}{\sigma^2 - \mu}$$

The above formulae can be derived from the formulae in Table 4P.1. The Student has to be conscious of some differences in the symbols used. The results given in [14] for these two parameters are as follows $p = 0.4703$ and $r = 3.837$.

<u>Solution</u>

Our calculations gave $\mu = 4.27475$ $\sigma^2 = 9.21134$ $p = 0.4641$ $r = 3.702$. And with the help of *Mathcad* a comparison was made which is shown in Fig. 4P.13 and documents the similarity of both results and insignificance of the differences in all four parameters. The *Red curve* shows parameters given in [14], the *blue curve*, the parameters given here. By the way, curiously, the formula denoted in [14] as (5), expressing the probability distribution, contains three errors. There is also another remark in this context, authors of [14] interpret the value $f(0)$ as a major estimate, as they called it, *the proportion of sterile couples* in the Jordanian population. Leaving aside a possible discussion of this attitude, we add both values from the *red* and from the *blue* curves. For the *red* curve $f_{red}(0) = 0.0583171$ and for the *blue* curve $f_{blue}(0) = 0.0553226$. The Student may easily check that the Jordanian census gave the value of $f_{census}(0) = 0.0941887$ therefore, looking at the paper [14] it is difficult to understand why this approach favors the *artificial* value of 0.058 instead of the real value 0.094 ?

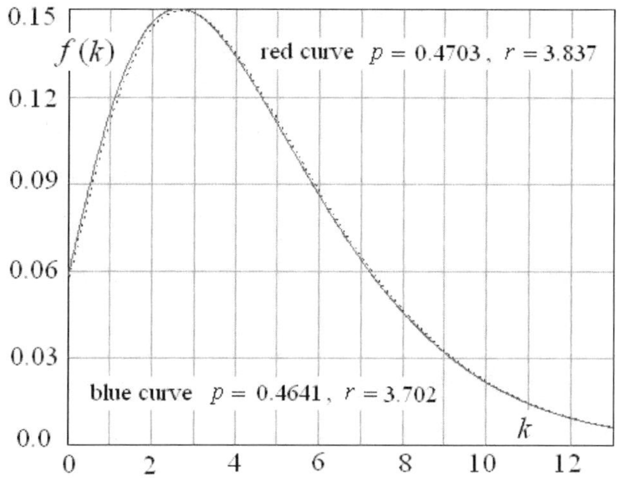

Fig. 4P.13 Two negative binomial distributions – formula (A)

***

<u>*Another generalization.*</u> Continuing the theoretical considerations of this Unit, we present one more version of the distribution belonging to the discussed here family of continuous negative binomials, based on interesting publication [19] accessible on the Internet. For the theoretical origin of this version we can consult a textbook by M.Fisz [7] but instead of providing a lengthy detailed procedure we will describe it briefly recommending that the Student with a deeper theoretical interest consult [7].

The starting point is the Poisson distribution, whose formula we provide here to help with specific symbolic notation:

$$\pi\left(Y_i = k\right) = e^{-\lambda}\,\frac{\lambda^y}{k!}$$

In the second important step it is assumed that the parameter λ of this distribution is a random variable and follows *gamma distribution* with parameters a and v given below:

$$f(\lambda) = \frac{a^r}{\Gamma(r)}\,\lambda^{r-1}\,e^{-a\lambda}$$

Then in the third step we can obtain (see [7], p.178) the final, desired formula of the negative binomial in the following form:

$$P(Y_i = k) = \frac{\Gamma(k+r)}{\Gamma(k+1)\,\Gamma(r)} \frac{(a)^r}{(a+1)^{k+r}} \qquad \text{(B)}$$

It is possible to justify (see: [7]), that the formula (B) is equivalent to the negative binomial distribution in the form presented earlier as formula (A). This point will be also discussed below. And now with all the presented theoretical background we propose to consider the following.

Problem 4.16 (see [19])

Examine the sample of 19 013 individuals presenting the number of car accidents from 1982-83 from the province of Quebec shown in Table 4P.15.

Table 4P.15 Car accident statistics, Quebec 1982-83

f(0) = 17 784	f(1) = 1 139	f(2) = 79	f(3) = 9	f(4) = 2
0.9353600168	0.0599063798	0.0041550518	0.0004733603	0.0001051912

Solution

Table 4P.16 Deriving empirical averages

y_i	0	1	2	3	4
$P(y_i)$	0.9353600168	0.0599063798	0.0041550518	0.0004733603	0,0001051912
$\sum y_i \cdot P(y_i)$	0,0700573291	0,0599063798	0,0083101036	0,0014200809	0,0004207648
$\sum_i y_i^2 \cdot P(y_i)$	0,0824698889	0,0599063798	0,0166202072	0,0042602427	0,0016830592

Numerical results given in Table 4P.16 allow to determine the mean value μ of the investigated statistics and its mean square (second column of Table 4P.16). The mean square serves to derive σ^2, i.e. the variance. So, we get: $\mu = 0.0700573291$ and $\sigma^2 = 0.077561859539374293\,19$. Note: the variance is greater than the mean.

In the next step, by using well known relations:

$$p = \frac{\mu}{\sigma^2} \qquad r = \frac{\mu^2}{\sigma^2 - \mu}$$

we obtained both parameters p and r which are necessary to determine version (A) of the negative binomial distribution. By using Word 7 calculator we obtained what follows:

$$p = 0.90324457814778645369804715708992 \quad \text{and}$$

$$r = 0.65400885508766416629108304554246$$

In the last step the formula of the distribution (A) was fed into *MathCad* to draw the diagram of this distribution and derive the initial values of this distribution, crucial for the investigated approximation.

Simultaneously we checked again the version of the negative binomial distribution (B) used for calculations in paper [19]. The authors of paper [19] gave the following values for both parameters of version (B) distribution $1/a = 9.9359$ and $r = 0.6960$. However, they do not provide the details of the procedure used to determine them. In the following step these values and formula (B) distribution were entered in *MathCad* and an appropriate diagram was drawn, together with the values of this distribution for the initial points.

As Fig. 4P.14 shows both diagrams are very similar to each other, with the scale of the diagram Fig. 4P.14, they are practically identical.

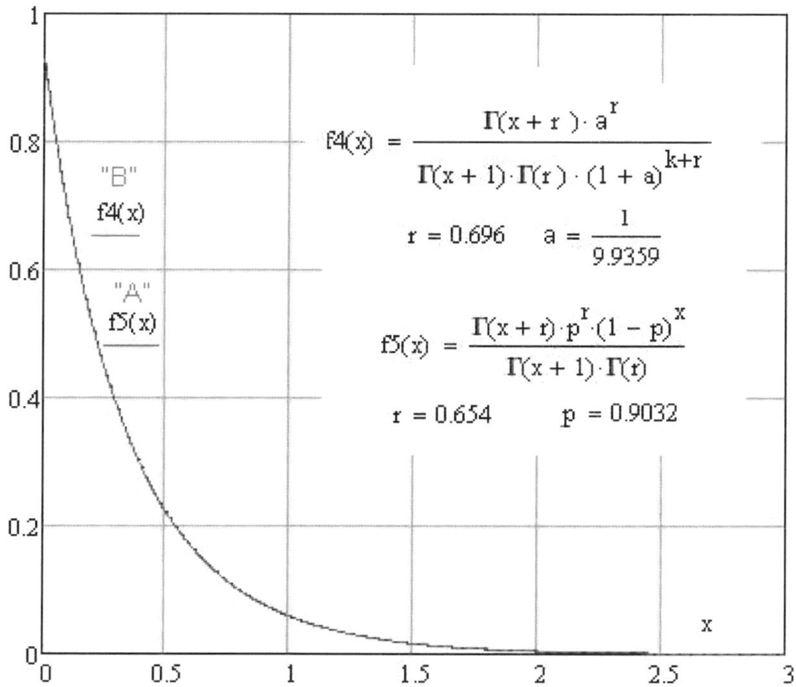

Fig. 4P.14 Negative binomials distributions (A) and (B) of [19]

The presented set of numerical values in Table 4P.17 shows numerical identity of both distributions i.e. (A) and (B) - supplied with the parameters derived in this book. For the sake of a numerical comparison we provide appropriate results based on paper [19]. We denote them as case (B) of [19]. Comparing the numerical results given in the two lowest rows of Table 4.P.17 clearly shows why in Fig.4P.14 both graphs seem to be identical.

Table 4P.17 Fitting by theoretical distributions (A) and (B)

y_i	0	1	2	3	4
(B)	0.9355837920	0.0592291902	0.0047415099	0.0004060427	0.0000359051
(A)	0.9355837920	0.0592291902	0.0047415099	0.0004060427	0.0000359051
(B) [19]	0.9354344679	0.0595344267	0.0046164645	0.0003793618	0.0000320532

Final remark: we resist the temptation to apply the test of the *goodness of fit* based on the *chi-square* distribution and establish quantitatively which case (B) or (B) of [19] is closer to empirical data. Nevertheless, in any case both approximations fit this statistics surprisingly well. To complete the theoretical proof that cases (A) and (B) are principally identical we provide the formulae given by M.Fisz [7] combining the parameters of both cases (A) and (B):

$$p = 1/(1+a) \quad \text{and} \quad q = a/(1+a)$$
$$a = (1-p)/p \quad \text{and} \quad 1/a = p/(1-p)$$

Ending this Unit we suggest to the Student one more approach to present the negative binomial distribution from the point of view of the Bernoulli trials.

Problem 4.17

Prove the following Theorem:

Performing n Bernoulli trials with the probability of success p until the occurrence of r successes, the probability of performing exactly n trials is given by the formula:

$$P(S_k = n) = \binom{n-1}{r-1} p^r (1-p)^{n-r}$$

Proof

Let A denote an event described in the Theorem, let A_1 denotes an event in which irrespective to the sequence r - 1 successes will occur in $n-1$ trials, and let A_2 denote such an event that success will occur in the n-th trial.

Moreover $A = A_1 \cap A_2$, and the events A_1 and A_2 are independent events, therefore we get the following results:

$$P(A_1) = \binom{n-1}{r-1} p^{r-1} q^{(n-1)-(r-1)} \quad \text{and} \quad P(A_2) = p$$

Therefore, in consequence the final step concludes the proof:

$$P(A) = \binom{n-1}{r-1} p^{r-1} q^{n-r} \cdot p = \binom{n-1}{r-1} p^r q^{n-r} .$$

References

[1] Yule, G.U.: An Introduction to the Theory of Statistics. Charles Griffin and Co, London (1911); 2-nd Edition translated into Polish by Z. Limanowski: Wstęp do Teorji Statystyki, Gebethner i Wolff, Warszawa 1921; p.1-446. Vi-th Edition of 1922 accessible by Internet, p.1-415. 14-th edition, co-author M.G. Kendall, 1950 translated into Polish as Wstęp do Teorii Statystyki, PWN, Warszawa (1966)

[2] Weinberg, G.H., Schumaker, J.A., Oltman, D.: Statistics – An Intuitive Approach, 4th edn., pp. 1–447. Brooks/Cole, Monterey (1981)

[3] Spiegel, M.R.: Schaum's Outline of Theory and Problems of Statistics, pp. 1–359. McGraw-Hill, New York (1972), 870 solved problems

[4] Uspensky, J.V.: Introduction to Mathematical Probability, p. s.411. McGraw-Hill, New York (1937)

[5] Feller, W.: An Introduction to Probability Theory and Its Applications, 3rd edn. I, Posthumous Edition, p. 509. John Wiley and Sons, New York (1971)

[6] Laudański, L.M.: Statystyka nie tylko dla licencjatów (in Polish: Statistics not only for undergraduates), part1, part2, 2nd edn. Publishing House of the Rzeszow TU, Rzeszów (2009)

[7] Fisz, M.: Rachunek Prawdopodobieństwa i Statystyka Matematyczna (in Polish: Probability and Mathematical Statistics). Posthumous 3rd edn., pp. 694. PWN, Warszawa (1967); there is also the English translation although the aAuthor of this book has no references to it

[8] Neyman, J.: Zasady rachunku prawdopodobieństwa i statystyki matematycznej, p. s.258. PWN, Warszawa (1969); English origin First Course In Probability and Statistics. Henry Holt & Co., New York (1950)

[9] Edwards, A.W.F.: Pascal's Arithmetical Triangle – The Story of a Mathematical Idea, pierwsze wydanie - w Anglii 1987 – Charles Griffin & Company Ltd, London, a od roku 2002 było drukowane jako Paperback, p. s.202. Johns Hopkins University Press, Baltimore

[10] A complete translation of the Ars Conjectandi is available as Jacob Bernoulli: The Art of Conjecturing, together with Letter to a Friend on Sets in Court Tennis, trans. by E.D. Sylla, p. 580 $57.60. Johns Hopkins University Press, Baltimore (2006)

[11] Pogorzelski, W.: Zarys Rachunku Prawdopodobieństwa i Teorii Błędów (in Polish: An outline of probability and the error theory). Towarzystwo Bratniej Pomocy Studentów PW (edited by the students organization later resolved by the communist government, Warsaw, pp.1–100 (1948)

[12] Parzen, E.: Modern Probability Theory and Its Applications, p. 464. J. Wiley & Sons, New York (1960)

[13] Hawkins, C.A., Weber, J.E.: Statistical Analysis. Applications to Business and Economics, p. 626. Harper & Row, New York (1980)
[14] Al-Saleh, M.F., AL-Batainah, F.K.: Estimation of the Proportion of Sterile Couples Using the Negative Binomial Distribution. Journal of Data Science 1, 261–274 (2003)
[15] Gosset, W.S.: An Explanation of Deviation from Poisson's Law in Practice, vol. 12(3/4), pp. 211–215 (November 1919); also: Paper in Student's Collected Papers, London, Biometrica Office, pp. 65–69 (1942)
[16] Gurland, J.: Some Applications of the Negative Binomial and Other Contagious Distributions. Pdf file No.1388, Wikipedia, pp. 1–12. Printed in A.J.P.H., vol. 49(10), pp. 1388–1399 (October 1959)
[17] Fisher, R.A.: The negative Binomial Distribution. Annuals of Eugenics 11, 182–187 (1941)
[18] Fisher, R.A.: Note on the Efficiency Fitting of the Negative Binomial. Biometrics 9, 197–199 (1953)
[19] Dionne, G., Vanasse, C.: A generalization of Automobile Insurance Rating Models: the negative Binomial Distribution with a Regression Component. ASTIN Bulletin 19(2), 199–212 (1989)

Unit 5
Normal Distribution. Binomial Heritage

Acquaintance with the normal distribution, tables of the normal distribution. Probabilistic paper. Sample means distribution and Monte Carlo simulation. Two theorems of de Moivre-Laplace. When does normal approximation fit binomial distribution data?

Instead of commencing this Unit with a numbered problem, we suggest following the below considerations which according to the promises made in Unit 4 apart from the solution to Problem 4.6 present the third approach to this problem. We commence by restating its main features. While solving Problem 4.6. we looked for such the lowest value r which guarantees that

$$\sum_{k=0}^{r} b(k) = 0.95$$

where $b\,(k)$ is the probability either given by the binomial distribution or by Poisson distribution

$$b(k;n,p) = \frac{n!}{k!\,(n-k)!}\,p^k\,q^{n-k} \qquad P(k;\lambda) = e^{-\lambda}\,\frac{\lambda^k}{k!}$$

It also has to be restated that in the considered problem we face $n = 90$ Bernoulli trials, with $p = 0.1$. Therefore, the mean value is equal to $n\,p = 9$ and the variance $n\,p\,q = 8.1$.

It is also to be mentioned that by resorting to the complement event there is also the following indirect formulation of the problem

$$\sum_{k=n-r}^{n} b(k) = 0.05$$

We will presently see that if we are looking for the solution with the help of De Moivre-Laplace's second theorem, a direct solution will produce the answer in a shorter time than an indirect one. In order to solve the problem by the applied method we have to apply the Normal Distribution Tables searching for such a value of the *z-variable* which corresponds to probability 0.45 (not 0.95). Doing

L.M. Laudański: Between Certainty and Uncertainty, ISRL 31, pp. 275–302.
springerlink.com

that we shall read the two closest values: $z_1 = 1.64$ which corresponds to 0.449497 and $z_1 = 1.65$ giving 0.450520. In the last step we have to move from these two derived *z-scored* variables to the real value variables – by using the formula:

$$k_i = n \cdot p + z_i \sqrt{n \cdot p \cdot (1-p)}$$

Substituting the above given values of z_i we will get two values of k_i 13.67 and 13.70, respectively . Both of them lead to the same answer. Moreover, the answer to both approximations received above is the same as the exact answer known from the solution of Problem 4.6: it is not enough to employ 13 secretaries, the above derived answers also state that required minimum number of secretarial staff should be 14 persons.

Closing the above considered problem we have to commence with the normal distribution problems from scratch. This Student who finds it difficult to follow the above given considerations may skip this passage and return to it later on.

Problem 5.1 (see [5], Problem 7.16, p.129)

Two students were informed that they received standard scores of + 0.8 and - 0.4 respectively on a multiple choice examination in English. If their marks were 88 and 64 respectively, find the mean and the standard deviation of the examination marks.

Solution: to solve the problem we have to know the mean and the variance of the normal distribution. Then the *z-scored* values can be obtained from:

$$X = \overline{X} + z \cdot \sigma_x$$

Data given in the problem allow to write two equations:

$$X_1 = \overline{X} + z_1 \cdot \sigma_x \qquad \text{and} \qquad X_2 = \overline{X} + z_2 \cdot \sigma_x$$

By substituting $X_1 = 88,\ z_1 = 0.8$ and $X_2 = 64,\ z_2 = -0.4$ the two above equations can be solved with respect to unknown $\overline{X}$ and σ_x -- giving the answer $\overline{X} = 72$ and $\sigma_x = 20$.

Problem 5.2 (see [5], Problem 7.20)

The mean length of 500 laurel leaves from a certain bush is 151 mm and the standard deviation is 15 mm. Assuming that the lengths are normally distributed find how many leaves measure between 120 and 155 mm, and how many more than 185 mm?

Solution can be obtained by using three methods, the first two methods apply *MathCad*, the third one uses a standard approach based on the Table values. Therefore, we have to consider the third one as the basic method, whereas the first two will serve as a specific *décor* of the Unit. Nevertheless, we apply the first, showing Fig. 5P.1 and Fig. 5P.2.

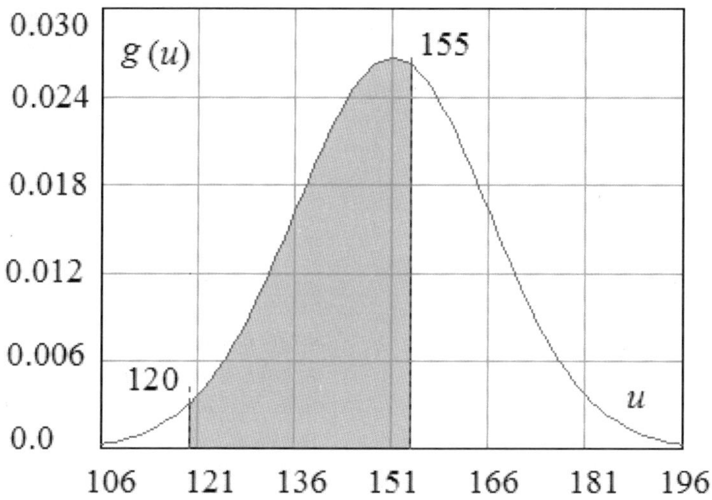

Fig. 5P.1 Normal distribution, $\mu = 151$, $\sigma_u = 15$

With the distribution illustrated by Fig. 5P.1 we have to derive the following integral formula to give the answer to the first question:

$$F(l,r) = \int_l^r g(u)\,du$$

here $g(u)$ is the probability density of the normal variable u with the average $\mu = 151$ mm and standard deviation $\sigma_u = 15$ mm, while the lower/upper bands of integration have to be $l = 120$ mm, and $r = 155$ mm. We recall the formula for the normal distribution in general case:

$$g(u) = \frac{1}{\sigma_u\sqrt{2\pi}}\exp\left(-(u-\mu)^2/2\sigma_u^2\right)$$

The answer to the first question obtained using *MathCad* is as follows: $F(120,155) = 0.5857543024471563$, which corresponds after appropriate rounding to 293 leaves.

This answer can be also derived by applying an approach which uses *z-scored* values and resorts to the normalized distribution as shown in Fig. 5P.2 .

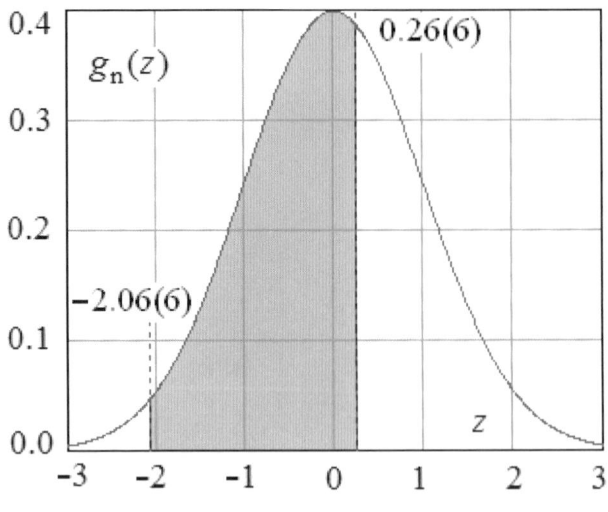

Fig. 5P.2 Normalized normal distribution, $\mu = 0$, $\sigma_z = 1$

To derive the appropriate *z-scores* we will use the formula (1.18) given at the beginning of Unit 5 in a formally modified fashion, by using the present symbols:

$$z_1 = \frac{u_1 - \mu}{\sigma_u} \quad \text{and} \quad z_2 = \frac{u_2 - \mu}{\sigma_u} \quad \text{moreover} \quad u_1 = l, \; u_2 = r$$

Therefore, by substituting numerical values the desired *z-scores* can be derived

$$z_1 = \frac{120 - 151}{15} \rightarrow z_1 = -\,2.06(6) \qquad z_2 = \frac{155 - 151}{15} \rightarrow z_2 = 0.26(6)$$

To use *MathCad, z-scores* must be given in a suitable fashion, we assumed $z_1 = -\,2.06666667$ and $z_2 = +\,0.26666667$ respectively.

This time we must also use a simplified integral formula applying normalized density $g(z)$ given by:

$$g(z) = 1/\sqrt{2\pi}\,\exp\left(-u^2/2\right),$$ nevertheless it will give us the result

$$F(z_1, z_2) = \int_{z1}^{z2} g(z)\,dz \rightarrow F(z_1, z_2) = 0.5857543038876679$$

which is practically identical with the above obtained result and leads to the same answer – indicating 293 leaves within the required limits.

In the third step we will solve this problem by using the Tables of the normal integrals. The second step above will give us the necessary references. To make use of the Tables, as we have already shown in Chapter 5, we have to determine *z-scores* - as they were obtained, i.e. $z_1 = -2.06(6)$ and $z_2 = +0.26(6)$ after small corrections. We remind the Student that our Tables use the *z-scored* ordinates rounded to two decimal places, therefore in fact they are $z_1 = -2.07$ and $z_2 = +0.27$. Then, in the next step first we have to read from the Table the area which lies between the above given $z_1 = -2.07$ with $z_2 = 0$ and ignore the negative sign and read the value 0.480774. Then, we find the area lying between $z_1 = 0$ and $z_2 = +0.27$ which is 0.106420. Their sum 0.587194 gives the answer of 294 leaves, slightly higher than the one obtained using the two more accurate approaches. In fact even this result could be corrected by using the method of linear interpolation – which, as it is easy to see, will give (0.480616 + 0.105136) = 0.585752 also giving 293 leaves. The derivation of this more accurate value is left for the Student.

The answer to the second question given in [5] indicates 5 leaves, although the answer to the first question given in [5] indicates a very rough result of 300 leaves.

Problem 5.3 (see 7.22 in [5], Modified)

The grades on a short quiz in biology were 0, 1, 2 …, 10 points, depending on the number answered correctly out of 10 questions. The mean grade was 6.7 and the standard deviation was 1.2. Assuming the grades to be normally distributed find the limiting values for the traditional grades A, B, and C [consider also grades including "+" and "-" assuming they correspond to 1/3-rd of the space dividing full marks]. Hint: assume passing grade at 50%, that grade A received the top 5%, and grade B the next 15%.

Solution

The essential part of the solution can be read from Fig. 5.P.3. Therefore, first we will follow what is shown there.

The *z-scores* corresponding to the limiting values are proposed as follows, grade C starts from $z = 0$ up to $z = 0.842$, grade B starts from, say $z = 0.843$ up to $z = 1.645$, and grade A lies above $z = 1.645$. In Problem 5.3 we are asked to determine the grade points from the scale (0, 10) corresponding to the above determined *z-score* limits. They follow the equation

$$x_i = 1.2 \cdot z_i + 6.7 \quad \text{therefore} \quad x_1 \approx 7.7 \quad \text{and} \quad x_2 \approx 8.7 \,.$$

A possible solution for the limiting values indicating fractional marks is left for the Student. Apparently in the context of the considered problem this fine partition has negligible applicability.

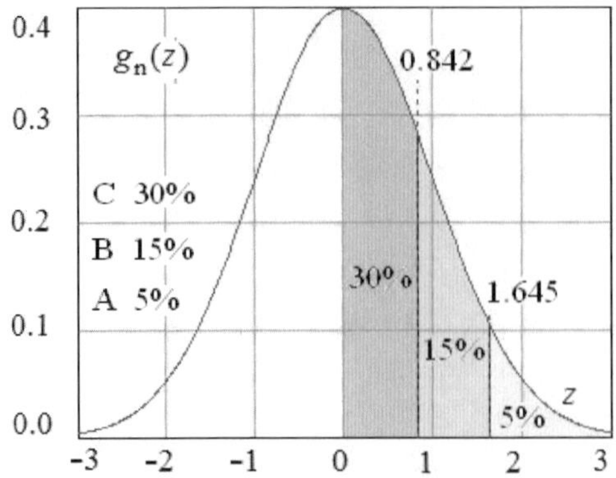

Fig. 5P.3 Grade distribution with a partition

Problem 5.4 ([1], 7.28 – No Answer)

Past experience has shown that the mean life of Whiz-Matic irons is 72 month with a standard deviation of 10 months. The product is sold with an unconditional four-year guarantee. The lives are normally distributed. (i) what fraction of the irons should be expected to be returned for failing to satisfy the guarantee? (ii) this year 100 irons were returned for not meeting the guarantee. Assuming that all four-year-old irons not still functioning were returned, how many irons were sold four years ago?

Solution

Numerical answers – supported by Fig. 5P.4
Regarding the first question, with the help of *MathCad* we can obtain a numerical result giving the fraction equal to $a = 0.008197535924596044$.

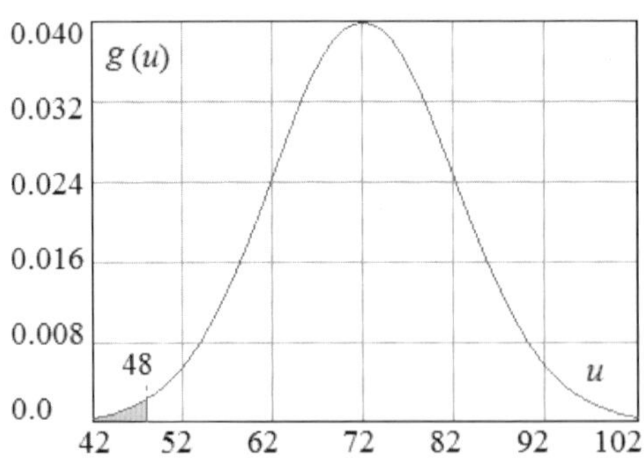

Fig. 5P.4 Distribution of irons with four-year guarantee

To answer the second question we have to perform the following calculations, first denote $b = 100 / a$ then we get a numerical result (by using the Windows 7 calculator) of $b = 12198.0970968528909490119541351 55$. This result means that about 12,200 irons were sold in the past four years.

Problem 5.5 ([1], 7.27 – Answers Enclosed)

Professor Trash has just finished grading final exams and now must determine final grades. She knows from experience that the scores will be approximately normal in distribution. The mean grade proves to be 74, and the standard deviation 9 (a) if the middle 35% are to be graded C, find the cut-off points for C [answer: 69.95 – 78.05] (b) if Laura's average is an 87, and if the top 10% received A's, can Laura expect an A? [answer: yes].

Solution

Problem 5.5 follows the above solved Problem 5.3 and we provide it to give our Student a chance to consider how teaching staff solves delicate problems of grading. We have in mind our final comments which we would like to propose for consideration. The starting points for this discussion are Fig. 3P.3 and Fig. 3P.5.

First let us explain the meaning of the *"middle 35%"*. We understand it in the way shown in Fig.5P.5. The position of both markers in the first question follows from the fact that each side of the mean value covers 17.5% of the distribution. The numerical value was not derived from the Table of normal distribution, but by using *MathCad* in the way presented twice before in the previous problems. For convenience we present the result obtained in this way as

$$\int_{74}^{78.084} h(u)\, du = 0.175005610187$$

In this way the position of both *markers* for the grade C was found to be (69.916, 78.084). To give the position of the third marker, of grade A, we get the following result:

$$\int_{74}^{85.52} h(u)\, du = 0.399727432$$

It is also justified to add that to determine the upper limits in both above integrals a method of trials an errors was used, as the simplest one for such purposes. The limits for the intermediate grade B were thus determined to give (78.084, 85.52).

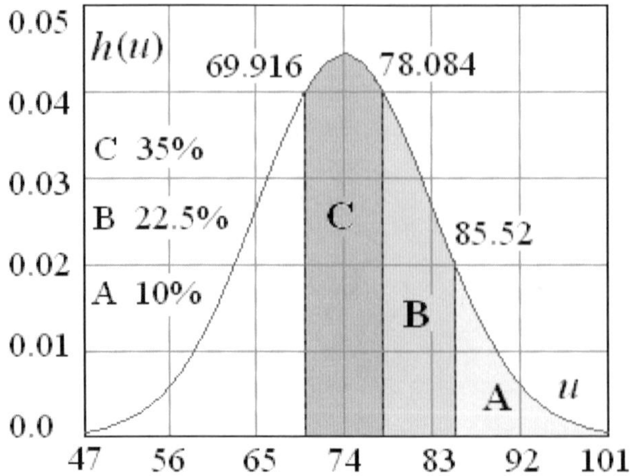

Fig. 5P.5 Final grades distribution with a partition

To obtain the percentage which covers grade B we moreover determined the value of the third integral

$$\int_{78.084}^{85.52} h(u)\, du = 0.22472182$$

To conclude these calculations, there is a remark summarizing two ways of distributing grading points, one shown in Fig. 5P.3 and the other in Fig. 5P.5. The so called *passing grade* is always a sensitive issue. In this matter the final decision is made by the instructor but sometimes important consequences follow from the policy of a particular university. Nevertheless the two examples depicted in Fig. 5P.3 and Fig. 5P.5 distinctly reflect two different attitudes in this respect which do not belong to the subject of Statistics. Returning to the second question: under the policy of professor Trash, if Laura gets 87 marks, she may get grade A but with the grading system suggested in problem 5.3, Laura should get not less than 89 marks to get grade A . Below we offer one more problem of a similar nature but not burdened with as much responsibility as academic grading philosophy.

Problem 5.6 ([1], 7.29, with a Single Answer; Modified)

Certain commodities are graded by weight and normally distributed; 20% are called standard, 50% large, 20% super, and 10% colossal. If the mean is 0.92 ounce with a standard deviation of 0.08 ounce, what are the limits of the weights of the all classes?

Solution

With the help offered by similar problems which we have already solved, this solution is given in Fig. 5P.6 together with numerical values for two classes of four shown in this figure, class C has limits (0.85267, 0.96195), and class B has limits (0.96195, 1.02252). Weinberg defines class B as (0.9616, 1.0224). The Student who wants to consider which boundaries are closer to ideal values ensuring the assumed percentage can personally try to find the answer.

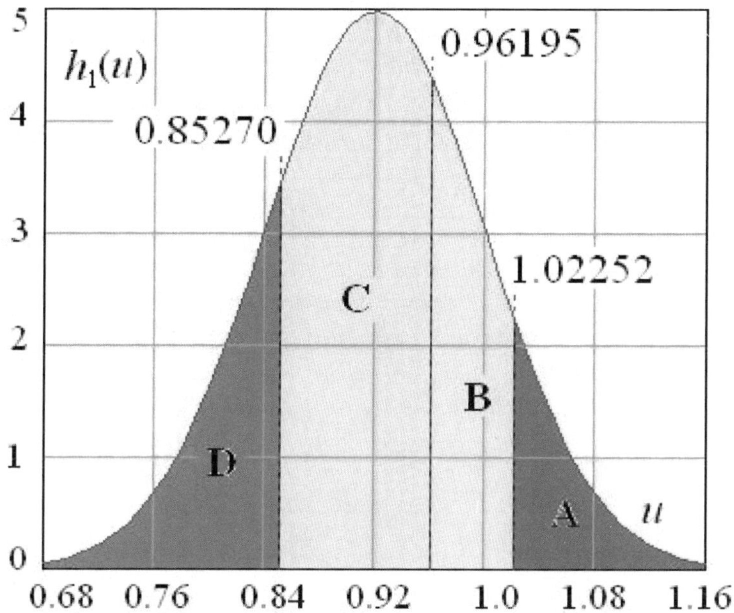

Fig. 5P.6 Weight distribution with complete partition

It should be understood that the accuracy in showing the limiting values for the particular classes given here is redundant for practical purposes. We may add that the MathCad package used to derive them, first served to determine the appropriate *z-scored* values: 0.841622, 0.524401, 1.281552 – they ensure the appropriate probabilities up to 6 decimal places. In the end we can add that 1 oz is equivalent approximately to 28.3 g.

Problem 5.7 (Following Problem 4.5)

An *unfair* coin $p = 0.4$ has been tossed $n = 11$ times. Investigate the case by drawing the diagram of both binomial and approximate normal distribution. Also provide numerical values for both these distributions.

Solution

This time we start with the numerical values given in Table 5P.1

Table 5P.1 Binomial distribution and normal approximation, $n = 11,\ p = 0.4$

Normal distribution	Binomial distr.	Normal distribution	Binomial distr.
f1(0) = 0.0062761744	f2(0) = 0.0036279706	f1(1) = 0.0274958023	f2(1) = 0.0266051174
f1(2) = 0.0824768952	f2(2) = 0.0886837248	f1(3) = 0.1693919340	f2(3) = 0.1773674496
f1(4) = 0.2382032395	f2(4) = 0.2364899328	f1(5) = 0.2293491395	f2(5) = 0.2207239373
f1(6) = 0.1511962712	f2(6) = 0.1471492915	f1(7) = 0.06824637075	f2(7) = 0 .0700710912
f1(8) = 0.02109174713	f2(8) = 0.0233570304	f1(9) = 0.0044631354	f2(9) = 0.0051904512
f1(10) = 0.0006466393	f2(10) =0.0006920602	f1(11) = 00006414734	f2(11) = 0.000041943

To see how close binomial values and their normal approximation are we suggest using Fig.5P.7.

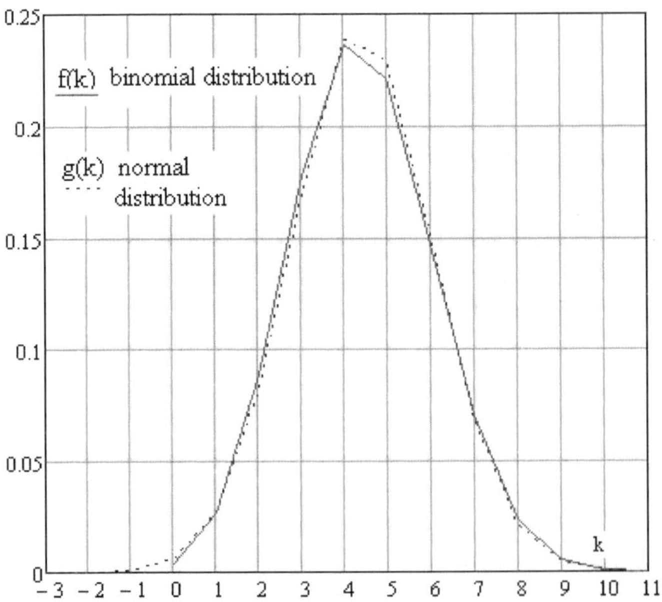

Fig. 5P.7 Goodness of approximation of the binomial and normal distributions

The Student is recommended to go back to Fig. 4P.4 and compare it with Fig. 5P.7. We suggest considering (i) to what extent normal approximation (Fig. 5P.7) is better than Poisson approximation (Fig. 4P.4) (ii) Fig. 5P.7 shows a certain insufficiency of *MathCad* drawing functions – they allow either to display diagrams for discrete or for continuous argument but not to combine the two.

Therefore, only the drawing of the binomial distribution can be accepted without reservations, whereas the drawing of the normal distribution is artificially distorted.

Problem 5.8 (Following Problem 4.7)

From Problem 4.7 we borrow the notation of the binomial distribution as $n = 12$, and $p = 0.8$ - we will also make use of Table 4P.7, however, its values must be complemented by the normal distribution approximation which follows given binomial terms.

Solution

Following the above stated we provide Table 5P.2 with numerical values of the binomial and normal distributions for the same rational arguments starting from 4, 5, up to 12 (skipping the less interesting lowest values)

Table 5P.2 Binomial distribution and normal approximation, $n = 12$, $p = 0.8$

Normal distribution	Binomial distr.	Normal distribution	Binomial distr.
f1(4) = 0.0000817563	f2(4) = 0.00051904512	f1(5) = 0.0011644367	f2(5) = 0.003321888768
f1(6) = 0.0098517997	f2(6) = 0.015502147584	f1(7) = 0.0495131102	f2(7) = 0.053150220288
f1(8) = 0.1478188424	f2(8) = 0.13287555072	f1(9) = 0.2621466691	f2(9) = 0.23622320128
f1(10) = 0.276161955	f2(10) = 0.283467841536	f1(11) = 0.1728177353	f2(11) = 0.206158430208
f1(12) = 0.0642418041	f2(12) = 0.068719476736		

Problem 5.9 (Following Problem 4.8)

Using the data from Problem 4.8, i.e. $n = 100$ and $p = 0.0162$ which determine the binomial distribution, derive the appropriate values of the normal approximation to this binomial distribution assuming arguments 0, 1, ..., 5

Solution

Table 5P.3 Binomial and normal distributions, $n = 100$, $p = 0.0162$

Normal distribution	Binomial distr.	Normal distribution	Binomial distr.
f1(0) = 0.1387164500	f2(0) = 0.195290817	f1(1) = 0.2801077566	f2(1) = 0.32158073
f1(2) = 0.3020124399	f2(2) = 0.26212196	f1(3) = 0.1738709616	f2(3) = 0.140999129
f1(4) = 0.0563448043	f2(4) = 0.056303627	f1(5) = 0.0087728173	f2(5) = 0.017801057

Comparing the similarity of the Poisson approximation and the binomial (see Table 4P.10) with the above given values approximating the same binomial by the normal distribution it is seen that the Poisson approximation fits the binomial very well, while the normal approximation does it much worse.

Problem 5.10 (Following Problem 4.10)

Solve Problem 4.10 to derive the probability of events concerning 90 Bernoulli trials, with $p = 0.65$ filling the range (66, 90) by applying the normal approximation.

Solution 1

This solution is based on the Tables of the normal distribution. We have to determine the appropriate *z-scored* range, assuming that $m = 58.5$, and $s = \sqrt{20.475}$. Considering that the right band lies well outside of the *three-sigma* limits we have to determine only the left band:

$z_1 = ((66 - 0.5) - 58.5)/\sqrt{20.475} \rightarrow z_1 \approx 1.547$ with upper rounding we get the following answer: 50 - 0.439429 = 0.060571, i.e. the result given by Weinberg [1] as 0.0606.

Solution 2

With the help of *MathCad* we have already derived the exact result obtained on the base of the binomial. For convenience we repeat it here as 0.05871866696. But now we want to obtain the result based upon the normal distribution for the limits (65.5, 90), which also using *MathCad* is 0.0609334390. It suggests that the distribution of errors in the approximations examined in this book and in this Unit is not so easy to follow: the above result may cause disappointment as there seems to be *much ado about nothing*. In fact both results differ only by 3.7%. Uncertainty of Statistical data usually exceeds 5%.

***

Further problems from this place onwards will be related mainly to the second De Moivre-Laplace theorem as it is for Problem 5.10, while the previous ones mainly illustrated the first De Moivre-Laplace theorem. We commence with an example using the idea given by Spiegel [5]. The subsequent problems were taken from Weinberg [1] generally with suitable modifications to fit the subject of Unit 5. Nevertheless all the solutions are the Author's own and were first presented in [9], although the versions cited here were duly revised, frequently modified and expanded.

Problem 5.11 (see [5], Problem 7.25)

A fair coin is tossed $n = 500$ times. Find the probability that the number of heads will not differ from 250 by more than 10 (options: more than 30).
The answer derived in [5] for the range from 239.5 to 260.5 shows probability of 0.6528.

Solution

Here as $n = 500$ *MathCad* cannot be used to determine the binomial distribution , therefore we can derive the exact solution only by calculating the necessary 11 probabilities with the Word 7 calculator. They are given in Table 5P.4. Then in the second approach we use *MathCad* to determine the normal distribution resorting to the second theorem of de Moivre-Laplace and apply the normal approximation. The third attempt uses Tables of the normal distribution (the answer given by Spiegel [5] was also obtained in this way).

Table 5P.4 Binomial results and normal approximations, $n = 500$, $p = 0.5$

k	Normal approx.	Binomial distribution
240 or 260	0.0239186832	0.023923296060921815475216087849104
241 or 259	0.025807364	0.025809365044977892213926069878701
242 or 258	0.0276233071	0.027622419614253198691763851647O4
243 or 257	0.029331437	0.029327507244762655401131990637598
244 or 256	0.0308970243	0.030890038368459026385618531122388
245 or 255	0.0322868452	0.032276938050308207162115689662577
246 or 254	0.0334703468	0.033457801637514604985119922211208
247 or 253	0.0344207602	0.034405998445055504721540324864967
248 or 252	0.0351161057	0.035099667768544526994152024963051
249 or 251	0.0355400375	0.035522555332020967078418916830075
250 or 250	0.0356824823	0.035664645553349050946732592497396

Total normal for the range (239.5 – 260.5): 0.65234551987 - *MathCad*
Total binomial: 0.651490045560032968996205628089 – Word 7 calculator

The binomial result is exact. In comparison with the estimate obtained by Spiegel who used Tables without interpolation, the difference is small and amounts to the marginal 0.2%, however, it saves a lot of effort. The third attempt commences with the *z-scored* limits of 239.5 and 260.5. Our solution uses the general formula of z-scored values and leads to the following results:

$$z_l = (239.5 - 250)/\sqrt{125} \approx -0.9391 \qquad z_r = (260.5 - 250)/\sqrt{125} \approx +0.9391$$

If we round the *z-score* to 0.94, it will allow us to read the following result from the Tables (in this book): 0.326391, if we then multiply this result by the "2", we will get the third answer of 0.652782. To close our considerations we will obtain the fourth result improved by linear interpolation. Reading from the Tables the area for 0.93 as 0.323814, and taking fraction 0.91 from the difference – will finally lead to the fourth answer of 0.65231814. This value has the same four decimal places as the value obtained with the help of *MathCad*.

Note: avoid the temptation of adding the results of the second column of Table 5P.4. However, the numerical methods to evaluate definite integrals apply the "trapezoid rule" and so the above mentioned values may be used to solve Problem 5.11 yet again.

Problem 5.12 (Weinberg [1] 10.5 p. 196)

The consulting psychologist for a large seaboard city believes that 45% of the city's gang members are White. The city crime commission decided to accept this claim if a random sample of 100 records of gang members on file with the Police department has from 40 to 50 White individuals (including this endpoints). What is probability that such a claim is wrong? Answer given in [1] : 0.2670

Solution

Let us first use *MathCad* to obtain the exact solution by resorting to the binomial distribution with the parameters $n = 100$, $p = 0.45$. They justify the mean $m =$ 45 and the standard deviation $s = \sqrt{24.75}$. Therefore, looking for an event corresponding to the appearance of the range of *successes* (40, 50) we will find that its probability expressed by the sum

$$b(k;n,p) = \frac{n!}{k!\,(n-k)!}\, p^k\, q^{n-k} \;; \qquad R = \sum_{40}^{50} b(k)$$

$R = 0.731169745575726$ was determined using *MathCad.* The answer to the question stated in the Problem indicates complement event probability ($1 - R$) whose numerical value can be rounded to 0.268830 . This result deserves to be called *the exact answer* as obtained by using the binomial distribution. The solution is supported by Fig. 5P.8 . It is seen that the distribution is almost symmetrical.

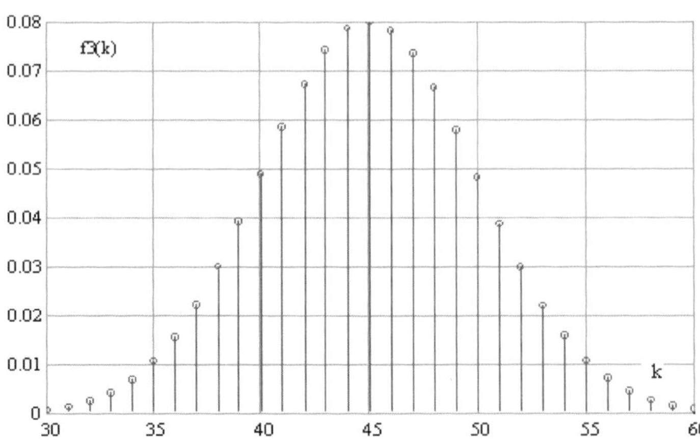

Fig. 5P.8 Central part of the binomial distribution $n = 100$, $p = 0.45$

It is also possible to derive a similar result by using a calculator available on the market, but in order to do it, it is necessary to calculate 11 probabilities and then summarize them. We advise such a solution to the Student as good exercise in using calculators. Continuing the examination of the problem we will use the normal approximation again by resorting to *MathCad* . The numerical results obtained in this way were paired with the appropriate result following the binomial distribution in order to see how close these values are. The Student's attention is drawn to the method of pairing seen in Table 5P.5.

Table 5P.5 Binomial distribution and normal approximation, $n = 100$, $p = 0.45$

Normal approximation	Binomial distribution	Normal approximation	Binomial distribution
---------------	f2(45) = 0.07998750025	f1(45) =0.0801904156	----------------
f1(44) = 0.0785866613	f2(44) = 0.078559152034	f1(46) = 0.0785866613	f2(46) = 0.0782486415
f1(43) = 0.073965289	f2(43) = 0.074118186325	f1(47) = 0.073965289	f2(47) = 0.073556750086
f1(42) = 0.066858993	f2(42) = 0.0671607321	f1(48) = 0.066858993	f2(48) = 0.066451836725
f1(41) = 0.0580422781	f2(41) = 0.0584336313	f1(49) = 0.0580422781	f2(49) = 0.057698440793
f1(40) = 0.0483299187	f2(40) = 0.04880290316	f1(50) = 0.0483929187	f2(50) = 0.04815197150

Returning now to answer the considered Problem, the *MathCad* result obtained for the range (39.5 – 50.5) indicates probability of 0.7310750194883805; the exact result obtained above gave 0.731169745575726.
Probabilities of the complementary events (under consideration) show that the exact result of 0.268830 and gives 0.268925 for the normal approximation. To finish our considerations, we have to resort to the Table data looking for the solution which is the easiest to get for the considered problem. Keeping in mind the symmetry of the limiting values with respect to the mean (value), it is enough to calculate a single *z-scored* limit:

$$z = (50.5 - 45) / \sigma = 5.5 / \sqrt{24.75} \cong 1.10554 \cong 1.11$$

The area read from Tables for this ordinate is 0.366500 - multiplying this value by two and resorting to the complementary event we finally get 0.2670, i.e. exactly the same result as given by Weinberg. However, it is possible to obtain in this way a significantly better result by resorting to the linear approximation. To do that, we have to read the area corresponding to *z-scored* value of 1.10 , in our Table we will see 0.364334, then the difference of 0.002166 should be multiplied by the value 0.554 - getting a finer approximation of the area as 0.365534. Complementary event gives the result of (1-2*0.365534) = 0.268932 - which is very close to the *MathCad* solution obtained for normal approximation – i.e. 0.268925. Comments are left for the Student.

Problem 5.13 (Weinberg [1], 10.7, p.196)

An advertising agency promises that 65% of the market can be captured if a certain campaign is adopted. After six months 900 people are asked if they use the product. The decision rule is to accept the claim if the outcome is within 20 units of the mean (inclusive). Suppose, in fact, the campaign is adopted, but only 60%

of the marked is captured. Find the probability of accepting the claim of the advertising company. Answers: 0.0475, 0.9525 given in [1] determine probability of a type-two-error, and the power of the test – i.e. the concepts lying outside of the scope discussed in this book.

Solution

Coming to this Problem we will first have to acknowledge that the binomial solution with $n = 900$ lies outside *MathCad* potential. Then let us note that the problem defines the binomial determined by $n = 900$, $p = 0.6$. Thus, the Student has to notice that the conditions which lead to accepting the claim of the advertising company refer to the 65% condition, which gives the value of 900*0.65 = 585 – basic to determine the so called *critical range* denoting successes of the advertising campaign as (565, 605). To clarify: we are asked for the probability of the appearance of the above given *critical range* but by applying the binomial distribution $n = 900$, $p = 0.6$. Now let us turn our attention to mathematical tools. In fact, the Word 7 calculator allows to determine the values of the probabilities under consideration – but it has to be done 40 times. Difficult, though achievable. To document our suggestion we provide a single value obtained in this way, showing the maximum value of this binomial as $b\,(\,k = 540\,) = 0.0271366261031992192934293975 76225$ and for the sake of instant comparison we provide here its normal approximation obtained by *MathCad* which equals 0.02714458399460666. As we see only the first three decimal places are identical and the next two are not so different (37 and 45). Although we consider the normal approximation as the only solution to the Problem with no further justification it should be obvious for the Student that the Poisson approximation cannot be expected to apply. It is enough to see that the mean 540, and the variance of about 216 are so different that they exclude such a possibility. Returning to the main subject, we have to determine the value of the following integral

$$g(u) = \frac{1}{\sigma_u \sqrt{2\pi}} \exp\left(-\left(u-\mu_u\right)^2 / 2\sigma_u^2\right) \qquad F(l,r) = \int_l^r g(u)\, du$$

Limiting values of the above given integral are as follows $l = 564.5$, $r = 605.5$ while the mean and the variance $\mu_u = 540$, $\sigma_u^2 = 216$ - then the answer obtained in this way gives $F\,(\,564.5, 605.5) = 0.04775166384427319$.

The above result has to be checked again using the Table data. It means that we have to determine the appropriate *z-scored* limiting values, due to the distant upper limit (outside 4 *sigma*) we consider only the lower one:

$$z_1 = (564.5 - 540)/\sigma \cong 1.66701 \cong 1.67$$

After rounding the ordinate to 1.67 the area read from the Tables shows 0.452540), therefore the answer is (0.5 – 0.452540) = 0.04746. Linear approximation can improve it giving 0.0477594. The last result coincides with the result of 0.0477517obtained using *MathCad*. To complement the solution we provide also Fig. 5P.9. In the end it will be noted that the first answer 0.0475 given by Weinberg (see above) is almost the same as our result. This result ensures that in testing hypothesis procedure under consideration, type one error expressing the probability of the *critical range* was correctly determined.

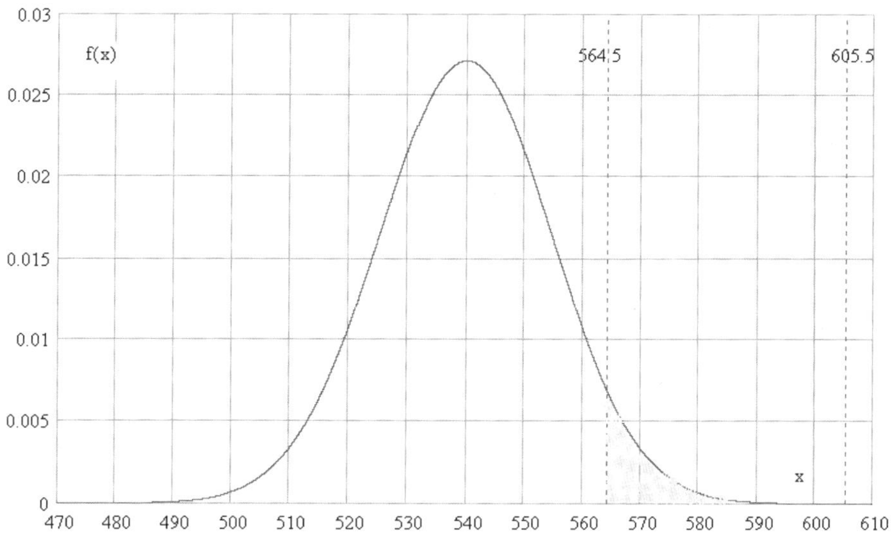

Fig. 5P.9 Normal distribution to approximate binomial $n = 900$, $p = 0.6$

Problem 5.14 (Weinberg [1], 10.9, s.196)

The developers of an achievement test for children between the ages of 9 and 11 believe they have advised a new test whose scores will be distributed with a mean of 65 and a standard deviation of 15. If the test results of 150 students selected at random have a mean that deviates less than 1.5 standard deviation from its mean, the hypothesis will be accepted and published in promotional literature. What interval of outcomes (*critical* region) would allow for acceptance of the hypothesis? Find the probability of rejecting true hypothesis.
Answers: (a) $(63.16 \div 66.83)$; (b) 0.1336.

Solution

If Problem is to be unique and less ambiguous, it has to be complemented by the following assumptions. Firstly, the mother distribution described in the problem is normal with the mean 65 and standard deviation 15. Secondly, the sample of 150 students is drawn from the mother distribution. With these additional assumptions the solution is presented below.

Let us consider what interval of outcomes (*critical region*) is determined in the Problem. We know that it is symmetrically distributed with respect to the mean

65 and the range of the interval is 50% greater than the span of a single standard deviation *of the sample*. The standard deviation of the sample follows from the theorem known from Chapter 5. The sample volume is N = 150, so, we recall Theorem 5.4 (from Chapter 5) which states:

$$\sigma_{\bar{X}} = \sigma / \sqrt{N}$$

In this formula $\sigma_{\bar{X}}$ denotes the sample standard deviation and σ the standard deviation of the paternal population. Therefore the value of the standard deviation of the mean results from $15 / \sqrt{150}$. It allows to determine both limiting values, lower $l = 65 - 15 / \sqrt{150}$, and upper $r = 65 + 15 / \sqrt{150}$.

Suitable numerical values rounded to three decimal places are l = 63.163 and r = 66.837. With these limits we can move on to the normal distribution and calculate the following integral by using *MathCad* :

$$F(l,r) = \int_l^r g(u)\, du\,.$$

It will lead to the numerical value 0.8663855974622837. Therefore, it gives the final numerical result as (1 – 0.8663855974622837) = 0.1336144025377163. Fig. 5P.10 explains this procedure in a visual manner.

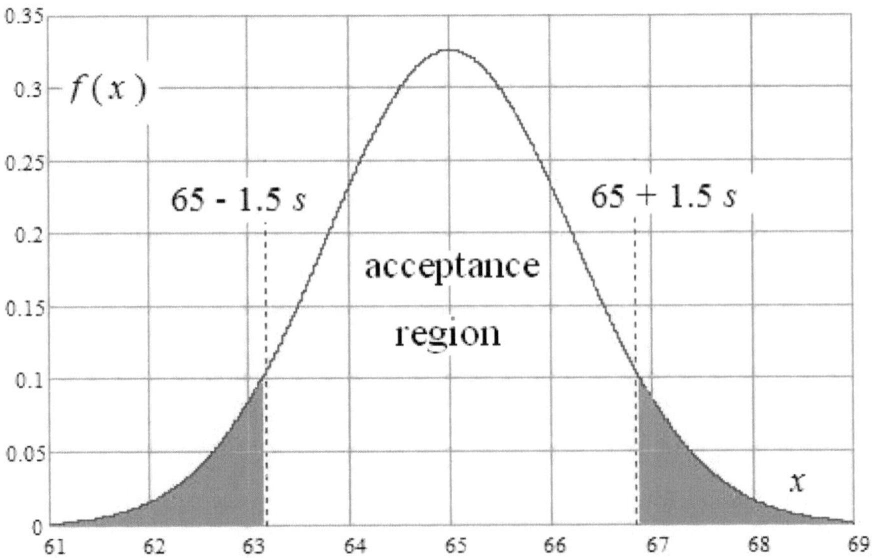

Fig. 5P.10 Normal distribution, $m = 65$, $s \approx 1.2247$, acceptance/rejection regions

In the last step of the proposed procedure we present the solution based on the Table data. From the point of view of the final goal we may use as a reference Fig. 5P.10 and determine a single *z-scored* value of the ordinate:

$$z_1 = \frac{(65 + 1.5 \cdot 15 / \sqrt{150}) - 65}{15 / \sqrt{150}} = 1.5$$

Note: this is the exact value! For this ordinate the appropriate area found from the Tables will be 0.433193 , it means that the left tail of the distribution contains (0.5 - 0.433193) = 0.066807. Multiplying this fraction by two the final result of 0.133614 will be obtained. It is interesting to note that with respect to the result obtained from *MathCad* – all six decimal places *are the same*. This successive example shows growing confidence in the method based on the Table data.

Problem 5.15 (Weinberg [1], 10.13, p.197)

The placement office of Hallowed Hall College contends that each year it places a mean of 175 graduating seniors in social work positions ($\sigma = 30$). The sophomore statistics class has been assigned to test this contention by contacting all social work graduates for the past five years. The decision rule to be used is as follows: Accept the hypothesis if the sample mean deviates less than 20 units from the mean.

a. What assumption(s) must the students make?
b. What is the probability of rejecting the contend if the hypothesis is true? What risk is involved?
c. If the true mean and standard deviation are 150 and 35 respectively, what is the probability that has been accepted the wrong hypothesis?
   Answers: (b) 0.1362, (c) 0.3724.

Solution(s)

In fact this problem can be considered similar to the previous one. Therefore, our Student is advised to try solving it by him/herself and only then to compare his/her own results with the results presented below.

Let us begin by listing the assumptions requested in on (a). Firstly, it is implicitly assumed that the parent population describing one year distribution of the employed is normal (175, 30). Secondly, the sample of $N = 5$ (years) is entirely drawn from this parent population. Therefore, the sample mean distribution is normal (175, $30 / \sqrt{5}$) . With these in mind we may go further.

Assuming that the contend is right, the *critical region* is given as (155, 195) - and with respect to question (b) about the probability of rejecting the true hypothesis – it is the probability of the tails of the normal distribution determined by (175, $30 / \sqrt{5}$) limited by the values 155 (the left tail) and 195 (the right tail). In this respect Fig. 5P.11 will be helpful.

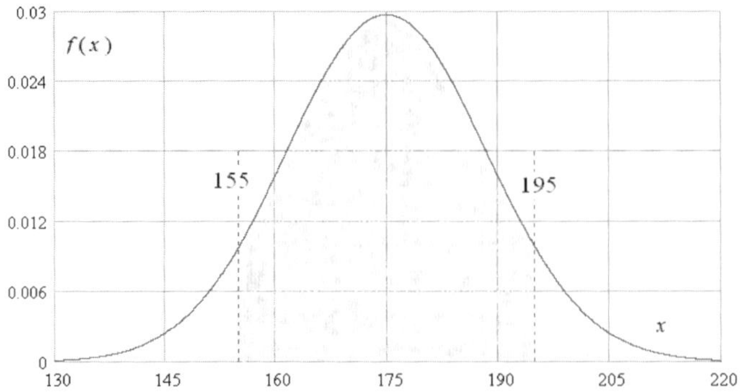

Fig. 5P.11 Normal distribution, $m = 175$, $s \approx 13.4164$, critical region α

Skipping the same details as in Problem 5.14 we provide the final answer obtained using *MathCad* as 0.13603712811414348 . Then in a subsequent step we also derive a solution using the Table data. So, we commence from determining the right limit *z-scored* value:

$$z_2 = \frac{195 - 175}{30 / \sqrt{5}} \approx 1.4907 \quad \rightarrow \quad z_2 \approx 1.49$$

The suitable area read from our tables of the normal distribution is 0.431888 (the area from $z_1 = 0$ to $z_2 \approx 1.49$), therefore the probability of the right tail is equal to (0.5 - 0.431888) = 0.068112. This value has to be multiplied by two to get the answer, i.e. 0.136224. Note, that the *refining* procedure in this case can be ignored as redundant.

To answer (c), we have to examine it in the following procedure. Assuming that the contend is wrong, the *critical region* will be determined by (130, 170) - and with respect to the value of the probability of accepting the false hypothesis – it is the probability of the occurrence of events from the complement set to the *critical region* (130, 170), still based on the normal distribution but now determined by the mean 175 and the standard deviation given by $35 / \sqrt{5}$ (following a suggestion in sub point c).

In solving this part of Problem 5.15, we have to consult Fig. 5P.12 and Fig. 5P.11. Probability of the *critical region* (130, 170) derived by *MathCad* is 0.37267660013929 . This probability determines the situation of accepting the wrong contend. Following the terminology used in the theory of testing statistical hypothesis (which is here used implicitly) such a situations corresponds to so called *type-two-error* , sometimes also called $\beta - error$, whereas the first problem examined above is called *type-one-error* – and corresponds to the rejection of the true hypothesis – also called $\alpha - error$.

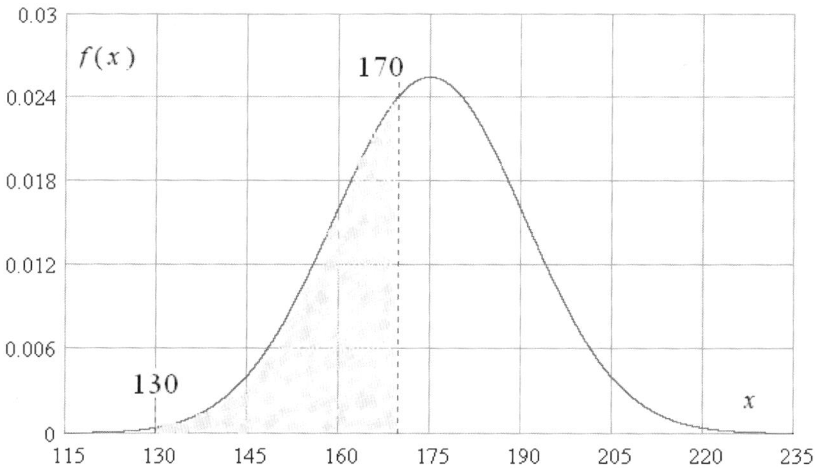

Fig. 5P.12 Normal distribution, $m = 175$, $s \approx 15.6525$, critical region β

As always, the last step requires resorting to the Table data. This time we have to deal with two *z-scored* values of two necessary ordinates, as is clearly suggested by the provided Fig. 5P.12, i.e. 130, and 170:

$$z_2 = \frac{130-175}{35/\sqrt{5}} \approx -2.874944543 \rightarrow z_2 \approx -2.87 \quad \text{the area is } 0.497948$$

$$z_1 = \frac{170-175}{35/\sqrt{5}} \approx -0.319438282 \rightarrow z_2 \approx -0.32 \quad \text{the area is } 0.125516$$

The critical region area covers (0.497948 - 0.125516) = 0.372432 .

Moreover this time the result can be *refined* by using the linear interpolation. For the first result it will give us 0.4979796, and for the second result we will get 0.1253028 - all in all, we will get the result 0.37267687. And this refinement is significant: the first six decimal places coincides with the solution obtained using *MathCad.*

Problem 5.16 (Weinberg [1], 10.15, p.197)

The manager of the produce department of a large supermarket suspects that the mean diameter of the "2-inch apples" is really 1 ¾ inches with a standard deviation of a half inch. Into what interval would the mean of 49 randomly selected apples have to fall to support the manager's belief, if the probability of a type-one-error is 0.05 and the acceptance interval is centered at the mean? Answer: (1.61 – 1.89) in.

Solution(s)

The main body of the solution can be outlined easily. Firstly, we must determine the probability distribution under the consideration. We have to determine a suspicious parent normal distribution suggested by the manager with the mean of 1.75 inches, and standard deviation of 0.5 inch. But in the testing procedure we will deal with the sample mean distribution which inherits from the parent population the mean of 1.75 inches and whose standard deviation will reflect the sample volume $N = 49$, that is will be $0.5 / 7$.

To move further, the helpful numerical details are to be found in Fig. 5P.13.

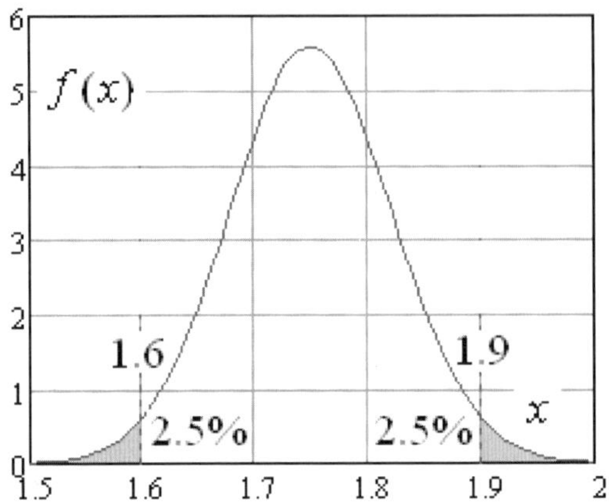

Fig. 5P.13 Normal distribution, $m = 1.75$, $s \approx 0.071$, critical region β

Problem 5.16 asks for the position of the symmetrical markers for the critical region β, in other words we have to resort to Table data but look for such ordinate which corresponds to the area given by the following result: (0.5 - 0.025), i.e. for 0.475. In a lucky case the area 0.475002 is determined for the ordinate + 1.96. This *z-scored* ordinate has to be transformed into the value in inches:

$$x_2 = 1.75 + 1.96 \cdot (0.5/7) = 1.89 \qquad x_1 = 1.75 - 1.96 \cdot (0.5/7) = 1.61$$

It is, therefore seen that the above determined values exactly follow the answer given by Weinberg. In reverse succession we may now recall a more accurate solution obtained using *MathCad* . In the beginning it is reasonable to answer the following question of where we expect an improvement on the obtained solution. It is obvious that this improvement may regard a more accurate value of the ordinate which corresponds to the area of 0.475 with higher precision than the above obtained value of 1.96. But it can also be predicted in the beginning that such a search will apply a method called "*trials and errors*" and from the practical point of view it may be described as an *overestimation* if not an *abuse* of the statistical tools. Therefore, we may leave it for the inquiring Student entirely as a sophisticated exercise.

Problem 5.17 (Weinberg [1] 10.23, p.198)

The president of Grey Goose Airlines maintains that, because of delays on the ground, the mean time that must be made up in the air is 12.5 minutes, (with standard deviation $s = 3$ minutes). A regulatory agency is willing to accept this contention if the mean of the times made up in the air of 64 randomly selected flights is between 12 and 13 minutes. What are the chances of a type-one-error being made?
Answer [1]: 0.1868

Solution

Let us first define the paternal population as the normal with the mean 12.5 minutes and standard deviation 3 minutes. In the second step let us define the sample mean distribution for the sample volume $N = 64$, with the same mean and standard deviation equal to $3 / 8$ minutes. This distribution obtained using *MathCad* is shown in Fig. 5P. 14.

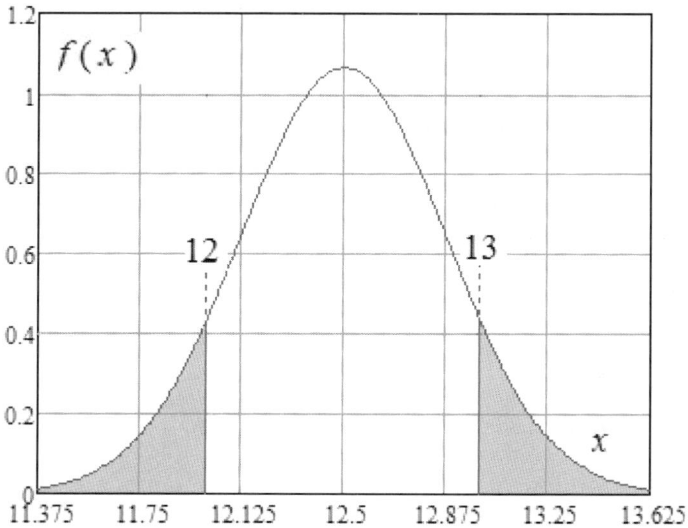

Fig. 5P.14 Normal distribution, $m = 12.5$, $s \approx 0.375$, critical regions α

The probability of rejecting the contention of the president of Grey Goose Airlines corresponds to the probability of the shadowed region in Fig. 5P.14. An apparently very accurate procedure to determine this probability is as follows:

$$2 \cdot \left(0.5 - \int_{12.5}^{13.0} \exp\left(-x^2/2\right) dx \right) / \sqrt{2\pi} \quad \rightarrow \quad 0.18242243945173564$$

It is always justified to determine this answer using the Table data. In order to do that we have to resort to *z-scored* values of both limits. The left limit corresponds to $z_1 = 0$, the right limit has to be calculated as:

$$z_2 = 8 \cdot (13 - 12.5)/3 = 1.3(3) \quad \rightarrow \quad z_2 \approx 1.33$$

The area corresponding to this ordinate is 0.408241 (and for the ordinate 1.34 the area is 0.409877), therefore this first approach gives us 0.183518. Let us see now which value gives a *refinement*.

It is easy to check that the linear interpolation leads to the area of 0.4087863(3), which finally gives the probability of 0.1824273(3), a result identical with the accurate result with respect to the first five decimal places. And also it is seen that this time Weinberg offers a less accurate answer.

***

Problems which were presented here starting from Problem 5.14 examine the normal sample means distribution. It is well known from teaching practice, that this distribution is frequently confused with the paternal population of normal or unknown distribution. Therefore, we decided to provide at the end of this Unit two more problems giving the Student a chance to practice this point. This passage will also be complemented by illustrative problems proposed by Weinberg but this time without answers.

Problem 5.18 (Weinberg [1], 8.8, p.157, No Answers)

The bus trip from Bathville to John City takes a mean of 85 minutes with a standard deviation of 15 minutes and is normally distributed. Bunny Hop Transit schedules eight buses daily between the two points. (i) what percent of the daily mean times should exceed an hour and half? (ii) one-fourth of the daily means are less than what time?

Solution

Paternal normal distribution is defined by averages (85, 15), while the sample means normal distribution drawn from this paternal distribution is determined by the averages (85, $15/\sqrt{8}$). This distribution is examined further in order to find the answers to two questions stated in the Problem. To help answering them we provide Fig. 5P.15 which shows the sample means distribution under consideration.

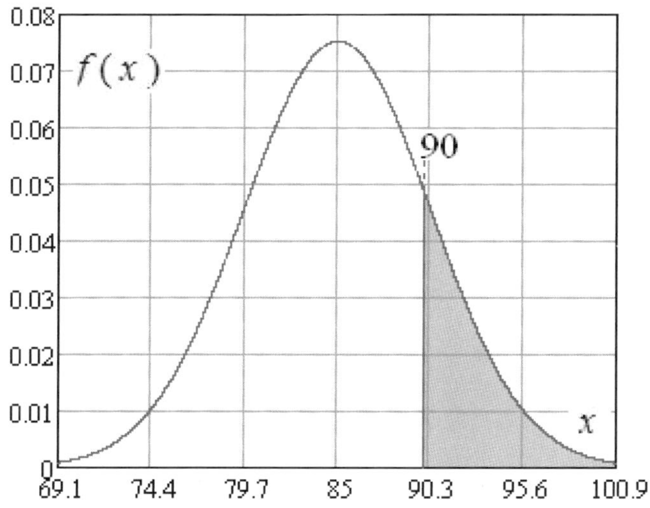

Fig. 5P.15 Sample means distribution, $m = 85$, $s \approx 5.3$

The first question is about the probability of the shadowed region in Fig.5P.15. To derive a numerical answer we first use *MathCad* getting

$$\left(0.5 - \int_{85}^{90} \exp\left(-x^2/2\right)\, dx\right) / \sqrt{2\pi} \quad \rightarrow \quad 0.172889293075801$$

Then we need to justify this answer by using the Table data. In order to do that, we have to resort to *z-scored* values of both limits. The left limit is simply $z_1 = 0$, the right limit has to be determined

$$z_2 = \sqrt{8}\cdot(90-85)/15 \approx 0.942809041 \quad \rightarrow \quad z_2 \approx 0.94$$

The area corresponding to this ordinate is 0.326391 (and for the ordinate 0.95 the area is 0.328944), therefore this first approach gives us 0.173609. Let us see now which value gives a *refinement*. It is easy to check that the linear interpolation leads to the shadowed area of 0.172891851. A result almost identical with the accurate result with respect to the first five decimal places.

Now let us have a look at the second question. Imagine that the shadowed area is exactly 0.25, and the question is which marker assesses this value. From the first answer it is clear that we are asked for a delay shorter than 5 minutes, but which exactly? We have again, two approaches: using *MathCad*, and using the Table data. Let us begin this time from the solution based upon the Table data.

There are usually two ordinates which give the area corresponding roughly to the desired one. Ordinate 0.67 gives an area of 0.248571, while the ordinate 0.68, an area of 0.251748. And now we have to derive the corresponding number of minutes to these two *z-scored* ordinates.

$$x_l = 85 + 0.67 \cdot \left(15/\sqrt{8}\right) = 88.5532116 \qquad x_h = 85 + 0.68 \cdot \left(15/\sqrt{8}\right) = 88.6062446$$

These two answers show the upper and the lower time limits. But using the linear interpolation we can derive the *z-scored* value quite closely related to the area 0.25 - the Student may check that it will be 0.67449795 which corresponds to $x = 85 + 0.67449795 \cdot \left(15/\sqrt{8}\right) = 88.5770656 \cdot$

Now let us resort to *MathCad* having in mind that the above approximation will be useful in determining the *MathCad* approximation by the method of trials and errors. Below we provide two of our results obtained in this way

$$88.57702 \to 0.250000124 \qquad 88.577021 \to 0.250000064207$$

It is obvious that our efforts in looking for the results which from purely mathematical point of view are nearer of an ideal answer satisfy only purely theoretical discoveries. The next problem is somewhat similar and somewhat different.

Problem 5.19 (Weinberg [1], 8.10, p.157 – No Answers)

Ruth's Poodle Palace schedules eight poodles for grooming each day. The times required to groom the dogs are normally distributed with a mean of 55 minutes and a standard deviation of 10 minutes. What percent of the time does Ruth work more than a 7-hour day (420 minutes) ? [Hint: this means the average μ per dog must be more than 420/8 minutes].

Solution

The paternal normal distribution has parameters $(55, 10)$, therefore with the sample volume of 8, the sample means normal distribution will be determined by $(55, 10 / \sqrt{8})$. To give the answer to the last question stated, it must be given an exact quantitative character. Guessing the intentionality (suggested by the hint) we should look for such a marker drawn in the sample means distribution which corresponds to 60 minutes. In conclusion: if there are seven such cases where each requires 60 minutes – it will fill a seven-hour working day. But according to the problem, there may be still be an eight dog waiting to be groomed. After solving this crucial point, the rest is rather trivial and evident from Fig. 5P.16. The Student may check the numerical result resorting to the Table data. With the refinement we got 0.078663 which is very close to the *MathCad* solution.

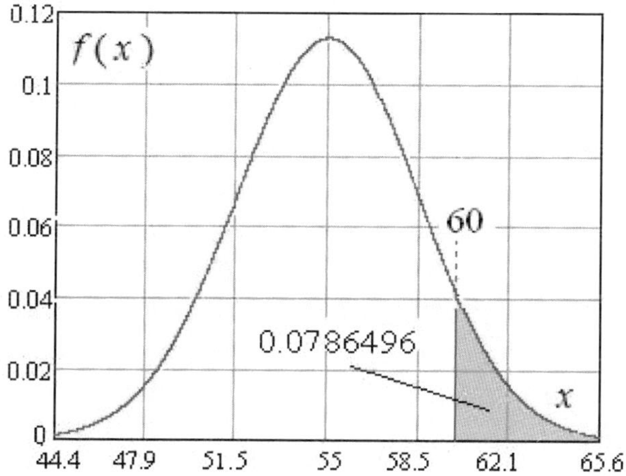

Fig. 5P.16 Sample means distribution, $m = 55$, $s \approx 3.54$

Problem 5.20

Assuming that the normal distribution is determined by its mean m and the standard deviation s, determine parameters of the sister binomial distribution (with the same mean and the same variance).

Solution

Following the symbols most often used in this Unit the two equations below state the problem:

$$m = n \cdot p$$

$$s^2 = n \cdot p \cdot (1 - p)$$

It means that we have to determine two unknown parameters of the binomial distribution i.e. n and p. In order to do that, we substitute $n = m/p$ in the second equation and obtain:

$s^2 = \dfrac{m}{p} \cdot p \cdot (1 - p)$ by ordering this result we shall arrive at the formula

$$m \cdot p = m - s^2$$

which indicates that the solution of the problem will correspond only to the case when $m > s^2$ and then – the final solution is given by what follows:

$$p = \frac{m - s^2}{m} \qquad n = \frac{m^2}{m - s^2} \quad .$$

Note: crucial condition $m > s^2$ selects such normal distributions which possess a property essential for the binomial distributions: their variance is always smaller than their mean.

References

[1] Weinberg, G.H., Schumaker, J.A., Oltman, D.: STATISTICS – An Intuitive Approach, 4th edn. Brooks/Cole Publishing Company, Monterey (1981)

[2] Neyman, J.: First Course in Probability and Statistics. HR&W, New York (1950) (Polish translation Zasady Rachunku Prawdopodobieństwa i Statystyki Matematycznej, PWN, Warszawa – 1969; Russian translation – Nauka, Moskwa - 1968)

[3] Fisz, M.: Rachunek Prawdopodobieństwa i Statystyka Matematyczna (in Polish: Probability and Mathematical Statistics). PWN, Warszawa – Wyd.3 (1967)

[4] Zubrzycki, S.: Wykłady z Rachunku Prawdopodobieństwa i Statystyki Matematycznej (in Polish: Letures on Probability and Mathematical Statistics). PWN, Warszawa (1966)

[5] Spiegel, M.R.: Schaum's Outline of Theory and Problems of Statistics, pp. 1–359. McGraw-Hill, New York (1972), 870 solved problems

[6] Czechowski, T.: Elementarny Wykład Rachunku Prawdopodobieństwa (in Polish: Introductory Course In Probability). PWN, Warszawa - Edition 1 – 1958, Edition 2 – 1968

[7] Stigler, S.M.: The History of Statistics – the Measurement of Uncertainty before 1900. The Balknap Press of Harvard UP, Cambridge (1986)

[8] de Moivre, A.: The Doctrine of Chances or A Method of Calculating the Probabilities of Events in Play, 3rd edn., Fuller, Clearer, and more Correct than the Former, London, A. Millar, pp. 1–378 (1756); Digitized by Google, Internet

[9] Laudański, L.M.: Statystyka nie tylko dla licencjatów (in Polish: Statistics not only for the undergraduates), part 2, 2nd edn. Publishing House of the Rzeszow TU (2009)

[10] Laudański, L.M., Dłubis, E.: Zbieżność i rozbieżność rozkładu dwumianowego względem rozkładu Gaussa w praktyce statystycznej (in Polish: The convergence and divergence between the binomial and Gaussian distribution as used in statistical practice). In: Proceedings of the IV-th International Scientific Conference in Jarosław: Development of the Rzeszow Province Due to European Integration. PWSZ Jarosław, pp. 263–272 (2000)

[11] Student: The probable error of a mean. Biometrica 6, 1–25 (March 1908), accessible via Internet

[12] Lloyd, E.: Handbook of Applicable Mathematics. Probability, vol. II, pp. 1–450. J. Wiley & Sons, New York (1980)

[13] Feller, W.: An Introduction to Probability Theory and Its Applications, 3rd edn., vol. I, Posthumous Edition, p. 509. John Wiley and Sons, New York (1971)

[14] Laudański, L.M.: Pre-stochastic Models of Computational Probability. Transactions of the Aviation Institute 159, 4/99, 30–36 (1999)

Error Function

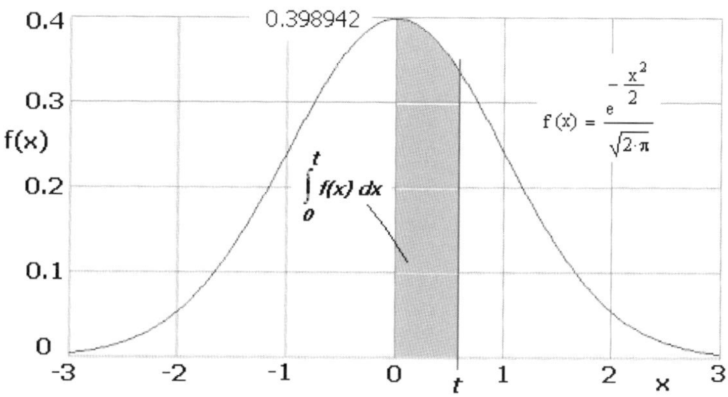

t	0.00	0.01	0.02	0.03	0.04	0.05	0.06	0.07	0.08	0.09
0.0	0.0000	0.3989	0.7978	1.1966	1.5953	1.9939	2.3922	2.7903	3.1881	3.5856
0.1	3.9828	4.3795	4.7758	5.1717	5.5670	5.9618	6.3559	6.7495	7.1424	7.5345
0.2	7.9260	8.3166	8.7064	9.0954	9.4835	9.8706	10.2568	10.6420	11.0261	11.4092
0.3	11.7911	12.1720	12.5516	12.9300	13.3072	13.6831	14.0576	14.4309	14.8027	15.1732
0.4	15.5422	15.9097	16.2757	16.6402	17.0031	17.3645	17.7242	18.0822	18.4386	18.7933
0.5	19.1462	19.4974	19.8468	20.1944	20.5401	20.8840	21.2260	21.5661	21.9043	22.2405
0.6	22.5747	22.9069	23.2371	23.5653	23.8914	24.2154	24.5373	24.8571	25.1748	25.4903
0.7	25.8036	26.1148	26.4238	26.7305	27.0350	27.3373	27.6373	27.9350	28.2305	28.5236
0.8	28.8145	29.1030	29.3892	29.6731	29.9546	30.2337	30.5105	30.7850	31.0570	31.3267
0.9	31.5940	31.8589	32.1214	32.3814	32.6391	32.8944	33.1472	33.3977	33.6457	33.8913
1.0	34.1345	34.3752	34.6136	34.8495	35.0830	35.3141	35.5428	35.7690	35.9929	36.2143
1.1	36.4334	36.6500	36.8643	37.0762	37.2857	37.4928	37.6976	37.9000	38.1000	38.2977
1.2	38.4930	38.6861	38.8768	39.0651	39.2512	39.4350	39.6165	39.7958	39.9727	40.1475
1.3	40.3200	40.4902	40.6682	40.8241	40.9877	41.1492	41.3085	41.4657	41.6207	41.7736
1.4	41.9243	42.0730	42.2196	42.3641	42.5066	42.6471	42.7855	42.9219	43.0563	43.1888
1.5	43.3193	43.4478	43.5745	43.6992	43.8220	43.9429	44.0620	44.1792	44.2947	44.4083
1.6	44.5201	44.6301	44.7384	44.8449	44.9497	45.0529	45.1543	45.2540	45.3521	45.4486
1.7	45.5435	45.6367	45.7284	45.8185	45.9070	45.9941	46.0796	46.1636	46.2462	46.3273
1.8	46.4070	46.4852	46.5621	46.6375	46.7116	46.7843	46.8557	46.9258	46.9946	47.0621
1.9	47.1283	47.1933	47.2571	47.3197	47.3810	47.4412	47.5002	47.5581	47.6148	47.6705
2.0	47.7250	47.7784	47.8308	47.8822	47.9325	47.9818	48.0301	48.0774	48.1237	48.1691
2.1	48.2136	48.2571	48.2997	48.3414	48.3823	48.4222	48.4614	48.4997	48.5371	48.5738
2.2	48.6097	48.6447	48.6791	48.7126	48.7455	48.7776	48.8089	48.8396	48.8696	48.8989
2.3	48.9276	48.9556	48.9830	49.0097	49.0358	49.0613	49.0863	49.1106	49.1344	49.1576
2.4	49.1802	49.2024	49.2240	49.2451	49.2656	49.2857	49.3053	49.3244	49.3431	49.3613
2.5	49.3790	49.3963	49.4132	49.4297	49.4457	49.4614	49.4766	49.4915	49.5060	49.5201
2.6	49.5339	49.5473	49.5604	49.5731	49.5855	49.5975	49.6093	49.6207	49.6319	49.6427
2.7	49.6533	49.6636	49.6736	49.6833	49.6928	49.7020	49.7110	49.7197	49.7282	49.7365
2.8	49.7445	49.7523	49.7599	49.7673	49.7744	49.7814	49.7882	49.7948	49.8012	49.8074
2.9	49.8134	49.8193	49.8250	49.8305	49.8359	49.8411	49.8462	49.8511	49.8559	49.8605
3.0	49.8650	49.8694	49.8736	49.8777	49.8817	49.8856	49.8893	49.8930	49.8965	49.8999
3.1	49.9032	49.9065	49.9096	49.9126	49.9155	49.9184	49.9211	49.9238	49.9264	49.9289
3.2	49.9313	49.9336	49.9359	49.9381	49.9402	49.9423	49.9443	49.9462	49.9481	49.9499
3.3	49.9517	49.9534	49.9550	49.9566	49.9581	49.9596	49.9610	49.9624	49.9638	49.9651
3.4	49.9663	49.9675	49.9687	49.9698	49.9709	49.9720	49.9730	49.9740	49.9749	49.9758

3.5	49.976737	4.0	49.996834	4.5	49.999664	5.0	49.999971	5.5	49.999998

References

[1] Arbuthnott, J.: An Argument for Divine Providence, Taken from the Constant Regularity Observ'd in the Births of Both Sexes. Phil. Transactions (1683-1775) 27, 186–190 (1710-1712)

[2] Bayes, T.: An Essay towards solving a Problem in the Doctrine of Chance. Philosophical Transactions of the Royal Society of London 53, 370–418 (1763)

[3] Benjamin, A.T., Quinn, J.J.: Proofs That Really Count. The Art of Combinatorial Proof. Mathematical Association of America, pp.194 (2003)

[4] Bernoulli, J.: Ars Conjectandi, A complete translation by Edith Dudley Sylla – The Art of Conjecturing, together with Letter to a Friend on Sets in Court Tennis. Johns Hopkins University Press, Baltimore, p. 580, $57.60 (2006) Hardcover, $57.60

[5] Booth, D.E.: Regression Methods and Problem Banks. COMAP, Inc. (1986)

[6] von Bortkiewicz, L.: Das Gesetz der kleinem Zahlen, p. 60, Teubner, Leipzig (1898)

[7] Boyer, C.B.: A History of Mathematics. Princeton UP (1985)

[8] Criqui, M.H.: University of California, San Diego, reported in the New York Times, December 28 (1994)

[9] Czechowski, T.: Elementarny Wykład Rachunku Prawdopodobieństwa. (In Polish Introductory Course to Probability. PWN, Warszawa), 1st edn. (1958), 2nd edn. (1968)

[10] Descartes, R.: La Geometrie (1637) Appendix to Discours de la method; Translated into English by Michael Mahoney. Dover, New York (1979), Hermann, R.: Internet offers pdf French Edition 82 pages, Paris (1886)

[11] Diamond, S.: The World of Probability. Statistics in Science, p.155. BASIC Books, New York (1970) (Russian translation Statystyka, Moskwa)

[12] Dionne, G., Vanasse, C.: A generalization of Automobile Insurance Rating Models: the negative Binomial Distribution with a Regression Component. ASTIN Bulletin 19(2), 199–212 (1989)

[13] Draper, N.R., John, J.A.: Influential Observations and Outliers in Regression. Technometrics 23, 21–26 (1981)

[14] Edwards, A.W.F.: Pascal's Arithmetical Triangle – The Story of a Mathematical Idea, 1st edn., p. 202. Charles Griffin & Company Ltd, since 2002 London as a paperback – Johns Hopkins University Press, Baltimore

[15] Feller, W.: An Introduction to Probability Theory and Its Applications, 3rd edn., Posthumous Edition, vol. I, p. 509. John Wiley and Sons, New York (1971)

[16] Fisher, R.A.: Frequency Distribution of the Values of the Correlation Coefficient in Samples from Indefinitely Large Population. Biometrica 10, 507–521 (1915)
[17] Fisher, R.A.: The Goodness of Fit of Regression Formulae and the Distribution of Regression Coefficients. J. Royal Statistical Society 85, 597–612 (1922)
[18] Fisher, R.A.: The design of experiments. Oliver & Boyd, Edinburg (1935)
[19] Fisher, R.A.: The negative Binomial Distribution. Annuals of Eugenics 11, 182–187 (1941)
[20] Fisher, R.A.: Note on the Efficiency Fitting of the Negative Binomial. Biometrics 9, 197–199 (1953)
[21] Fisz, M.: Rachunek Prawdopodobieństwa i Statystyka Matematyczna (In Polish: Probability and Mathematical Statistics), Posthumous Third Edition, p. 694. PWN, Warszawa (1967); (there is also the English translation although the Author of this book has no references to it)
[22] Flaschmeyer, J.: Kombinatoryka – podstawowy wykład w ujęciu mnogościowym. (In Polish: Combinatorics, basic theory by using theory of set). PWN, Warszawa (1974); Translated from German "Kombinatorik". VEB, Berlin (1969)
[23] Galilei, G.: Dialog o dwu najważniejszych układach świata Ptolemeuszowym i Kopernikowym, pp. 314–316. PWN, Warszawa (1962); Translated from Italian into Polish by Edward Ligocki; (Dialogo sopra i due massimi sistemi del mondo Tolemaico e Copernico, Firenze (1632); Leyden (1638)
[24] Galton, F.: Finger Prints. MacMillan and Co., London, p. s.247 (1892); see also Hereditary Genius, 1st edn., p. 423 (1869); Both books pdf copies offers Internet
[25] Galton, F.: Regression towards mediocrity in hereditary stature. Journal of the Anthropological Institute of Great Britain and Ireland 15, 246–263 (1886)
[26] Gauss, C.F.: Theoria motus corporum coelestium in sectionibus conicis Solem ambienitum. Hamburg (1809); English translation - Theory of the Motion of the Heavenly Bodies Moving About the Sun in Conic Sections by Charles Henry Davis, p. 416. Little, Brown, and Company, Boston (1857); accessible via Internet
[27] Gerstenkorn T., Śródka T.: Kombinatoryka i rachunek prawdopodobieństwa. (In Polish: Combinatorics and Probabilisty), 3rd edn. PWN, Warszawa (1976)
[28] Good, I.J. : Some Statistical Applications of Poisson's Work. Statistical Science 1(2), 157–170 (1986)
[29] Gosset, W.S.: An Explanation of Deviation from Poisson's Law in Practice 12(3/4), 211–215 (1919); Also: Paper in "Student's" Collected Papers. London, Biometrica Office, 65–69 (1942)
[30] Gumbel, E.J.: Statistics of Extremes. Columbia University Press, New York (1958, 1962); Russian translation of B. Ю.Тамарский; (W.Yu. Tamarski), p. 450, Mir, Moscow (1965), Lit.: 647 entries
[31] Gumbel, E.J.: Bortkiewicz, Ladislaus von. International Encyclopedia of Statistics 1, 24–27 (1978); Reprinted from the International Encyclopedia of Social Sciences (1968)
[32] Gurland, J.: Some Applications of the Negative Binomial and Other Contagious Distributions. Pdf file No.1388, Wikipedia, pp. 1-12; Printed in A.J.P.H. 49(10), 1388–1399 (1959)
[33] Haight, F.A.: Handbook of the Poisson Distribution. Wiley, New York (1978)
[34] Hawkins, C.A., Weber, J.E.: Statistical Analysis. Applications to Business and Economy, pp. 1–626. Harper & Row, New York (1980)
[35] Houbolt, J.C.: Atmospheric Turbulence. AIAA Journal 11(4), 421–437 (1973)

[36] Hubble, E.P.: A relation between distance and radial velocity among extra-galactic nebulae. In: Proceedings of the National Academy of Sciences 15, 168–173 (1929)

[37] Jacobs, H.R.: Geometry, 2nd edn., pp. 1–668. W. H. Freeman & Co., New York (1987); 1st edn. (1974)

[38] Kemp, A.W., Kemp, C.D.: Weldon's Dice Data Revised. The American Statistician 45(3), 216–222 (1991)

[39] Kendal, M.G.: Ronald Aylmer Fisher. Biometrika, vol. 50, No.1-2, pages 1-17 (1963)

[40] Kingman, J.F.C.: Poisson Processes. Oxford UP (1993); translated into Polish. PWN, Warszawa (2002)

[41] Marquis de Laplace, P.S.: A Philosophical Essay on Probabilities. Transl. from French, p. 196. Dover (1951)

[42] Laudański, L.M.: Pre-stochastic Models of Computational Probability. Transactions of the Aviation Institute 159(4/99), 30–36 (1999)

[43] Laudański, L.M.: Statystyka Ogólna z Elementami Statystyki Matematycznej. (In Polish: General Statistics and Probability). Wydawnictwa PWSZ, Jarosław (2000)

[44] Laudański, L.M., Dłubis, E.: Zbieżność i rozbieżność rozkładu dwumianowego względem rozkładu Gaussa w praktyce statystycznej. (In Polish: The convergence and divergence between the binomial and Gaussian distribution as used in statistical practice). Proceedings of the IV-th International Scientific Conference in Jarosław: Development of the Rzeszow Province Due to European Integration, pp. 263–272. PWSZ, Jarosław (2000)

[45] Laudański, L.M.: Dylematy jakości nauczania w epistolografii św. Pawła. (In Polish: Quality Dillemmas in Letters of St. Paul), Conference Proceedings Materiały Konferencji Naukowej nt. Dylematy jakości kształcenia w uczelniach wyższych, pp. 111–121. Politechnika Rzeszowska, Rzeszów (2008)

[46] Laudański, L.M.: Statystyka nie tylko dla Licencjatów (In Polish: Statistics not only for undergraduates), 2nd edn., vol. 1, 2. Publishing House of the Rzeszow TU, Rzeszów (2009)

[47] Lehman, E.L.: Jerzy Neyman 1894-1981. A Biographical Memoir. National Academy of Sciences, Washington, p. 28 (1994)

[48] Lewis, C.S.: Out of the Silent Planet, AVON Book Division, , p. 159. The Hearst Corporation, New York (1949); Polish translation by Andrzej Polkowski: Z Milczącej Planety, p. 160, Wydawnictwo M, Kraków (1989)

[49] Lloyd, E.: Probability. Handbook of Applicable Mathematics. vol. II, , pp.1– 450. J.Wiley & Sons, New York (1980)

[50] Леонид Ефимович Майстров: Теория Вероятностей – Исторический Очерк, Наука, Москва (1967); Also known as the English edition: Maistrov, L. E.: Probability Theory - A Historical Sketch. Academic Press, New York (1974)

[51] Makać, W. and Urbanek-Krzysztofiak, D.: Metody Opisu Statystycznego. (In Polish: General Statistics, Outline). Wydawnictwa Uniwersytetu Gdańskiego (1995, 2001)

[52] Al-Saleh, M.F., AL-Batainah, F.K.: Estimation of the Proportion of Sterile Couples Using the Negative Binomial Distribution. Journal of Data Science 1, 261–274 (2003)

[53] de Moivre, A.: The Doctrine of Chances or A Method of Calculating the Probabilities of Events in Play, 3rd edn., Fuller, Clearer, and more Correct than the Former, pp. 1-378. A. Millar, London (1756); Digitized by Google, Internet
[54] de Montmort, P.: Essay d'analyse sur les jeux de hazard, 2nd edn., Jacque Quillau, Paris, p. 468 (1713) (pdf copy accessible by Internet)
[55] Neyman, J., Pearson: Egon Sharpe: Joint Statistical Papers, p. 300. Cambridge University Press (1967)
[56] Neyman, J.: First Course in Probability and Statistics. Henry Holt & Co., New York (1950); Polish translation: Zasady Rachunku Prawdopodobieństwa i Statystyki Matematycznej. PWN, Warszawa (1969); Russian translation – Nauka, Moskwa (1968)
[57] O'Flynn, M.: Probabilities, Random Variables, and Random Processes, p. 523. Harper & Row, Cambridge (1982)
[58] O'Connor, J.J., Robertson, E.F.: Walter Frank Raphael Weldon – Wikipedia
[59] Papoulis, A.: Probability, Random Variables, and Stochastic Processes, p. 583. McGraw-Hill, New York (1962)
[60] Parzen, E.: Modern Probability Theory and Its Applications, p. 464 . J. Wiley & Sons, New York (1960)
[61] Pascal, B.: Traite du triangle arithmetique. Desprez, Paris (1665)
[62] Pearson, K.: Cloudiness: Note on a Novel Case of Frequency. Proceedings of the Royal Society 62, 287 (1897)
[63] Pearson, K.: On the Criterion that a Given System of Deviations from the Probable in the Case of the Correlated Systems of Variables Is Such That it Can be Reasonably Supposed to Have Arisen from Random Sampling. Philosophical Magazine 50, 157–175 (1900)
[64] Pearson, K., Lee, A.: On the Laws of Inheritance on Man. Biometrica 2(4), 357–462 (1903)
[65] Pearson, K.: Francis Galton. Nature 85, 440–445 (1911)
[66] Pearson, K.: The Grammar of Science. Dover Publications (1992, 2004)
[67] Pepys, S.: The Diary of Samuel Pepys. Translated into Polish and selected by Maria Dąbrowska. Dziennik Samuela Pepysa, vol. 1 (1660-1665), vol. 2 (1666-1669), pp. 435–519, PIW, Warsaw (1952)
[68] Pogorzelski, W.: Zarys Rachunku Prawdopodobieństwa i Teorii Błędów. (In Polish: An Outline of Probability and the Error Theory). Towarzystwo Bratniej Pomocy Studentów PW (edited by the students organization soon confiscated by the communist government), Warsaw, pp. 1– 100 (1948)
[69] Press, H.: The Application of the Statistical Theory of Extreme Values to Gust-load Problems. NACA Report 991, Washington, p. 16 (1949)
[70] Reichmann, W.J.: Use and Abuse of Statistics. Penguin Books, Middlesex, pp. 345 (1961, 1976); Polish translation by Robert Bartoszyński (1933-1998); Drogi i bezdroża statystyki, pp. 1–395. PWN, Warszawa (1968)
[71] Romanowski, M.: On the Normal Law of Errors. National Research Council of Canada. Report APH-1178, Ottawa, pp.1–29 (February 1964)
[72] Rubin, E., Schell, E.D.: Questions and Answers. The American Statistician 14 (4), 27–30 (1960)
[73] Smoluk, A.: Mathematics – a Universal Science. Didactics of Mathematics, vol. 6, pp. 5–9. Wroclaw University of Economics, Wrocław (2005)

[74] Soper, H.E.: On the probable error of the correlation coefficient to a second approximation. Biometrica 9, pages 91–115 (1913); Tables of Poisson's exponential binomial limit. Biometrika 10, pages 25–35 (1914)
[75] Spiegel, M.R.: Schaum's Outline of Theory and Problems of Statistics, pp.1–359. McGraw-Hill, New York (1972) 870 solved problems
[76] Stigler, S.M.: Poisson on the Poisson Distribution. Statistics & Probability Lectures 1, 33–35 (1982)
[77] Stigler, S.M.: The History of Statistics – the Measurement of Uncertainty before 1900. The Balknap Press of Harvard UP, Cambridge (1986)
[78] Stigler, S.M.: The Dark Ages of Probability in England: The Seventeenth Century Work of Richard Cumberland and Thomas Strode. International Statistical Review / Revue Internationale de Statistique 56(1), 75–88 (1988)
[79] Stigler, S.M.: Isaac Newton as a Probabilist. Statistical Science 21(3), 400–403 (2006)
[80] Stigler, S.M.: Chance is 350 Years Old 20(4), 33–36 (2007)
[81] Stigler, S.: Karl Pearson's Theoretical Errors, and the Advanced They Inspired. Statistical Science 23(2), 261–271 (2008)
[82] Stigler, S.: Karl Pearson and the Rule of Three. To appear in Biometrika, circa 14 p.
[83] Student: The probable error of a mean. Biometrika, vol. 6, pages 1-25 (1908); Accessible in Internet
[84] Tennant-Smith, J.: BASIC Statistics, p. 160. Butterworths, London (1985)
[85] Brown, L.: The New Shorter Oxford English Dictionary on Historical Principles, vol. 1, p. 1620. Clarendon Press, Oxford (1993)
[86] Honderich, T.: The Oxford Companion to Philosophy, Oxford (1995)
[87] Uspensky, J.V.: Introduction to Mathematical Probability, p. 411. McGraw-Hill, New York (1937)
[88] von Brzeski, J.G.: Application of Lobatchevsky's Formula on the Angle of Parallelism to Geometry of Space and to the Cosmological Redshift. Russian Journal of Mathematical Physics 14, 366 (2007)
[89] von Brzeski, J.G.: Expansion of the Universe – Mistake of Edwin Hubble? Cosmological Redshift and Related Electromagnetic Phenomena in Static Lobatchevskian (Hyperbolic) Universe. Acta Physica Polonica 39, 1501 (2007)
[90] von Collani, C.: Biography of Karl Pearson, 26 pages, `http://encyclopedia.stochastikon.com`
[91] Weinberg, G.H., Schumaker, J.A., Oltman, D.: Statistics – An Intuitive Approach, 4th edn., pp. 1–447. Brooks/Cole, Monterey (1981)
[92] Wieleitner, H.: Geschichte der Mathematik. Part I from Descartes to about 1800. Leipzig, vol.2 (1911-1921); The reference based on the Russian translation: История Математики од Декарта до середины XIX столетия. Наука, Москва (1956)
[93] Wilenkin, N. J.: Kombinatoryka, translated into Polish PWN, Warszawa (1972); Russian original Russian, Nauka, Moscow (1962)
[94] Willerman, L., Schultz, R., Rutledge, J.N., Bigler, E.: In vivo brain size and intelligence. Intelligence 15, 223–228 (1991)
[95] Williams, R.H., Zumbo, Roos, Zimmerman: On the Intellectual Versatility of Karl Pearson. Human Nature Review 3, 296–301 (2003)
[96] Williams, R.H.: George Udny Yule: Statistical Scientist. Human Nature Review 4, 31–37 (2004)

[97] Yates, F., Mather, K.: Ronald Aylmer Fisher. Biographical Memoirs of Fellows of the Royal Society of London, 9, pages 91–120 (1963)

[98] Yule, G.U.: An Introduction to the Theory of Statistics. Charles Griffin and Co., London (1911); 2nd edn., Translated into Polish by Z. Limanowski: Wstęp do Teorji Statystyki, Gebethner i Wolff, Warszawa, pp.1–446 (1921); 6th edn., accessible by Internet, pp.1–415 (1922); 14th edn., co-author M.G. Kendall (1950); Translated into Polish as Wstęp do Teorii Statystyki. PWN, Warszawa (1966)

[99] Zubrzycki, S.: Wykłady z rachunku prawdopodobieństwa i Statystyki Matematycznej. (In Polish: Lectures on Probability and Mathematical Statistics), p. 334. PWN, Warszawa (1966)

[100] Stigler Law, see Internet

[101] A Guide to the Microfilm Edition of: The Emil J. Gumbel Collection contains 10 pages biography written by Arthur Brenner, p.32. Leo Baeck Institute, New York. University Publications of America. No year of publication (approximately 1990)

Index

T

U

V

W

Y

Z

Printed by Printforce, the Netherlands